PROGRESS IN CLINICAL AND BIOLOGICAL RESEARCH

Series Editors
Nathan Back
George J. Brewer

Vincent P. Eijsvoogel
Robert Grover
Kurt Hirschhorn

Seymour S. Kety
Sidney Udenfriend
Jonathan W. Uhr

RECENT TITLES

Vol 50: **Rights and Responsibilities in Modern Medicine: The Second Volume in a Series on Ethics, Humanism, and Medicine,** Marc D. Basson, *Editor*

Vol 51: **The Function of Red Blood Cells: Erythrocyte Pathobiology,** Donald F. H. Wallach, *Editor*

Vol 52: **Conduction Velocity Distributions: A Population Approach to Electrophysiology of Nerve,** Leslie J. Dorfman, Kenneth L. Cummins, and Larry J. Leifer, *Editors*

Vol 53: **Cancer Among Black Populations,** Curtis Mettlin and Gerald P. Murphy, *Editors*

Vol 54: **Connective Tissue Research: Chemistry, Biology, and Physiology,** Zdenek Deyl and Milan Adam, *Editors*

Vol 55: **The Red Cell: Fifth Ann Arbor Conference,** George J. Brewer, *Editor*

Vol 56: **Erythrocyte Membranes 2: Recent Clinical and Experimental Advances,** Walter C. Kruckeberg, John W. Eaton, and George J. Brewer, *Editors*

Vol 57: **Progress in Cancer Control,** Curtis Mettlin and Gerald P. Murphy, *Editors*

Vol 58: **The Lymphocyte,** Kenneth W. Sell and William V. Miller, *Editors*

Vol 59: **Eleventh International Congress of Anatomy,** Enrique Acosta Vidrio, *Editor-in-Chief.* Published in 3 volumes:
Part A: **Glial and Neuronal Cell Biology,** Sergey Fedoroff, *Editor*
Part B: **Advances in the Morphology of Cells and Tissues,** Miguel A. Galina, *Editor*
Part C: **Biological Rhythms in Structure and Function,** Heinz von Mayersbach, Lawrence E. Scheving, and John E. Pauly, *Editors*

Vol 60: **Advances in Hemoglobin Analysis,** Samir M. Hanash and George J. Brewer, *Editors*

Vol 61: **Nutrition and Child Health: Perspectives for the 1980s,** Reginald C. Tsang and Buford Lee Nichols, Jr., *Editors*

Vol 62: **Pathophysiological Effects of Endotoxins at the Cellular Level,** Jeannine A. Majde and Robert J. Person, *Editors*

Vol 63: **Membrane Transport and Neuroreceptors,** Dale Oxender, Arthur Blume, Ivan Diamond, and C. Fred Fox, *Editors*

Vol 64: **Bacteriophage Assembly,** Michael S. DuBow, *Editor*

Vol 65: **Apheresis: Development, Applications, and Collection Procedures,** C. Harold Mielke, Jr., *Editor*

Vol 66: **Control of Cellular Division and Development,** Dennis Cunningham, Eugene Goldwasser, James Watson, and C. Fred Fox, *Editors.* Published in 2 Volumes.

Vol 67: **Nutrition in the 1980s: Constraints on Our Knowledge,** Nancy Selvey and Philip L. White, *Editors*

Vol 68: **The Role of Peptides and Amino Acids as Neurotransmitters,** J. Barry Lombardini and Alexander D. Kenny, *Editors*

Vol 69: **Twin Research 3, Proceedings of the Third International Congress on Twin Studies,** Luigi Gedda, Paolo Parisi, and Walter E. Nance, *Editors.* Published in 3 volumes:
Part A: **Twin Biology and Multiple Pregnancy**

	Part B: **Intelligence, Personality, and Development**
	Part C: **Epidemiological and Clinical Studies**
Vol 70:	**Reproductive Immunology,** Norbert Gleicher, *Editor*
Vol 71:	**Psychopharmacology of Clonidine,** Harbans Lal and Stuart Fielding, *Editors*
Vol 72:	**Hemophilia and Hemostasis,** Doris Ménaché, D. MacN. Surgenor, and Harlan D. Anderson, *Editors*
Vol 73:	**Membrane Biophysics: Structure and Function in Epithelia,** Mumtaz A. Dinno and Arthur B. Callahan, *Editors*
Vol 74:	**Physiopathology of Endocrine Diseases and Mechanisms of Hormone Action,** Roberto J. Soto, Alejandro De Nicola, and Jorge Blaquier, *Editors*
Vol 75:	**The Prostatic Cell: Structure and Function,** Gerald P. Murphy, Avery A. Sandberg, and James P. Karr, *Editors.* Published in 2 volumes:
	Part A: **Morphologic, Secretory, and Biochemical Aspects**
	Part B: **Prolactin, Carcinogenesis, and Clinical Aspects**
Vol 76:	**Troubling Problems in Medical Ethics: The Third Volume in a Series on Ethics, Humanism, and Medicine,** Marc D. Basson, Rachel E. Lipson, and Doreen L. Ganos, *Editors*
Vol 77:	**Nutrition in Health and Disease and International Development: Symposia From the XII International Congress of Nutrition,** Alfred E. Harper and George K. Davis, *Editors*
Vol 78:	**Female Incontinence,** Norman R. Zinner and Arthur M. Sterling, *Editors*
Vol 79:	**Proteins in the Nervous System: Structure and Function,** Bernard Haber, Joe Dan Coulter, and Jose Regino Perez-Polo, *Editors*
Vol 80:	**Mechanism and Control of Ciliary Movement,** Charles J. Brokaw and Pedro Verdugo, *Editors*
Vol 81:	**Physiology and Biology of Horseshoe Crabs: Studies on Normal and Environmentally Stressed Animals,** Joseph Bonaventura, Celia Bonaventura, and Shirley Tesh, *Editors*
Vol 82:	**Clinical, Structural, and Biochemical Advances in Hereditary Eye Disorders,** Donna L. Daentl, *Editor*
Vol 83:	**Issues in Cancer Screening and Communications,** Curtis Mettlin and Gerald P. Murphy, *Editors*
Vol 84:	**Progress in Dermatoglyphic Research,** Christos S. Bartsocas, *Editor*
Vol 85:	**Embryonic Development,** Max M. Burger and Rudolf Weber, *Editors.* Published in 2 volumes:
	Part A: **Genetic Aspects**
	Part B: **Cellular Aspects**
Vol 86:	**The Interaction of Acoustical and Electromagnetic Fields With Biological Systems,** Shiro Takashima and Elliot Postow, *Editors*
Vol 87:	**Physiopathology of Hypophysial Disturbances and Diseases of Reproduction,** Alejandro De Nicola, Jorge Blaquier, and Roberto J. Soto, *Editors*
Vol 88:	**Cytapheresis and Plasma Exchange: Clinical Indications,** W.R. Vogler, *Editor*
Vol 89:	**Interaction of Platelets and Tumor Cells,** G.A. Jamieson, *Editor*

See pages following the index for previous titles in this series.

INTERACTION OF PLATELETS AND TUMOR CELLS

The Chesapeake Conference

Organizing Committee

G. A. Jamieson, *Chairman*
E. Bastida
James F. Haggerty
A. Ordinas
Alice R. Scipio

INTERACTION OF PLATELETS AND TUMOR CELLS

Papers from the proceedings of
The Chesapeake Conference
held at St. Mary's College
St. Mary's, Maryland
June 26–30, 1981

Editor

G. A. Jamieson

Associate Director
American Red Cross Blood Services
Bethesda, Maryland
and
Adjunct Professor
Schools of Medicine and Dentistry
Georgetown University
Washington, D.C.

Assistant Editor

Alice R. Scipio

American Red Cross Blood Services
Bethesda, Maryland

ALAN R. LISS, INC., NEW YORK

Address all Inquiries to the Publisher
Alan R. Liss, Inc., 150 Fifth Avenue, New York, NY 10011

Copyright © 1982 Alan R. Liss, Inc.

Printed in the United States of America.

Under the conditions stated below the owner of copyright for this book hereby grants permission to users to make photocopy reproductions of any part or all of its contents for personal or internal organizational use, or for personal or internal use of specific clients. This consent is given on the condition that the copier pay the stated per-copy fee through the Copyright Clearance Center, Incorporated, 21 Congress Street, Salem, MA 01970, as listed in the most current issue of "Permissions to Photocopy" (Publisher's Fee List, distributed by CCC, Inc.) for copying beyond that permitted by sections 107 or 108 of the US copyright law. This consent does not extend to other kinds of copying, such as copying for general distribution, for advertising or promotional purposes, for creating new collective works, or for resale.

Library of Congress Cataloging in Publication Data

Main entry under title:

Interaction of platelets and tumor cells.

(Progress in clinical and biological research; v. 89)
Papers from the Chesapeake Conference held at Saint Mary's College of Maryland, June 26-30, 1981.
Includes bibliographical references and index.
1. Cancer cells—Congresses. 2. Blood platelets—Congresses. 3. Cell interaction—Congresses. 4. Metastasis—Congresses. I. Jamieson, G. A. (Graham A.), 1929-   . II. Scipio, Alice R. III. Chesapeake Conference (1981: Saint Mary's College of Maryland)   IV. Series.
RC267.I5255   616.99'407   82-6530
                                        AACR2

ISBN 0-8451-0089-0

Contents

Contributors xi
Participants xvii
Introductory Remarks
G. A. Jamieson xix

SESSION I

PATHOLOGY AND PHARMACOLOGY

A Classification of Disorders of Hemostasis and Thrombosis in Patients With Malignancy
Alvin H. Schmaier and H. James Day 1
Discussion 1 20
Intra-Tumoral Platelet Consumption: The Kasabach-Merritt Syndrome as a Model for Malignant Platelet Destruction
Raymond P. Warrell, Jr. 21
Antithrombotic Effects of Drugs Which Suppress Platelet Function: Their Potential in Preventing Growth of Tumour Cells
Alexander G. G. Turpie 31
Platelet Cancer Cell Interaction in Metastasis Formation. Platelet Aggregation Inhibitors: A Possible Approach to Metastasis Prevention
Helmuth Gastpar, Julian L. Ambrus, and Clara M. Ambrus 63
Discussion 2 81
Platelet Aggregation Inhibitors and Metastatic Spread of Neoplastic Cells
Julian L. Ambrus, Clara M. Ambrus, and Helmuth Gastpar 83
Antimetastatic and Antitumor Effect of Platelet Aggregation Inhibitors
Julian L. Ambrus, Clara M. Ambrus, Helmuth Gastpar, E. Huberman, R. Montagna, W. Biddle, S. Leong, and J. Horoszewicz 97
Discussion 3 109
The Biologic Basis for Anticoagulant Treatment of Cancer
Leo R. Zacharski 113
Discussion 4 128
Antimetastatic Action of RX-RA 69, a New Potent PDE-Inhibitor in the Lewis Lung Carcinoma of the Mouse
Rosemarie Lichtner and Walter Haarmann 131

SESSION II

MODELS FOR PLATELET TUMOR CELL INTERACTION

Blood Platelets and Tumour Dissemination
P. Hilgard 143

Discussion 5 156

Animal Models for the Study of Platelet-Tumor Cell Interactions
M. B. Donati, D. Rotillio, F. Delaini, R. Giavazzi, A. Mantovani, and A. Poggi 159

Discussion 6 175

In Vivo Models for Studies of Human Tumor Metastasis
Nabil Hanna 177

Discussion 7 188

Cultured Human Tumor Cells for Cancer Research: Assessment of Variation and Stability of Cultural Characteristics
Jørgen J. Fogh, Nicholas Dracopoli, James D. Loveless, and Helle Fogh ... 191

Discussion 8 222

Positive and Negative Aggregation Responses to Cultured Human Tumor Cell Lines Among Different Normal Individuals
Eva Bastida, Antonio Ordinas, and G.A. Jamieson 225

Discussion 9 230

SESSION III

PLATELET FACTORS IN CELL PROLIFERATION

A Mitogenic Factor for Transformed Cells From Human Platelets
Allan Lipton, Nancy Kepner, Cheryl Rogers, Elizabeth Witkoski, and Kim Leitzel 233

Discussion 10 247

Effect of Platelet Growth Factor(s) on Growth of Human Tumor Colonies
Dale H. Cowan and Joyce Graham 249

Discussion 11 267

Human Platelet Chemotaxis Can Be Induced by Low Molecular Substance(s) Derived From the Interaction of Plasma and Collagen
Rosalin Wu Lowenhaupt 269

Discussion 12 279

SESSION IV

INTERACTION OF PLATELETS AND TUMOR CELLS WITH BLOOD VESSELS

Role of the Vascular Endothelium
John C. Hoak, William M. Parks, Glenna L. Fry, Abigail F.A. Brotherton, and Robert L. Czervionke 281

Discussion 13 294

Control of Tumor Growth and Metastasis With Prostacyclin and Thromboxane Synthetase Inhibitors: Evidence for a New Antitumor and Antimetastatic Agent (Bay g 6575)
Kenneth V. Honn, Jay Meyer, Gregory Neagos, Thomas Henderson, Christine Westley, and Vaneerat Ratanatharathorn 295

Discussion 14 329

Tumor Cell Interactions With Vascular Endothelial Cells and Their Extracellular Matrix
Randall H. Kramer, Kathryn G. Vogel, and Garth L. Nicolson 333

Discussion 15 349

Interactions of Tumor Cells With Whole Basement Membrane in the Presence or Absence of Endothelium
Calvin M. Foltz, Raimondo G. Russo, Gene P. Siegal, Victor P. Terranova, and Lance A. Liotta 353

The Use of a Perfusion Model for Studying Aggregation and Attachment of Platelets and Tumor Cells at Subendothelial Surfaces
Antonio Ordinas, J. Michael Marcum, Manley McGill, Eva Bastida, and G.A. Jamieson 373

SESSION V

BIOCHEMICAL ASPECTS

Interaction of Platelets and Tumor Cells
Manfred Steiner 383

Mechanisms of Platelet Aggregation by Human Tumor Cell Lines
G.A. Jamieson, Eva Bastida, and Antonio Ordinas 405

Isolation and Partial Purification of an Activity From Human Pulmonary Epidermoid Carcinoma Cells in Culture That Clots Heparinized Plasma
S. F. Mohammad, H. Y. K. Chuang, J. Szakacs, and R. G. Mason 415

Plasma Membrane Vesicles as Mediators of Interactions Between Tumor Cells and Components of the Hemostatic and Immune Systems
Gabriel J. Gasic and Tatiana B. Gasic 429

Discussion 16 444

Platelet Aggregating Material (PAM) of Two Virally-Transformed Tumors: SV3T3 Mouse Fibroblast and PW20 Rat Renal Sarcoma. Role of Cell Surface Sialylation
Simon Karpatkin, Edward Pearlstein, Peter L. Salk, and Ganesa Yogeeswaran 445

Discussion 17 476

Inhibition of the Platelet-Aggregating Activity of Two Human Adenocarcinomas of the Colon and an Anaplastic Murine Tumor With a Specific Thrombin Inhibitor: Dansylarginine N-(3-Ethyl-1,5-Pentanediyl) Amide
Edward Pearlstein, Cynthia Ambrogio, Gabriel J. Gasic, and Simon Karpatkin 479

Discussion 18 501

Subject Index 503

Contributors

Cynthia Ambrogio [479]
Department of Medicine, New York University Medical Center, New York, NY

Clara M. Ambrus, M.D., Ph.D. [63, 83, 97]
Roswell Park Memorial Institute and State University of New York at Buffalo, Department of Pediatrics, Buffalo, NY

Julian L. Ambrus, M.D., Ph.D. [63, 83, 97]
Roswell Park Memorial Institute and State University of New York at Buffalo, Department of Internal Medicine, Experimental Pathology and Pediatrics, Buffalo, NY

Eva Bastida, Ph.D. [225, 373, 405]
Servicio de Hemoterapia y Hemostasis, Hospital Clinico y Provincial, Facultad de Medicina, Universidad de Barcelona, Barcelona 36, Spain

W. Biddle [97]
Associated Biomedic Systems, Buffalo, NY

Abigail F.A. Brotherton, Ph.D. [281]
Division of Hematology-Oncology, Department of Medicine, University of Iowa Hospitals, Iowa City, IA

H.Y.K. Chuang, Ph.D. [415]
Department of Pathology, College of Medicine, University of Utah, Salt Lake City, UT

Dale H. Cowan, M.D., J.D. [249]
Division of Hematology/Oncology, Saint Luke's Hospital, Case Western Reserve University, Cleveland, OH

Robert L. Czervionke, Ph.D. [281]
Division of Hematology-Oncology, Department of Medicine, University of Iowa Hospitals, Iowa City, IA

H. James Day, M.D. [1]
Department of Medicine and Pathology, Clinical Laboratories, Temple University Health Science Center, Philadelphia, PA

F. Delaini, Biol. Sci. D. [159]
Istituto di Richerche Farmacologiche, "Mario Negri," 20157 Milano, Italy

M.B. Donati, M.D., Ph.D. [159]
Istituto di Richerche Farmacologiche, "Mario Negri," 20157 Milano, Italy

Nicholas Dracopoli, Ph.D. [191]
Human Cell Laboratory, Sloan-Kettering Institute for Cancer Research, Rye, NY

Helle Fogh [191]
Human Cell Laboratory, Sloan-Kettering Institute for Cancer Research, Rye, NY

The bold face number in brackets following each contributor's name indicates the opening page of that author's paper.

Jørgen J. Fogh, M.D. [191]
  Human Cell Laboratory, Sloan-Kettering Institute for Cancer Research, Rye, NY
Calvin M. Foltz, Ph.D. [353]
  Laboratory of Chemistry, National Institute of Arthritis, Diabetes, and Digestive and Kidney Diseases, National Institutes of Health, Bethesda, MD
Glenna L. Fry, B.S. [281]
  Division of Hematology–Oncology, Department of Medicine, University of Iowa Hospitals, Iowa City, IA
Gabriel J. Gasic, M.D. [429, 479]
  Department of Pathology, University of Pennsylvania, School of Medicine, Philadelphia, PA
Tatiana B. Gasic, M.S. [429]
  Department of Pathology, University of Pennsylvania, School of Medicine, Philadelphia, PA
Helmuth Gastpar, M.D., D.Sc. [63, 83, 97]
  University of Munich Medical School, 8000 Munich 2, West Germany
R. Giavazzi, Biol. Sci. D. [159]
  Istituto di Richerche Farmacologiche, "Mario Negri," 20157 Milano, Italy
Joyce Graham, B.S., M.D. [249]
  Division of Hematology/Oncology, Saint Luke's Hospital, Cleveland, OH
Walter Haarmann, M.D. [131]
  Dr. Karl Thomae GmbH, Biologische Forschung, Birkendorfer Strasse 65, D-7950 Biberach, West Germany
Nabil Hanna, Ph.D. [177]
  Cancer Metastasis and Treatment Laboratory, NCI Frederick Cancer Research Center, Frederick, MD
Thomas Henderson [295]
  Departments of Radiology and Radiology Oncology, Wayne State University, School of Medicine, Gordon H. Scott Hall of Basic Sciences, Detroit, MI
P. Hilgard, M.D. [143]
  Bristol-Meyers International Corporation, Clinical Research (EMEA), Chaussee de la Hulpe, 185, 1170 Brussels, Belgium
John C. Hoak, M.D. [281]
  Division of Hematology–Oncology, Department of Medicine, University of Iowa Hospitals, Iowa City, IA
Kenneth V. Honn, Ph.D. [295]
  Departments of Radiology and Radiation Oncology, Wayne State University, School of Medicine, Gordon H. Scott Hall of Basic Sciences, Detroit, MI
J. Horoszewicz, Ph.D., D.Sc. [97]
  Roswell Park Memorial Institute, Buffalo, NY
E. Huberman, Ph.D. [97]
  Oakridge National Laboratory, Oakridge, TN
G.A. Jamieson, Ph.D., D.Sc. [xix, 225, 373, 405]
  American Red Cross Blood Services, Bethesda, MD
Simon Karpatkin, M.D. [445, 479]
  Department of Medicine, New York University Medical School, New York, NY

Nancy Kepner [233]
The Milton S. Hershey Medical Center of the Pennsylvania State University, Department of Medicine, Hershey, PA

Randall H. Kramer, Ph.D. [333]
Departments of Anatomy and Oral Medicine, University of California, San Francisco, CA

Kim Leitzel [233]
The Milton S. Hershey Medical Center of the Pennsylvania State University, Department of Medicine, Hershey, PA

S. Leong, Ph.D. [97]
Roswell Park Memorial Institute, Buffalo, NY

Rosemarie Lichtner, Ph.D. [131]
Dr. Karl Thomae GmbH, Biologische Forschung, Birkendorfer Strasse 65, D-7950 Biberach, West Germany

Lance A. Liotta, M.D., Ph.D. [353]
Laboratory of Pathophysiology, National Cancer Institute, National Institutes of Health, Bethesda, MD

Allan Lipton, M.D. [233]
The Milton S. Hershey Medical Center of the Pennsylvania State University, Department of Medicine/Oncology Division, Hershey, PA

James D. Loveless [191]
Human Cell Laboratory, Sloan-Kettering Institute for Cancer Research, Rye, NY

Rosalin Wu Lowenhaupt, M.D. [269]
Department of Physiology, University of Cincinnati, College of Medicine, Cincinnati, OH

Manley McGill, Ph.D. [373]
University of Cincinnati Medical Center, The Paul I. Hoxworth Blood Center, Cincinnati, OH

A. Mantovani, M.D. [159]
Istituto di Richerche Farmacologiche, "Mario Negri," 20157 Milano, Italy

J. Michael Marcum, Ph.D. [373]
Coconut Grove, FL

R.G. Mason, M.D., Ph.D. [415]
Department of Pathology, College of Medicine, University of Utah, Salt Lake City, UT

Jay Meyer [295]
Departments of Radiology and Radiology Oncology, Wayne State University, School of Medicine, Gordon H. Scott Hall of Basic Sciences, Detroit, MI

S.F. Mohammad, Ph.D. [415]
Department of Pathology, College of Medicine, University of Utah, Salt Lake City, UT

R. Montagna, Ph.D. [97]
Associated Biomedic Systems, Buffalo, NY

Gregory Neagos [295]
Departments of Radiology and Radiology Oncology, Wayne State University, School of Medicine, Gordon H. Scott Hall of Basic Sciences, Detroit, MI

Garth L. Nicolson, Ph.D. [333]
Department of Tumor Biology, The University of Texas System Cancer Center, M.D. Anderson Hospital and Tumor Institute, Houston, TX

Antonio Ordinas, M.D. [225, 373, 405]
Servicio de Hemoterapia y Hemostasis, Hospital Clinico y Provincial, Facultad de Medicina, Universidad de Barcelona, Barcelona 36, Spain

William M. Parks, M.D. [281]
Division of Hematology–Oncology, Department of Medicine, University of Iowa Hospitals, Iowa City, IA

Edward Pearlstein, Ph.D. [445, 479]
Department of Medicine, New York University Medical Center, New York, NY

A. Poggi, Biol. Sci. D. [159]
Istituto di Richerche Farmacologiche, "Mario Negri," 20157 Milano, Italy

Vaneerat Ratanatharathorn, M.D. [295]
Departments of Radiology and Radiology Oncology, Wayne State University, School of Medicine, Gordon H. Scott Hall of Basic Sciences, Detroit, MI

Cheryl Rogers [233]
The Milton S. Hershey Medical Center of the Pennsylvania State University, Department of Medicine, Hershey, PA

D. Rotillio, Chem. D. [159]
Istituto di Richerche Farmacologiche, "Mario Negri," 20157 Milano, Italy

Raimondo G. Russo, M.D. [353]
Laboratory of Pathophysiology, National Cancer Institute, National Institutes of Health, Bethesda, MD

Peter L. Salk, M.D. [445]
Autoimmune and Neoplastic Disease Laboratory, The Salk Institute for Biological Studies, San Diego, CA

Alvin H. Schmaier, M.D. [1]
Department of Medicine, Coagulation Laboratory, Temple University Health Science Center, Philadelphia, PA

Gene P. Siegal, M.D., Ph.D. [353]
Laboratory of Pathophysiology, National Cancer Institute, National Institutes of Health, Bethesda, MD

Manfred Steiner, M.D., Ph.D. [383]
Division of Hematologic Research, The Memorial Hospital, Pawtucket, and Brown University, Providence, RI

J. Szakacs, M.D. [415]
Department of Pathology, Saint Joseph's Hospital, Tampa, FL

Victor P. Terranova, D.M.D., Ph.D. [353]
Laboratory of Developmental Biology and Anomalies, National Institute of Dental Research, National Institutes of Health, Bethesda, MD

Alexander G. G. Turpie, M.B., F.R.C.P., F.A.C.P. [31]
McMaster University, Hamilton Civic Hospital, Hamilton, Ontario, Canada

Kathryn G. Vogel, Ph.D. [333]
Department of Biology, University of New Mexico, Albuquerque, NM

Raymond P. Warrell, Jr., M.D. [21]
Developmental Chemotherapy Service, Memorial Sloan-Kettering Cancer Center, New York, NY

Christine Westley [295]
Departments of Radiology and Radiology Oncology, Wayne State University, School of Medicine, Gordon H. Scott Hall of Basic Sciences, Detroit, MI

Elizabeth Witkoski [233]
The Milton S. Hershey Medical Center of the Pennsylvania State University, Department of Medicine, Hershey, PA

Ganesa Yogeeswaran, Ph.D. [445]
Department of Cancer Biology, The Salk Institute for Biological Studies, San Diego, CA

Leo R. Zacharski, M.D. [113]
Dartmouth Medical School, Hanover, New Hampshire, and Veterans Administration, Medical and Regional Office Center, White River Junction, VT

Participants

Julian L. Ambrus, M.D., Ph.D.
Roswell Park Memorial Institute and State University of New York at Buffalo, Departments of Internal Medicine, Experimental Pathology and Pediatrics, Buffalo, New York

Eva Bastida, Ph.D.
Servicio de Hemoterapia y Hemostasis, Hospital Clinico y Provincial, Facultad de Medicina, Universidad de Barcelona, Barcelona 36, Spain

Dale H. Cowan, M.D., J.D.
Division of Hematology/Oncology, Saint Luke's Hospital, Case Western Research University, Cleveland, Ohio

H. James Day, M.D.
Department of Medicine and Pathology, Clinical Laboratories, Temple University Health Science Center, Philadelphia, Pennsylvania

M.B. Donati, M.D., Ph.D.
Istituto di Richerche Farmacologiche, "Mario Negri," 20157 Milano - Via Eritrea, 62, Italy

Jørgen J. Fogh, M.D.
Human Cell Laboratory, Sloan-Kettering Institute for Cancer Research, Rye, New York

Gabriel J. Gasic, M.D.
Department of Pathology, School of Medicine, University of Pennsylvania, Philadelphia, Pennsylvania

Helmuth Gastpar, M.D., D. Sc.
University of Munich Medical School, 8000 Munich 2, West Germany

Nabil Hanna, Ph.D.
Cancer Metastasis and Treatment Laboratory, NCI Frederick Cancer Research Center, Frederick, Maryland

P. Hilgard, M.D.
Bristol-Meyers International Corporation, Clinical Research (EMEA), Chaussee de la Hulpe, 185, 1170 Brussels, Belgium

John C. Hoak, M.D.
Division of Hematology-Oncology, Department of Medicine, University of Iowa Hospitals, Iowa City, Iowa

Kenneth V. Honn, Ph.D.
Departments of Radiology and Radiation Oncology, School of Medicine, Wayne State University, Detroit, Michigan

G.A. Jamieson, Ph.D., D.Sc.
American Red Cross Blood Services, Bethesda, Maryland

Simon Karpatkin, M.D.
Department of Medicine, New York University Medical School, New York, New York

Randall H. Kramer, Ph.D.
Departments of Anatomy and Oral Medicine, University of California, San Francisco, California

Rosemarie Lichtner, Ph.D.
Dr. Karl Thomae GmbH, Biologische Forschung, Birkendorfer Strasse 65, D-7950 Biberach, West Germany

Lance A. Liotta, M.D., Ph.D.
National Cancer Institute, Laboratory of Pathophysiology, Bethesda, Maryland

Allan Lipton, M.D.
The Milton S. Hershey Medical Center of The Pennsylvania State University Department of Medicine/Oncology Division, Hershey, Pennsylvania

Rosalin Wu Lowenhaupt, M.D.
Department of Physiology, University of Cincinnati, College of Medicine, Cincinnati, Ohio

S.F. Mohammad, Ph.D.
Department of Pathology, College of Medicine, University of Utah, Salt Lake City, Utah

Antonio Ordinas, M.D.
Servicio de Hemoterapia y Hemostasia, Hospital Clinico y Provincial, Facultad de Medicina, Universidad de Barcelona, Barcelona 36, Spain

Edward Pearlstein, Ph.D.
Department of Medicine, New York University Medical School, New York, New York

Manfred Steiner, M.D., Ph.D.
Division of Hematologic Research, The Memorial Hospital of Pawtucket, and Brown University, Providence, Rhode Island

E.V. Sugarbaker, M.D.
Miami Cancer Institute, Miami, Florida

Alexander G.G. Turpie, M.B., F.R.C.P., F.A.C.P.
Department of Medicine, McMaster University and Hamilton Civic Hospitals, Hamilton, Ontario, Canada

Raymond P. Warrell, Jr., M.D.
Developmental Chemotherapy Service, Memorial Sloan-Kettering Cancer Center, New York, New York

Leo R. Zacharski, M.D.
Dartmouth Medical School, Hanover, New Hampshire, and Veterans Administration, Medical and Regional Office, White River Junction, Vermont

INTRODUCTORY REMARKS

G. A. Jamieson
American Red Cross Blood Services
Blood Services Laboratories
9312 Old Georgetown Road
Bethesda, Maryland 20814

The hematogenous spread of cancer involves the seeding of cells from the primary tumor, their transport in the circulation, their attachment at a distant site and their extravascular migration and growth: platelets may be involved in three of these stages, namely, transport, attachment and growth.

Qualitative observations of platelet-tumor cell interaction were first made postmortem over a century ago and have been confirmed numerous times in a wide variety of studies in human and animal pathology. About twenty years ago the Gasics made the important quantitative observation that the incidence of tumor metastasis in mice was inversely related to the platelet count and showed that there was a rough correlation between the ability of various tumor cells to aggregate platelets in vitro, the number of lung metastases produced in mice and the beneficial effect of thrombocytopenia in reducing metastasis. At the present time there has been a marked increase in interest on the possible role of platelets in the spread of tumors. Intensive studies have been carried out to elucidate the mechanisms of tumor cell-induced aggregation and the possible role of plasma components, and new model systems have been described for the evaluation of tumor cell attachment and the extravasation of tumor cells. Other studies have shown that platelets contain growth factors for tumor cells which may differ from the growth factors which have been described for normal smooth muscle cells.

The Chespeake Conference on the "Interaction of Platelets and Tumor Cells" was intended to bring together investigators in this rapidly developing area so as to promote and exchange information between scientists carrying out basic studies on tumor cell biology, clinicians involved in clinical trials and oncologists involved in cancer treatment. The superb weather and the pleasant surroundings for the Chesapeake Conference helped to stimulate this type of interaction leading to a number of collaborative projects.

We are particularly grateful to Baxter Travenol Laboratories, Inc. and The Becton, Dickinson Foundation for providing substantial funding for this conference. Additional funds were generously provided by Abbott Laboratories, Bayer/Cutter Miles Organization, Chrono-Log corporation, CIBA-GEIGY Corporation, Electro-Nucleonics, Inc., Endo Laboratories, Inc. (A subsidiary of E. I. Du Pont de Nemours), General Diagnostics (Division of Warner-Lambert), Haemonetics Research Institute, McNeil Pharmaceutical, Organon Pharmaceuticals (A Division of Organon, Inc.) and Syntex Research.

I am also grateful to the members of the Organizing Committee for their assistance and, in particular, to my assistant editor, Ms. Alice R. Scipio, for her many contributions to the smooth running of the conference and to the preparation of this volume.

Our thanks to Alan R. Liss, Inc., for assisting with the rapid publication of these proceedings.

A CLASSIFICATION OF DISORDERS OF HEMOSTASIS AND THROMBOSIS IN PATIENTS WITH MALIGNANCY

Alvin H. Schmaier and H. James Day

Dept. of Medicine, Hematology/Oncology Section, Dept. of Pathology, Clinical Laboratories and Thrombosis Research Center, Temple University Health Science Center, Philadelphia, PA. 19140

The patient with a malignancy may manifest a multitude of problems that impinge on the hemostatic system. Hemostatic problems may run the gamut from bleeding to thrombosis. These abnormalities may be the first sign of an underlying malignancy pointing the physician towards the diagnosis. Hemostatic defects may also occur as the result of therapy for the malignancy, thus complicating patient management. This communication attempts to provide a clinically-oriented framework in approaching the cancer patient with a hemostatic or thrombotic disorder. Disorders of hemostasis and thrombosis occurring in the patient with malignancy will be classified according to their pathogenetic defect. This classification does not pretend to be complete; rather it is a working hypothesis which is flexible enough to allow growth and evolution as improved understanding of the workings of the hemostatic system occurs. Although present knowledge is greater concerning those specific defects which allow hemorrhage, equally specific information as to the pathogenesis of thrombosis is beginning to be recognized.

It is appropriate at this time to subclassify prehemorrhagic and hypercoagulable states into defects of either plasma proteins, platelets and endothelial cell function (Table 1).

Table 1

DISORDER OF HEMOSTASIS AND THROMBOSIS

Abnormalities in plasma proteins

Abnormalities of platelets

Abnormalities of endothelial cell function

This classification is both pathogenetic and clinical in orientation, and is based upon the framework of the evolution of clinical coagulation testing. Using present clinical coagulation screening tests, disorders of hemostasis can be classified as to an abnormality of the activated partial thromboplastin time or prothrombin time (plasma protein disorders) or bleeding time (platelet or endothelial cell disorders). With present techniques, disorders of thrombosis have no equally recognized laboratory screening tests. However, some disorders can be recognized by specific testing for abnormalities in plasma proteins and platelets which allows one to classify thrombotic disorders in this manner.

BLEEDING DISORDERS IN THE CANCER PATIENTS.

Abnormalities of Plasma Proteins that are Associated with Hemorrhage.

Clinically recognized prolongations of the activated partial thromboplastin time (APTT) and prothrombin time (PT) in patients with malignancy have been shown to be due to a wide variety of defects. With the advent of one-stage factor assays and widely available factor deficiency plasmas, specific clotting factor deficiencies have been recognized in patients with malignancy (Table 2).

Table 2

SPECIFIC COAGULATION FACTOR DEFECTS IN
PATIENTS WITH MALIGNANCIES

True Deficiency
 Factor V
 Prekallikrein
 Factors X, IX, VII

Factor Inhibitors
 Factor VIII
 Factor VII
 Fibrinogen

Abnormal Factor
 Fibrinogen

Defects in plasma coagulation factors in patients with malignancy have been recognized to be true deficiencies, secondary to acquired inhibitors, or presumed synthesis of an abnormal molecule. Factor V deficiency has been recognized in chronic myelogenous leukemia in patients with a large myeloid-megakaryocytic cell mass (Hasegawa et al, 1980). Factor V deficiency in these patients was a true deficiency that corrected with fresh frozen plasma and cells. Plasma prekallikrein deficiency has been reported in a patient with chronic lymphocytic leukemia (Waddell et al, 1980). This patient had less than 1% plasma prekallikrein by coagulant and radioimmunoassay. Acquired Factor X deficiency has been recognized in patients with primary or immunoglobulin-type amyloidosis (Howell, 1963). Usually this factor X deficiency is an isolated abnormality; however, some patients have been described with combined Factor X and IX deficiency (McPherson et al, 1977; Griepp et al, 1979) or Factor X and VII deficiency (Furie et al, 1981).

Inhibitors to specific clotting factors have been recognized in patients with malignancy. Specific factor VIII inhibitors have been recognized in patients with macroglobulinemia (Castaldi and Penny, 1970), myeloma (Glueck and Hung, 1974) and lymphoma (Wenz and Friedman, 1974). In addition in one myeloma patient, macroglobulin coating of lymphocytes was reported to adsorb plasma Factor

VIII allowing for a hemorrhagic syndrome which cleared after splenectomy (Brody et al, 1979). Patients with myeloma have also been recognized to have depression of more than one clotting factor (Perkins et al, 1970). The presence of these pan-coagulation inhibitors, in our hands, has often been a clue to an underlying plasma cell dyscrasia. When looked for, these patients often have a lupus-like anticoagulant, as assessed by the tissue thromboplastin inhibition assay (Schleider et al, 1976). The bleeding in these patients is often difficult to ascribe to this pan-inhibitor since these patients often have other reasons for hemorrhage (to be enumerated below). A Factor VII deficiency due to an apparent immunoglobulin inhibitor has been reported in a patient with bronchogenic carcinoma (Campbell et al, 1980). We had the opportunity to study another patient with a myeloproliferative disorder who had an apparent factor VII deficiency. On plasma mixing experiments, patient plasma inhibited normal plasma factor VII coagulant activity in factor VII deficient plasma.

Dysfibrinogenemias have been recognized in patients with malignancy. Although the defect appears to be one of delayed polymerization of the fibrin monomer, the genesis of this defect apparently can have different forms. Gralnick et al (1978) found four of seven patients with malignant hepatoma who had increased carbohydrate content on the fibrinogen molecule. Enzymatic cleavage of the sialic acid from the abnormal fibrinogen restored fibrinogen function to normal. Coleman et al (1972) studied a group of myeloma patients in whom they found that purified IgA and IgG bound to certain regions of the growing fibrin chain interfering with clot formation and retraction. The inhibitory activity of these intact, whole immunoglobulins resided in the Fab portion of the molecule.

The ubiquitous effects of myeloma proteins on hemostasis (Table 3) can appear as an apparent dysfibrinogenemia. Recently, Khoory et al (1980) described a patient with an IgA kappa type M protein who had a prolonged thrombin time, but normal reptilase time. The thrombin time was corrected by the addition of protamine sulfate and Platelet Factor 4 (PF_4). This material was isolated on a PF_4-Sepharose affinity column, suggesting that it was a proteoglycan. Electrophoretic analysis suggested that the material was of the heparan-sulfate type and functionally it had one-tenth the specific activity of commercially available heparin.

In addition, it was indistinguishable from heparin in its ability to act as a cofactor in the antithrombin III inhibition of thrombin.

Table 3

HEMOSTATIC DEFECTS ASSOCIATED WITH MYELOMA

Abnormalities in Plasma Proteins
    Factor X, IX, VII Deficiency
    Factor VIII Inhibitors
    Pan-Inhibitors
    Dysfibrinogenemia
    Heparin-like Anticoagulant

Abnormalities in Platelets
    Thrombocytopenia

Abnormalities in Platelets that are Associated with Hemorrhage.

Thrombocytopenia is the most common cause of bleeding in patients with both solid tumors and hematologic malignancies. Thrombocytopenia can result from marrow replacement by tumor or the result of cytotoxic chemotherapy. Thrombocytopenia may also result secondary to hypersplenism due to the malignant process infiltrating the spleen (Marymount and Gross, 1963; Slichter and Harker, 1974). Recently, immune thrombocytopenia has been recognized in the cancer patient. Immune thrombocytopenia has been a recognized problem in patients with lymphomas, occurring often as the initial manifestation of the disorder (Fink and Al-Mondhiry, 1976). The fact that non-lymphoid tumors can be associated with immune thrombocytopenia has been recently reported (Kim and Boggs, 1979; Spivack et al, 1979). Immune thrombocytopenia associated with carcinomas may antedate recognition of the underlying malignancy. The response to steroids in all these conditions have been variable.

Apart from thrombocytopenia, malignancies of hematopoietic tissues have been associated with abnormal platelet function (Table 4). These functional abnormalities of platelets can occur with low, normal, or high platelet counts. Actually, thrombocytosis itself is a common accompaniment of cancer (Levin and Conley, 1964). Thrombocytosis

---

Table 4

ABNORMALITIES IN PLATELETS LEADING TO
HEMORRHAGE IN PATIENTS WITH MALIGNANCIES.

Thrombocytopenia
    Marrow Infiltration
    Chemotherapy
    Hypersplenism
    Immune

Thrombocytopathia
    Aggregation/Secretion
    Platelet Coagulant Activity
    Thromboxane Synthesis
    Storage Pool Disorder
    Antibiotics
    DIC

---

in patients with myeloproliferative disorders paradoxically has been associated with bleeding and thrombosis (Zucker and Mielke, 1972; Walsh et al, 1977). In general, the finding of abnormal platelet function has been described as a discriminatory test between primary thrombocythemia and secondary reactive thrombocytosis (Ginsberg, 1975). A number of large patient population studies have shown that abnormal epinephrine-induced platelet aggregation occurred in platelets from patients with myeloproliferative disorders (Zucker and Mielke, 1972; Spaet et al, 1969; Adams et al, 1974; Boneu et al, 1980). However, variability in abnormalities of the bleeding time, ADP- and collagen-induced aggregation also occurred in this population. The study of Kaywin et al (1978) discriminates between patients with essential thrombocythemia who don't respond to epinephrine from those who do. They found that patients who did not respond to epinephrine had a deficiency of the α-adrenergic receptor on their platelets; patients who did respond had normal ^{3}H-dihydroergocryptine, an α-adrenergic antagonist, binding to platelets. Other authors have suggested alternative mechanisms of abnormal platelet function in patients with myeloproliferative disorders. Walsh et al (1977) showed that two such patients with an associated hemorrhagic defect had a deficiency of either platelet contact forming activity or collagen-induced coagulant activity. Semeraro et al (1979) showed that five patients with bleeding tendency had markedly reduced factor X activating activity. Russell et al (1979) studied the platelets of two preleukemic patients who had a hemorrhagic tendency and found normal antigenic amounts but low functional activity of their platelet thromboxane A_2. Likewise, abnormalities of platelet function associated with normal or reduced intracellular levels of adenine nucleotides and diminished release of storage pool nucleotides have been reported in acute leukemias and preleukemic states (Cowan et al, 1975; Sultan and Caen, 1972).

Patients with malignancies undergoing chemotherapy often develop granulocytopenia and fever which is treated with a variety of antibiotics. In addition to intrinsic platelet function defects, certain antibiotics have the propensity to produce acquired defects in platelet function. These include the natural and synthetic penicillins. High-dose penicillins, especially the semi-synthetic penicillins, carbenicillin and ticarcillin, will cause in vivo prolongation of the bleeding time, that increases the risk for hemorrhage in these often already thrombocytopenic patients (Brown et al, 1974). The mechanism of this functional defect appears to be impairment of binding of the platelet agonist (ADP, epinephrine and von Willebrand's factor) to the platelets surface membrane (Shattil et al, 1980).

Disseminated intravascular coagulation (DIC) occurring in cancer patients can cause abnormalities in both plasma coagulation proteins and platelets. The mechanism has been recently reviewed by Colman et al (1979). Clinical manifestations of tumor associated DIC in one large series range from hemorrhage in three quarters of the individuals to venous or arterial thrombosis in the other remaining quarter (Al-Mondhiry, 1975). One can theoretically interpret the diversity of manifestations of tumor associated DIC as to the amount of thromboplastin and plasminogen activator injected into the vasculature and the rate of clearance of each. The best examples of acute hemorrhagic DIC associated with malignancy are those associated with acute promyelocytic leukemia and following surgical manipulation in prostate cancer.

In acute promyelocytic leukemia on presentation or with induction chemotherapy, presumed massive amounts of leukocytes' lysosomal granular material with clot-promoting activity requiring Factor VII or granulocytic proteases are released into the circulation (Gralnick and Abrell, 1973; Gouault-Heilmann et al, 1973; Egbring et al, 1977). This massive release of thromboplastic material results in depletion of coagulation factors and coupled with concurrent thrombocytopenia due to the leukemia leads to a severe, hemorrhagic coagulopathy. Heparin therapy given to these patients appears to be beneficial (Collins et al, 1978; Drapkin et al, 1978). Surgical manipulation of carcinoma of the prostate also results in acute DIC (Tagnon et al, 1952; Pellman et al, 1966). However, the release of thromboplastic material from the surgical assault on the prostate is probably more related to the nature of the glandular material than carcinomatous involvement. Studies have shown that coagulation parameters will be altered in patients with either prostatic cancer or benign prostatic hypertrophy (Friedman et al, 1969; Betkurer et al, 1979). Besides leukemia and prostatic cancer, hemorrhagic DIC has also been associated with adenocarcinomas of gastric, pancreatic, lung, biliary, ovarian and colonic origin, fibrosarcomas, neuroblastomas, lymphomas, melanomas, and Wilms tumor (Goodnight, 1974; Susens et al, 1976).

One variant of acute DIC associated with malignancy is the concurrent presence of microangiopathic hemolytic anemias. This syndrome which usually occurs with metastatic adenocarcinomas may be associated with either acute hemorrhage (Goodnight, 1974; Frick, 1956) or with thrombosis (Brain et al, 1970; Cohan et al, 1972). The genesis of the intravascular hemolysis has been proposed to be due to either the laying down of fibrin, due to intravascular coagulation on small arterioles or tumor emboli itself (Cohan et al, 1972), both of which tend to shear passing red cells.

Abnormalities in Endothelium Cell-Platelet Interactions.

The bleeding time serves as a screening parameter for platelets and endothelial integrity. Although platelet abnormalities most commonly will prolong the bleeding time in patients with cancer, reports now exist of acquired von Willebrand's disease in patients with malignancy. Three patients, one with chronic lymphocytic leukemia, one with

poorly differentiated lymphocytic lymphoma, and one with
hairy cell leukemia (Wautier et al, 1976; Joist et al, 1978;
Roussi et al, 1980) have been described who had decreased
FVIIR:Ag, qualitatively less anodic FVIIIR:Ag
and decreased FVIIIR:RCO activity. Treatment of the underlying malignancy improved or corrected these abnormalities
in each of these patients.

THROMBOTIC DISORDERS IN PATIENTS WITH CANCER.

Thrombosis has been considered a common hemostatic
abnormality associated with cancer. However, there is little
clear information as to the incidence of malignancy in
patients presenting with thrombosis (Table 5)

Table 5

ASSOCIATION OF THROMBOSIS AND CANCER.

Incidence of Occult Malignancy in Thrombosis

Study	Incidence
Ackerman and Estes (1951)‡	5.8%
Anylan et al (1956)	0%
Lieberman et al (1961)	2.2%
Pineo et al (1974)	0.5%

Incidence of Thrombosis in Patient with Malignancy.

Study	Incidence
Pineo et al (1974)	22.9%*
Negus et al (1980)	22.0%

‡Dates in parentheses represent bibliography references.
*Value represents incidence after control group (non-
malignant patients who developed thrombosis) subtracted
out.

All studies have been retrospective relying on the clinical
diagnosis of thrombosis. Ackerman and Estes (1951) followed
up 88 cases of idiopathic thrombophlebitis after five and
ten years; 5.8 percent of these patients turned out to have
an occult neoplasm. Anylan et al (1956) studied 301 consecutive cases of venous thromboembolism. In 14.5 percent of
these patients neoplasm was clinically evident at the time
of thrombosis diagnosis; followup studies on the rest of

these patients revealed no occult malignancies. In a study of 1,400 patients with venous thrombosis, Lieberman et al (1961) noted a 2.2 percent incidence of occult cancers. In another study (60) of 200 nonsurgical patients presenting with venous thrombosis, Pineo et al (1974) found only one patient to have an occult malignancy. These workers also studied the incidence of post-operative thrombosis diagnosed by ^{125}I-fibrinogen scans in 164 patients undergoing abdomino-thoracic surgery for malignant and non-malignant conditions (Table 5). In the non-cancer group, 10.4% had post-operative thrombosis; in the cancer group, 33.3% had thrombosis. This difference was highly significant (p <0.005). This increased risk for the development of post-operative venous thrombosis in patients with malignancy was echoed recently in a trial (Negus et al, 1980) of low-dose intravenous heparin prophylaxis. In the control group, 8 of 24 patients (33.3%) with malignancy developed post-operative deep venous thrombosis, diagnosed by ^{125}I-fibrinogen scan. In the heparin group, DVT developed in 1 of 17 patients (6%) with malignant disease (p <0.025). The data taken as a whole leads to the conclusion that presentation of thrombosis alone does not warrant investigation for occult malignancy. On the other hand, patients with known malignancy may appear to be at increased risk for thromboembolism.

Alteration of Plasma Coagulation Proteins

Activation of coagulation factors would appear to be an important factor in the pathogenesis of venous thrombosis. Experimentally, it has been shown that there is a marked difference in thrombogenicity between activated and nonactivated clotting factors, especially in the presence of venous stasis (Wessler and Yin, 1968). Clinically, the best example of activation of coagulation factors associated with malignancy is chronic DIC.

Chronic DIC is recognized clinically as a compensated intravascular coagulation with mildly depressed platelets and elevated fibrin degradation products, but with normal fibrinogen and coagulation factors (Sun et al, 1974). Clinically the consequence of chronic DIC associated with malignancy can manifest itself as migratory thrombophlebitis (Trousseau, 1865), marantic endocarditis (Reagen and Okazaki, 1974) or microangiopathic hemolytic anemias (Frick, 1956; Brain et al, 1970; Cohan et al, 1972). These three hypercoagulable manifestations of cancer has been recently reviewed by Sack et al (1977) and have been grouped together under the eponym, Trousseau's syndrome. Eighty percent of the tumors associated with this syndrome have been mucin-secreting adenocarcinomas. Experimentally purified mucin has been shown to contain a factor X activator (Pineo et al, 1973). In addition, a serine protease-like factor X activator has also been isolated from adenocarcinomas (Gordon et al, 1975). The variant of migratory thrombophlebitis may be an early diagnostic clue to an occult malignancy. Characteristically, these patients are notoriously coumadin resistant, which serves as another clue as to the diagnosis. Heparin therapy either intravenously or subcutaneously administered is only a temporizing measure. Control of the patient's recurrent thrombophlebitis depends on treating the underlying malignancy.

Alternatively, patients with cancer may have reduced mechanisms to clear the products of activated coagulation. Although no specific defects have yet been described, one must postulate that this may be at least a contributory mechanism. One recent large study of 73 cancer patients substratified into those with and without DIC showed that cancer patients without DIC but with hepatic metastasis did not have decreased functional or immunologic antithrombin III levels (Kies et al, 1980). No study in cancer patients has yet presented data on fibronectin levels, which recently has been invoked as an important opsonic protein involved in reticuloendothelial clearance function of activated products of coagulation (Niehans et al, 1980; Blumenstock, et al, 1977).

Besides, activation of coagulation proteins, cancer patients may have decreased fibrinolytic activity allowing for thrombosis. This idea has to be weighed against recent evidence invoking activation of the fibrinolytic system as an initiation of tumor metastasis (Malone et al, 1979). Two studies suggest that patients with malignancy do have decreased fibrinolytic capacity. In one study by Ogston and Dawson (1973), fibrinolytic activity in 40 patients with lymphoma assessed by the euglobulin clot lysis time was reduced. In a further study by Rennie and Ogston (1975), patients with disseminated malignant disease had a lower mean fibrinolytic activity again assessed by the euglobulin lysis time than normal controls or patients with localized cancer. Venous occlusion stimulated fibrinolytic activity in all groups tested. Although these studies are suggestive of a relationship between decreased fibrinolysis and thrombosis in cancer patients, no specific fibrinolytic defect in cancer patients has yet been described.

Alterations in Platelets and Endothelium

Patients with myeloproliferative syndromes may have high platelet counts with bleeding or thrombosis. The potential platelet functional defects leading to bleeding has already been discussed. Wu (1978) suggests that bleeding or thrombotic complications due to platelets can be differentiated according to platelet function. He describes six patients with myeloproliferative disorders and thromboembolism, but no bleeding, whose platelets when tested at a concentration of 300,000/mm^3 had increased spontaneous aggregation and circulating platelet aggregates. All of these patients platelets responded "normally" to fixed doses of ADP, epinephrine and collagen and they had normal bleeding times. Walsh et al (1977) described two patients with myeloproliferative disorders and thromboembolism. Both of these patients had significant elevation of platelet coagulant activities. Interestingly, one of Walsh's patients with thrombosis also had a prolonged bleeding time and absent ADP-, epinephrine- and collagen-induced platelet aggregation.

Endothelial disorders in cancer patients leading to thrombosis has yet to be clearly described. Recently, thrombotic thrombocytopenic purpura occurring in a patient with disseminated Hodgkins disease has been recognized (Crain and Chondhury, 1980). This report, of course, may be a chance occurrence between a patient with a malignancy and a disorder of presumed endothelial injury derivation. Paradoxically, the lupus-like anticoagulant which gives a prolonged activated partial thromboplastin time but generally is not associated with a bleeding tendency has been associated with a risk for thrombosis (Mueh et al, 1980). Schleider et al (1976) reported three patients with underlying malignancy and the lupus-like anticoagulant. Cooper et al (1974) purified an IgM molecule from serum in a patient with chronic lymphocytic leukemia and Hodgkins disease at autopsy who had a prolonged prothrombin time and multiple serologic abnormalities. These authors showed that this IgM reacted with a number of phospholipids. Recently, Thiagarajan et al (1980) purified an IgM from a patient with macroglobulinemia and a lupus-like anticoagulant that reacted with the phospholipid component contained within the routine coagulation assay. This work strongly suggests that the lupus anticoagulant is a phospholipid inhibitor that interferes with the coagulation assay, but does not induce bleeding. However, this phospholipid inhibitor may interfere with endothelial prostaglandin synthesis that may predispose to thrombosis. Recently, the IgG fraction from a patient with recurrent arterial thrombosis and a lupus anticoagulant was shown to inhibit secretion of prostacyclin from rat aorta rings and production of 6-keto-PGF$_{1\alpha}$ by cultured bovine endothelial cells (Carreras et al, 1981). The IgG fraction of plasma also contained the lupus anticoagulant.

A summary of the specific hemostatic and thrombotic disorders associated with malignancy is given in Table 6.

Table 6

DISORDERS OF HEMOSTASIS AND THROMBOSIS ASSOCIATED WITH MALIGNANCY.

HEMOSTASIS	THROMBOSIS
Abnormalities of Plasma Proteins	
Clotting Factor Deficiencies V, VIII, X, IX, VII	DIC
Factor Inhibitors	
Abnormal Molecules	
Anticoagulants	
Pan Inhibitors	
Abnormalities of Platelets	
Thrombocytopenia	Hyperactive Platelets
Thrombocytopathia	
Abnormalities of Endothelial Function	
von Willebrand's	Lupus Anticoagulant

Malignancy has many manifestations on the hemostatic system. By using a clinical classification as proposed above, the physician may sort out the varied manners in which abnormal hemostasis and thrombosis can occur in the cancer patient to provide specific, adjunctive therapy in the management of the cancer patient.

ACKNOWLEDGEMENTS

We appreciate Dr. Robert W. Colman's critical review of this manuscript and Ms. Terry Cruice for her superb typing.

This work was supported in part by a NHLBI Clinical Investigator's Award (AHS) and a Temple University Biomedical Research Grant SO7 RRO5417 (AHS).

REFERENCES

Ackerman RF, Estes JE (1951). Prognosis in idiopathic thrombophlebitis. Ann Int Md 34:902.

Adams S, Schultz L, Goldberg L (1974). Platelet function abnormalities in the myeloproliferative disorders. Scand J Haemat 13:215

Al-Mondhiry H (1975). Disseminated intravascular coagulation. Experience in a major cancer center. Thromb Diath Haemorr 34:181.

Anlyan WB, Shingleton WW, DeLaughter, Jr GD (1956). Significance of idiopathic venous thrombosis and hidden cancer. J A M A 161:964.

Betkurer B, Guinan P, Verma A, Morillo E, Ablin RJ, Sparkuhl A, Cochin A (1979). Effect of transurethral resection on coagulation in carcinoma of prostate. Urology XIII:142.

Blumenstock F, Saba T, Weber P, et al (1977). Isolation and biochemical characterization of alpha-2-opsonic glycoprotein from rat serum. J Biol Chem 252:7156.

Boneu B, Nowel C, Sie P, Caranobe C, Combes D, Laurent G, Pris J, Bierme R (1980). Platelets in myeloproliferative disorders. Scan J Haematol 25:214.

Brain MC, Azzopardi JG, Baker LRI, Pineo GF, Roberts PD, Dacie JV (1970). Microangiopathic hemolytic anemia and mucin-forming adenocarcinoma. Brit J Haemat 18:183.

Brody JI, Haidar ME, Rossman RE (1979). A hemorrhagic syndrome in Waldenstrom's macroglobulinemia secondary to immunoadsorption of factor VIII. New Engl J Med 300:408.

Brown CH, Natelson EA, Bradshaw MW, Williams TW, Alfrey CP (1974). The hemostatic defect produced by carbenicillin. New Engl J Med 291:265.

Campbell E, Salahattin S, Mattson J, Walker L, Estry S, Mueller L, Schwartz M, Hampton S (1980). Factor VII inhibitor. Amer J Med 68 962.

Carrecas LO, Machin SJ, Deman R, Defreyn G, Vermylen J, Spitz B, Assche AV (1981). Arterial thrombosis, intrauterine death and "Lupus" anticoagulant: Detection of immunoglobulin interfering with prostacyclin formation. Lancet 1:244.

Castaldi PA, Penny R (1970). A macroglobulin with inhibitory activity against coagulation factor VIII. Blood 35:370.

Cohan M, Pittman G, Hoffman GC (1972). Hemolytic anemia, tumor cell emboli and intravascular coagulation. Arch Path 93:305.

Coleman M, Vigliano EM, Weksler M, Nachman RL (1972). Inhibition of fibrin monomer polymerization by lambda myeloma globulins. Blood 39:210.

Collins AJ, Bloomfield CD, Peterson BA, McKenna RW, Edson JR (1978). Management of coagulopathy during daunorubicin-prednisone remission induction. Arch Int Med 138:1677.

Colman RW, Robboy SJ, Minna JD (1979). Disseminated intravascular coagulation: A reappraisal. Ann Rev Med 30:359.

Cooper MR, Cohen JH, Huntley CC, Waite BM, Spees L, Spurr CL (1974). A monoclonal IgM with antibody-like specificity for phospholipids in a patient with lymphoma. Blood 43:493.

Cowan DM, Graham, Jr RC, Baunach D (1975). The platelet defect in leukemia. J Clin Invest 56:188.

Crain SM, Chondhury AM (1980). Thrombotic thrombocytopenic purpura in a splenectomized patient with Hodgkins disease. Amer J Med Sci 280:35.

Drapkin RL, Gee TS, Dowling MD, Arlin Z, McKenzie S, Kemplin S, Clarkson B (1978). Prophylactic heparin therapy in acute promyelocytic leukemia. Cancer 41:2484.

Egbring R, Schmidt W, Fuchs H, Haveman K (1977). Demonstration of granulocytic proteases in plasma of patients with acute leukemia and septicemia with coagulation defects. Blood 49:219.

Fink K, Al-Mondhiry H (1976). Idiopathic thrombocytopenic purpura in lymphoma. Cancer 37:1999.

Frick PG (1956). Acute hemorrhagic syndrome with hypofibrinogenemia in metastatic cancer. Acta Haemat 16:11

Friedman NJ, Hoag MS, Robinson AJ (1969). Hemorrhagic syndrome following transurethral prostatic resection for benign adenoma. Arch Int Med 124:341.

Furie B, Voo L, Keith PW, McAdam K, Furie BC (1981). Mechanism of Factor X deficiency in systemic amyloidosis. New Engl J Med 304:827.

Ginsberg AD (1975). Platelet function in patients with high platelet counts. Ann Int Med 82:506.

Glueck HI, Hung R (1974). A circulating anticoagulant in γ1A-multiple myeloma: Its modification by penicillin. J Clin Invest 44:1866.

Goodnight SH (1974). Bleeding and intravascular coagulation in malignancy: A review. Ann N Y Acad Sci 230:271.

Gordon SG, Franks JL, Lewis B (1975). Cancer procoagulant A: A factor X activating procoagulant from malignant tissue. Thromb Res 6:127.

Gouault-Heilmann M, Chardon E, Sultan C (1973). The procoagulant factor of leukemia promyelocytes: Demonstration of immunologic cross-reactivity with human brain tissue factor. Br J Haemost 30:151.

Gralnick HR, Abrell E (1973). Studies of the procoagulant and fibrinolytic activity of promyelocytes in acute promyelocytic leukemia. Brit J Haemat 24:89.

Gralnick HR, Givelber H, Abrams E (1978). Dysfibrinogenemia associated with Hepatoma. New Engl J Med 299:221.

Griepp PR, Kyle RA, Bowie EJW (1979). Factor X deficiency in primary amyloidosis. New Engl J Med 301:1050.

Hasegawa DK, Bennett AJ, Coccia PF, Ramsay NK, Nesbit ME, Krivit W, Edson JR (1980). Factor V deficiency in Philadelphia-positive chronic myelogenous leukemia. Blood 56:585.

Howell M (1963). Acquired factor X deficiency associated with systematized amyloidosis - A report of a case. Blood 21:739.

Joist JH, Cowan JF, Zimmerman TS (1978). Acquired von Willebrand's disease. New Engl J Med 298:988.

Kaywin P, McDonough M, Insel PA, Shattil SJ (1978). Platelet function in essential thrombocythemia. New Engl J Med 299:505.

Khoory MS, Nesheim ME, Bowie EJW, Mann K (1980). Circulating heparin sulfate proteoglycan anticoagulant from a patient with a plasma cell disorder. J Clin Invest 65:666.

Kies MS, Posch JJ, Giolma JP, Rubin RN (1980). Hemostatic function in cancer patients. Cancer 46:831.

Kim HD, Boggs DR (1979). A syndrome resembling idiopathic thrombocytopenic purpura in 10 patients with diverse forms of cancer. Amer J Med 67:371.

Levin T, Conley CL (1964). Thrombocytosis associated with malignant disease. Arch Int Med 114:497.

Lieberman JS, Borrero J, Urdaneta E, et al (1961). Thrombophlebitis and cancer. J A M A 177:72.

Malone JM, Wangensteen SL, MOore WS, Keown K (1979). The fibrinolytic system. A key to tumor metastasis. Ann Surg 190:342.
Marymount JH, Gross S (1963). Patterns of metastatic cancer in the spleen. Am J Clin Pathol 40:58.
McPherson RA, Onstad JW, Ugoretz RJ, Wolf PL (1977). Coagulation in amyloidosis: Combined deficiency of Factors IX and X. Amer J Hemat 3:225.
Mueh JR, Herbst KD, Rapaport SI (1980). Thrombosis in patients with the Lupus anticoagulant. Ann Int Med 92:156.
Negus D, Cox SJ, Friedgood A, Peel ALG (1980). Ultra-low dose intravenous heparin in the prevention of postoperative deep vein thrombosis. Lancet 1:891.
Niehans GD, Schumacker PR, Saba TM (1980). Reticuloendothelial clearance of blood-borne particulates. Relevance to experimental lung microembolization and vascular injury. Ann Surg 191:479.
Ogston D, Dawson AA (1973). The fibrinolytic system in malignant lymphomas. Acta Haemost 49:89.
Pellman CM, Ridlon HC, Phillips LL (1966). Manifestation and management of hypofibrinogenemia and fibrinolysis in patients with carcinoma of the prostate. J of Urology 96:375.
Perkins HA, Mackenzie MR, Fudenberg HH (1970). Hemostatic defects in dysproteinemias. Blood 35:695.
Pineo GF, Regoeczi E, Hatton MWC, Brain MC (1973). The activation of coagulation by extracts of mucus: A possible pathway of intravascular coagulation accompanying adenocarcinomas. J Lab Clin Med 82:255.
Pineo GF, Brain MC, Gallus AS, Hirsh J, Hatton MWC, Regoeczi E (1974). Tumors, mucous production and hypercoagulability. Ann N Y Acad Sci 230:262.
Reagan JJ, Okazaki H (1974). The thrombotic syndrome associated with carcinoma: A clinical and neuropathologic study. Arch Neurol 31:390.
Rennie JAW, Ogston D (1975). Fibrinolytic activity in malignant disease. J Clin Path 28:872.
Roussi JH, Houbouyan LL, Alterescu R, Franc B, Goguel AF (1980). Brit J Haemost 46:503.
Russell NH, Keenan JP, Bellingham AJ (1979). Thrombocytopathy in preleukemia: Association with a defect of thromboxane A_2 activity. Brit J Haemat 41:417.
Sack G, Levin J, Bell WR (1977). Trousseau's syndrome and other manifestation of chronic disseminated coagulopathy in patients with neoplasma: Clinical, pathophysiologic and therapeutic features. Medicine 56:1.

Schleider MA, Nachman RL, Jaffe EA, Coleman MA (1976). Clinical study of the Lupus Anticoagulant. Blood 48:499.

Semeraro N, Cortellazzo S, Colucci M, Barbui T (1979). A hitherto undescribed defect of platelet coagulant activity in polycythaemia vera and essential thrombocythaemia. Thromb Res 16:795.

Shattil SJ, Bennett JS, McDonough M, Turbull J (1980). Carbenicillin and Penicillin G inhibit platelet function in vitro by impairing the interaction of agonists with the platelet surface. J Clin Invest 65:329.

Slichter SJ, Harker LA (1974). Hemostasis in malignancy Ann N Y Acad Sci 230:252.

Spaet TH, Lejnieks I, Gaynor E, Goldstein ML (1969). Defective platelets in essential thrombocythemia. Arch Int Med 124:135.

Spivack M, Brenner SM, Markham MJ, Snyder EL, Berkowitz D (1979). Presumed immune thrombocytopenia and carcinoma: Report of three cases and review of the literature. Amer J Med Sci 278:153.

Sultan Y, Caen JP (1972). Platelet dysfunction in preleukemic states and in various types of leukemia. Ann N Y Acad Sci 201:300.

Sun NCJ, Bowie EJW, Kazmier FJ, Elveback LR, Owen CA (1974). Blood coagulation studies in patients with cancer. Mayo Clin Proc 49:636.

Susens GP, Hendrickson C, Barto DA, Sams BJ (1976). Disseminated intravascular coagulation syndrome with metastatic melanoma: Remission after treatment with 5-(3,3-Dimethyl-1-Triazeno) Imidazole-4-carboxamide (DTIC). Ann Int Med 84:175.

Tagnon HJ, Whitmore, Jr WF, Shulman NR (1952). Fibrinolysis in metastatic cancer of the prostate. Cancer 5:9.

Thiagarajan P, Shapiro SS, DeMarco L (1980). Monoclonal immunoglobulin M coagulation inhibitor with phospholipid specificity. J Clin Invest 66:397.

Trousseau A (1865). Clinique Médicale de l'Hôtel-Dieu de Paris. Vol. 3, Fifth edition. Paris, Bailliere et Fils p. 94.

Waddell CC, Brown JA, Udden MM (1980). Plasma prekallikrein (Fletcher Factor) deficiency in a patient with chronic lymphocytic leukemia. S Med J 73:1653.

Walsh PN, Murphy S, Barry WE (1977). The role of platelets in the pathogenesis of thrombosis and hemorrhage in patients with throbocytosis. Thrombos Haemostas 38:105.

Wautier JL, Levy-Toledano S, Caen JP (1976). Acquired von Willebrand's syndrome and thrombopathy in a patient with chronic lymphocytic leukemia. Scan J Haemat 16:128.
Wenz B, Friedman G (1974). Acquired factor VIII inhibitor in a patient with malignant lymphoma. Amer J Med Sci 268:295.
Wessler S, Yin ET (1968). Experimental hypercoagulable state induced by factor X: Comparison of the non-activated and activated forms. J Lab Clin Med 72:256.
Wu KK (1978). Platelet hyperaggregability and thrombosis in patients with thrombocythemia. Ann Int Med 88:7.
Zucker S, Mielke CH (1972). Classification of thrombocytosis based on platelet function tests: correlation with hemorrhagic and thrombotic complications. J Lab Clin Med 80:385.

DISCUSSION

Dr. J. L. Ambrus (Roswell Park Memorial Institute, Buffalo). I would like to summarize very briefly some of our findings in relation to Dr. Day's interesting paper. We have found that tumor cells have very high thromboplastic activity, they have platelet aggregating substances and they have very high plasminogen activator activities. We found that some of the fibrinogen degradation products have specific action on the membrane of red cells in that they affect red cell deformability and membrane fluidity.

When one has thromboembolic disorders the preference is for heparin therapy. On the other hand, we have seen patients who had done extremely poorly with heparin and we found hyperaggregable platelets present. Platelet aggregating ability increases 1-2 fold after heparin.

Dr. Day also pointed out that the incidence of thrombocytosis-related vascular problems is relatively rare. On the other hand, the incidence of hyperaggregable platelets is extremely high. We must point out that the incidence of thromboembolic activity was not quite as high as you would expect in laboratory mice.

INTRA-TUMORAL PLATELET CONSUMPTION: THE KASABACH-MERRITT SYNDROME AS A MODEL FOR MALIGNANT PLATELET DESTRUCTION

Raymond P. Warrell, Jr.

Developmental Chemotherapy Service
Memorial Sloan-Kettering Cancer Center
1275 York Avenue
New York, New York 10021

   Human neoplasia is associated with a wide variety of coagulation disorders including recurrent thrombosis, non-bacterial thrombotic endocarditis and disseminated intravascular coagulation. Moreover, cancer patients are known to have shortened platelet survival even in the absence of factors such as hemorrhage or sepsis (Slichter and Harker, 1974; Abrahamsen, 1972). Although in most instances platelet destruction appears to be related to a consumptive coagulopathy, platelet and fibrinogen survival in some cancer patients exhibit considerable independence. This independent relation has been recently reported in both canine and human neoplasia (Williams et al, 1980; O'Donnell et al, 1981). Thus, a shortened platelet survival time is not necessarily associated with an increase in fibrinogen turnover.

   The mechanism for this accelerated platelet destruction has never been satisfactorily explained although several possibilities have been proposed. Autoimmune thrombocytopenia is associated with several malignant diseases (Kim and Boggs, 1979); however, it seems unlikely that autoimmune sensitization occurs in more than a small percentage of cases (principally patients with malignant lymphoma). Inapparent intravascular coagulation could account for increased platelet destruction, especially since tumor cells are known to release a variety of procoagulant substances (Gordon et al, 1979; Dvorak et al, 1981) Such intravascular coagulation might not be detected owing to the insensitivity of routinely measured clotting tests. This mechanism, however, implies increased fibrinogen turnover which, as noted, is a common but not essential feature of malignant platelet consumption.

Selective consumption of platelets could occur by two other mechanisms. The vascular supply of many tumors is extremely disordered and microscopic examination of these abnormal vessels reveals multiple gaps in their endothelial lining. Conceivably, platelets could be damaged and removed by exposure to poorly endothelialized surfaces; such a process would be exacerbated by inefficient tumor perfusion. Alternatively, tumor cells are now known to shed substances from their cell membranes which cause the direct aggregation of human platelets (Pearlstein et al, 1979; Hara et al, 1980). Exposure of platelets to such materials could likewise produce aggregation without activation of clotting reactions. All of these factors are closely related and each factor probably contributes to platelet and fibrinogen consumption, to some degree, in a given individual.

Tumor cells are known to elaborate procoagulants and platelet aggregating materials, and fibrin deposition around growing tumors is a well-recognized phenomenon (Hiramoto et al, 1960). Several investigators have reported localization of antifibrin antibodies or radiolabeled fibrinogen at sites of tumors in humans (Dewey et al, 1963; DeNardo et al, 1979). Although the significance of these findings has recently been challenged (Petersson, 1979), these data suggest that coagulation reactions and platelet consumption occur at the site of the tumor, that is a _localized_ rather than a "disseminated" consumption.

The importance of fibrin deposition around tumors has been addressed at length in recent publications (Dvorak et al, 1979; Zarcharski et al, 1979); however, localized platelet consumption may represent a more important biologic process. Platelets undergoing aggregation are known to secrete factors which stimulate the growth of vascular smooth muscle cells (Ross et al, 1974). The release of these factors has been implicated in the pathogenesis of vascular disease, especially atherosclerosis. It is quite possible that release of these factors at the periphery of growing tumors accelerates neovascularization of the tumor. Platelets are also known to secrete growth factors for virally transformed (Kepner and Lipton, 1981) and neoplastic cells (Hara et al, 1980). Moreover, platelets appear to facilitate tumor cell metastasis (Gasic et al, 1973). Thus, localized intratumoral platelet aggregation could affect the growth, metastasis, and neovascularization of malignant tumors.

THE KASABACH-MERRITT SYNDROME

The association of thrombocytopenia with hemangiomata was first noted 40 years ago. Most cases appear in childhood, particularly young infants, and the syndrome is quite rare in adults. Many patients have an associated consumptive coagulopathy (Shim, 1969) although this is not an invariable finding. Furthermore, eradication of the tumor by surgery or radiation causes the coagulopathy to disappear and the platelet count to normalize. Thus, despite the evidence for intravascular coagulation in many of these cases, it is widely held that the thrombocytopenia is secondary to intra-tumoral platelet entrapment.

Attempts to document such localized consumption using radiolabeled platelets have produced conflicting results (Blix and Aas, 1961; Brizel and Raccuglia, 1965; Propp and Scharfman, 1966). Some patients who did manifest a local increase in platelet-associated radioactivity had recently received radiation to their tumor (Brizel and Raccuglia, 1965; Propp and Scharfman, 1966). Few studies allowed for an increased blood content within these vascular tumors and all previous studies used allogeneic platelets. To some extent, these conflicting data have resulted from the inherent limitations of the most commonly used isotope for platelet labeling, chromium-51. The newly-developed isotope chelate, indium-111 oxine, offers several major advantages over ^{51}Cr as a platelet label: a higher yield of gamma photons, a higher labeling efficiency, and tight intracellular binding with minimal in vivo reutilization (Joist et al, 1978).

In a manner analgous to previously cited studies which employed radiolabeled fibrinogen as an oncophilic agent, we are using ^{111}Indium-labeled platelets to examine whether intra-tumoral consumption of platelets occurs in patients with neoplastic disease. In the course of our studies in cancer patients, we recently studied an adult patient with the Kasabach-Merritt syndrome. This patient had multiple hemangiomata involving bone and soft tissue. In particular, there were large lesions in the right hemi-thorax and mediastinum along with two large abdominal masses. The patient has marked thrombocytopenic (plts = 52,000/mm^3), a severe consumptive coagulopathy with prolongation of her clotting times, hypofibrinogenemia (45 µg/dl, control = 150-400 µg/dl), and elevated fibrin degradation products (96 µg/ml, control < 10 µg/ml). Using a minor modification of a previously published method (Hawker et al, 1980; Warrell et al, 1981), we injected the patient with 419 µCi of autologous In-labeled platelets obtained from 2 units of whole blood. At 3 hrs, 24 hrs, and 48 hrs following injection, serial gamma camera and rectilinear camera imaging showed prompt accumulation of platelet-associated radioactivity within the intra-thoracic masses. Using computerized subtraction of the sequential images, we showed that the accumulation actually increased over time. This indicates that platelet sequestration occurred as an active process and was not observed secondary to an increased tumor blood content. Another striking observation was the lack of any platelet accumulation in other lesions, particularly the large abdominal masses. We performed a qualitative assessment of tumor blood flow using radionuclide and radiographic contrast angiography. Each of the abdominal and intra-thoracic masses was relatively hypovascular. Besides showing that platelet localization was not due to whole blood content, this indicates a striking heterogeneity of these masses with respect to platelet-trapping ability, even within the same individual.

We then attempted to determine whether platelet consumption in this patient occurred secondary to generation of thrombin or by some other mechanism. The patient was admitted to the hospital and received heparin in a continuous IV infusion for 5 days. Levels of clotting factors gradually increased (fibrinogen, 45→140 mg/dl; Factor VIII:c, 50%→96% of control); however the platelet count failed to increase. These data suggest that platelets were removed by a process that did not require the generation of thrombin since inhibition of the coagulopathy did not correct the thrombocytopenia.

The findings in this patient with Kasabach-Merritt syndrome illustrate several features which we believe may apply more generally to patients with other neoplastic diseases. First: in at least some individuals, thrombocytopenia results from localized intra-tumoral platelet consumption rather than from "disseminated" coagulation reactions.

Second: it appears that tumors are heterogeneous in platelet-trapping ability despite similar histologic appearance. Whether these differences represent clonal evolution from the initially transformed cell or if it results from local changes in tumor blood vasculature is unknown. Third: despite pronounced activation of intrinsic clotting reactions, platelet consumption in this disorder is not simply a function of thrombin-induced aggregation. A similar conclusion was reached by Hoak et al who showed that heparin did not improve the thrombocytopenia produced in an animal model of this disorder (Hoak et al, 1971).

INDIUM-PLATELET STUDIES IN CANCER

Our current investigations will evaluate whether the phenomenon of intra-tumoral platelet consumption occurs more generally in cancer patients and, if so, in which diseases and with what frequency. These studies have only recently begun and any conclusions are premature. We have observed possible tumor localization of ^{111}In-labeled platelets in one patient with non-small cell lung cancer. This case illustrates both the possibilities and limitations of this technology. This patient had a large mass in the left superior mediastinum, enlargement of a right supraclavicular lymph node, and multiple miliary nodules in both lung fields. The platelet scan (Figure 2) showed unusual accumulation of radioactivity in the superior mediastinum above the cardiac blood pool and also in the right neck. (Normal platelet distribution is seen in Figure 1 for comparison.) There are several reservations, however. First, ^{111}In-labeled platelets will accumulate not only at sites of thrombosis (Goodwin et al, 1978) but also in areas of atherosclerotic vascular disease (Davis et al, 1978). Therefore, the platelet radioactivity seen in Figure 2 could have been due to platelet deposition upon atherosclerotic plaques in the ascending aorta and right common carotid artery. Second, because serial scans were not performed, no evaluation can be made of platelet consumption over time. Third, no correction was made for tumor blood content, therefore the apparent localization could have resulted from an increased local amount of whole blood.

These reservations illustrate the important objections to be addressed if this technology can be presumed to demonstrate intra-tumoral platelet localization. Proper patient

Figure 1   Figure 2

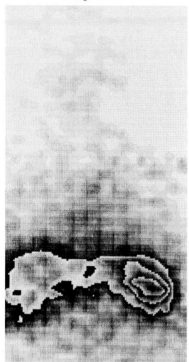 

Figure 1 shows the distribution of ^{111}In-labeled platelets in a normal subject. Activity in the cardiac blood pool is faintly seen above the liver-spleen activity at the center of the image. This scan can be compared with the image obtained from a patient with lung cancer in Figure 2. Increased platelet accumulation is noted in the superior mediastinum above the cardiac pool and in the right neck; both areas were clinically involved with tumor. (Scans obtained at 48 hrs using dual opposing rectilinear cameras.)

---

selection is vital. Individuals undergoing the examination should have clearly localized disease, preferably distant from major arteries (especially in elderly patients). Serial imaging in multiple views must be performed using detectors interfaced with computers to permit digital quantitation of radioemission and image subtraction. It would be desirable to have at least a qualitative assessment of tumor vascularity to be confident that platelet localization was not due to increased tumor blood content.

SUMMARY

The etiology of accelerated platelet destruction in patients with neoplastic disease is undoubtedly multifactorial. I suggest that the florid intra-tumoral consumption seen in patients with the Kasabach-Merritt syndrome may occur to a more subtle degree in many other patients with cancer. This proposal of localized platelet consumption has important theoretical implications as previously noted. Furthermore, there are several therapeutic implications as well. Cancer patients who exhibit this phenomenon may benefit from the use of relatively non-toxic agents which inhibit platelet aggregation. If such localized platelet consumption represents an important biologic process, then those agent(s) which optimize platelet survival in cancer patients should undergo more extensive clinical testing as anti-metastatic/anti-tumor compounds. Indium platelet scanning might also be used to select which cancer patients, if any, would benefit from the use of chemotherapy-loaded platelets (Ahn, et al, 1980).

The current studies using ^{111}In-labeled platelets will only determine whether intra-tumoral platelet consumption occurs to a significant degree in vivo. The results will not allow a choice between alternative mechanisms nor will they demonstrate that such consumption represents a significant factor in tumor biology. These important questions however are undergoing active study by the many other investigators featured in these Proceedings. Hopefully we may expect that these studies will prove beneficial to the increasing numbers of individuals afflicted with cancer.

References

Abrahamsen AF (1972). The effects of acetylsalicylic acid and dipyridamole on platelet economy in metastatic cancer, Scand J Haemat 9:562-565.
Ahn YS, Byrnes JJ, Harrington WJ, Collin A, Pall LM (1980). Treatment of malignancies of the mononuclear phagocyte

system with vinca-laden platelets, Clin Res 28:525 (Abstr)
Blix S, Aas K (1961). Giant hemangioma, thrombocytopenia, fibrinogenopenia, and fibrinolytic activity, Acta Med Scand 169:63-70.
Brizel H, Raccuglia G (1965). Giant hemangioma with thrombocytopenia, Blood 26:751-756.
Davis HH, Siegal BA, Joist JH, Heaton WA, Mathias CJ, Sherman LA, Welch MJ (1978). Scintigraphic detection of atherosclerotic lesions and venous thrombi in man by Indium-111-labelled autologous platelets, Lancet i:1185-1187.
DeNardo SJ, DeNardo GL, Swanson MA, Trelford JD, Krohn KA (1979). Efficacy of highly iodinated fibrinogen in patients with cancer, J Nuc Med 20:645 (Abstr.)
Dewey WC, Bale WF, Rose RG, Marrack D (1963). Localization of antifibrin antibodies in human tumors, Acta Un Int Cancer 19:185-196.
Dvorak HF, Dvorak AM, Manseau EJ, Wiberg L, Churchill WH (1979). Fibrin gel investment associated with line 1 and line 10 solid tumor growth, angiogenesis, and fibroplasia in guinea pigs. Role of cellular immunity, myofibroblasts, microvascular damage and infarction in line 1 tumor regression, J Nat Cancer Inst 6:1459-1467.
Dvorak HF, Quay SC, Orenstein NS, Dvorak AM, Hahn P, Bitzeram, Carvalho AC (1981). Tumor shedding and coagulation, Science 212:923-924.
Gasic GJ, Gasic TB, Galanti N, Johnson T, Murphy S (1973). Platelet-tumor cell interactions in mice. The role of platelets in the spread of malignant disease, Int J Cancer 11:704-718.
Goodwin DA, Bushberg JT, Doherty PW, Lipton MJ, Conley FK, Diamanti CI, Meares CF (1978). Indium-111 labeled autologous platelets for location of vascular thrombi in humans, J Nuc Med 19:626-634.
Gordon SG, Franks JJ, Lewis BJ (1979). Comparison of procoagulant activities in extracts of normal and malignant human tissue, J. Nat Cancer Inst 62:773-776.
Hara Y, Steiner M, Baldini MG (1980). Platelets as a source of growth-promoting factor(s) for tumor cells, Cancer Res 40:1212-1216.
Hawker RJ, Hawker LM, Wilkinson AR (1980). Indium (^{111}In)-labelled human platelets: optimal method, Clin Science 58:243-248.
Hiramoto R, Bernecky J, Jurandowski J, Pressman D (1960). Fibrin in human tumors, Cancer Res 20:592-593.
Hoak JC, Warner ED, Cheng HE, Fry GL, Hankenson RR (1971).

Hemangioma with thrombocytopenia and microangiopathic anemia (Kasabach-Merritt syndrome): an animal model, J Lab Clin Mcd 77:941-950.

Joist JH, Baker RK, Thakur ML, Welch MJ (1978). Indium-111-labeled human platelets: uptake and loss of label and in vitro function of labeled platelets, J Lab Clin Med 92: 829-836.

Kepner N, Lipton A (1981). A mitogenic factor for transformed fibroblasts from human platelets. Cancer Res 41: 430-432.

Kim HD, Boggs DR (1979). A syndrome resembling idiopathic thrombocytopenic purpura in 10 patients with diverse forms of cancer, Am J Med 67:371-377.

O'Donnell MR, Slichter SJ, Weiden PL, Storb R (1981). Platelet and fibrinogen kinetics in canine tumors, Cancer Res 41:1379-1383.

Pearlstein E, Karpatkin S, Salk PL, Yogeeswaran G (1979). Metastatic potential correlates with platelet aggregation ability in vitro and sialic acid content of ten variant cell lines derived from a PW-20 Wistar-Furth renal sarcoma, Blood 54:255 (Abstr).

Petersson H-I (1979). Vascular and extra-vascular spaces in tumors: tumor vascular permeability. In Petersson H-I (ed): "Tumor blood circulation: angiogenesis, vascular morphology and blood flow of experimental and human tumors", Boca Raton: CRC Press pp77-86.

Propp RP, Scharfman WB (1966). Hemangioma-thrombocytopenia syndrome associated with microangiopathic hemolytic anemia, Blood 28:623-633.

Ross R, Glomset J, Kariya B, Harker L (1974). A platelet-dependent serum factor that stimulates the proliferation of arterial smooth muscle cells in vitro, Proc Nat Acad Sci 71:1207-1210.

Shim WKT (1969). Hemangiomas of infancy complicated by thrombocytopenia, Am J Surg 116:896-906.

Slichter SJ, Harker LA (1974). Hemostasis in malignancy, Proc NY Acad Sci, 230:252-262.

Warrell RP, Kempin SJ, Benua RS, Reiman RE, Young CW (1981). Intra-tumoral consumption of Indium-111 labeled platelets in a patient with hemangiomatosis and intravascular coagulation (submitted for publication).

Williams CKO, Pineo GF, Gallus AS, McCulloch (1980). The relevance of platelet and fibrinogen kinetics and coagulation studies to extent of disease and performance status in patients with adenocarcinoma, Med Pediatr Oncol 8:367-374.

Zacharski LR, Henderson WG, Rickles FR, Forman WB, Cornell CJ, Forcier RJ, Harrower HW, Johnson RO (1979). Cancer 44:732-741.

ANTITHROMBOTIC EFFECTS OF DRUGS WHICH SUPPRESS PLATELET FUNCTION: THEIR POTENTIAL IN PREVENTING GROWTH OF TUMOUR CELLS

Alexander G. G. Turpie

Department of Medicine, McMaster University and Hamilton Civic Hospitals
Hamilton, Ontario, Canada

The potential value of drugs that suppress platelet function in the treatment of malignant disease is based on the premise that the spread of malignant cells and the growth of metastases is facilitated by the presence of a platelet-fibrin matrix. Although there is experimental evidence that antithrombotic drugs may be effective in limiting the spread of experimental neoplasms, their potential use in humans with malignant disease has not been adequately evaluated. There is, however, good evidence that a number of drugs that suppress platelet function are effective antithrombotic agents.

The use of drugs that suppress platelet function for the prevention and treatment of thromboembolism is based on the important role that platelets play in the genesis of thrombosis. The antithrombotic affect of these drugs is influenced both by their mode of action on platelets and by the nature of the thrombogenic stimulus which may vary from one thromboembolic condition to another. For these reasons, it is difficult to predict which drugs will be clinically effective on the basis of in vitro tests of platelet function or from experimental models of thrombosis in animals. Therefore, assessment of the efficacy of antiplatelet drugs in thromboembolic disease depends upon the results obtained from well-designed clinical trials. Since the drugs have different modes of action on platelet function, the results obtained from studies with any one drug cannot be used to conclude that similar results would be obtained from other antiplatelet drugs or to predict that the drug would have the same effect for all thromboembolic disorders.

Many compounds have been shown to inhibit platelet function in vitro, and four drugs have been evaluated for antithrombotic effects in clinical trials in humans. These are aspirin, dipyridamole, hydroxychloroquine and sulphinpyrazone, and of these, aspirin, dipyridamole and sulphinpyrazone are the most important. Dextran, a glucose polymer introduced as a volume expander, also inhibits platelet function and has antithrombotic properties, but these properties may not be due to its antiplatelet effect alone. Each of these drugs was initially introduced as a therapeutic agent because of a pharmacological effect other than inhibition of platelet function and it is not known if their antithrombotic effects are due only to the action on platelets or to some other mechanisms.

An understanding of the effects of antiplatelet drugs involves knowledge of the factors that initiate thrombosis and of the different types of thrombosis. Three basic types of thrombi are recognized. In the initial stages of formation, arterial thrombi are mainly composed of platelets and fibrin with a few red and white blood cells whereas venous thrombi are composed predominantly of fibrin and red cells. These are known as white and red thrombi respectively. Mixed thrombi composed of components of both white and red thrombi also occur (Mustard and Packham, 1979). The composition of thrombi is influenced by different mechanisms depending upon the initiating factors which include the site, the patterns and velocity of blood flow and whether the thrombi formed in response to single injury or to repeated injury of the vessel wall. In general plateletrich thrombi form in high flow systems while coagulation thrombi occur in regions of stasis. Antiplatelet drugs are more likely to be effective in preventing the formation of platelet-rich thrombi than red thrombi which are more responsive to anticoagulants.

The key role of platelets in the genesis of thrombosis has recently been defined. When a vessel wall is damaged, platelets adhere to collagen, basement membrane and to the microfibils in the subendothelium, (Baumgartner et al, 1976) release the contents of their granules (Hovig, 1963) and form thromboxane A_2 from arachidonate liberated from membrane phospholipids by phospholipases that are stimulated when platelets adhere to collagen (Smith et al, 1974; Bell et al, 1979). In regions where blood flow is disturbed, thromboxane A_2 and adenosine

diphosphate released from platelets stimulate platelets to
adhere to one another to form a platelet aggregate. The
blood coagulation system is activated in and around the
platelet aggregate and the thrombin that is generated causes
further platelet aggregation, release of platelet granule
contents and the formation of thromboxane A_2. The thrombin
also converts fibrinogen to fibrin which stabilizes the
platelet mass (Mustard and Packham, 1979). Thrombi are not
static structures and may undergo dissolution and reforma-
tion. The conversion of plasminogen to plasmin by activators (Majno and Joris, 1978) results in digestion of fibrin
and under the force of blood flow, platelet emboli or
platelet-fibrin emboli may break off and embolize distally
to the microcirculation. Moncada et al (1976)
demonstrated that prostaglandin precursors PGG_2 and PGH_2
were converted by vessel wall microsomes into a prostaglandin with potent antiplatelet activity. The substance was
named prostacyclin, later prostaglandin I_2 (PGI_2). The
possibility that PGI_2 has a role in limiting thrombus formation is under intense investigation. A number of cell types
in vessel walls can form PGI_2 when stimulated (Moncada and
Vane, 1979) and in regions where blood flow is slow or
arrested, the formation of PGI_2 may limit the size of
thrombi (Kelton et al, 1978). However, in large arteries
where blood flow is rapid, PGI_2 is less likely to provide
a protective effect(Dejana et al, 1980). The interactions
between platelets and tumour cells are less well understood
but might be different than those involving thrombosis.

MODE OF ACTION OF ANTIPLATELET DRUGS

ASPIRIN

Aspirin inhibits platelet adhesion to collagen under
conditions of stasis, in systems of low flow rate and at
low hematocrit levels (Cazenave et al, 1978; Davies et al,
1979). However, aspirin has no effect on platelet adhesion
at physiological rates of shear and at normal hematocrit
levels (Malmgren et al, 1979). The effect of aspirin on
platelet aggregation in-vitro is influenced by its concentration, the duration of incubation with platelets and the
aggregating simulus used (Weiss and Aledort, 1967). With
adrenalin or adenosine diphosphate (ADP), aspirin (5-10 μM)
inhibits the second phase of platelet aggregation.
Following 15 minutes incubation with platelets, aspirin

(40 uM) inhibits collagen-induced platelet aggregation (Zucker and Peterson, 1970). With shorter incubation periods, higher concentrations of aspirin are required to produce equivalent degrees of inhibition of collagen-induced platelet aggregation. Unlike other non-steroidal antiinflammatory drugs, repeated washing of platelets exposed to aspirin does not influence its inhibitory effect. In addition, aspirin inhibits the release reaction induced by antigen-antibody complexes gamma globulin-coated surfaces and low concentrations of thrombin. In humans, inhibition of platelet function can be measured ex-vivo at doses as low as 50 mg. of aspirin. The antiplatelet effect of aspirin is maintained over long periods indicating that the aspirin effect is not overcome by enzyme induction. The effect of aspirin on platelet function requires the acetate radical. In vitro, sodium salicylate can inhibit platelet function but the concentration required is 100 times higher than aspirin (Zucker and Rothwell, 1978). Salicylate does not alter platelet function in vivo and there is evidence that it may block the inhibitory effect of aspirin on platelets by limiting the accessibility of platelet cyclo-oxygenase to aspirin (Vargaftig, 1978).

The effect of aspirin on bleeding time has been assessed in many studies (Mielke et al, 1973; Hirsh et al, 1973; Buchanan et al, 1977). Each study has reported that aspirin prolongs the bleeding time and has been observed with all standard methods of assessing bleeding time, namely, the Ivy, Duke and template methods. Prolonged bleeding times have been reported with doses ranging from 300 to 3600 mg. (Mielke and Britten, 1970).

Aspirin has no effect on reduced platelet survival seen in some patients with diseases associated with atherosclerosis or thromboembolism (Ritchie and Harker, 1977) but it potentiates the effects of dipyridamole.

High doses of aspirin prolong the prothrombin time after a lag phase of 2 to 3 days. Doses of aspirin between 100 and 300 mg. per day have no effect but 1300 to 2000 mg. have been shown to decrease coagulation factors II, VII, IX and X (Loew and Vinazzer, 1976). The mechanism of action is unclear but is corrected by the administration of vitamin K (Goldsweig et al, 1976).

The mechanism by which aspirin inhibits platelet function has recently been defined. Aspirin inhibits the synthesis of cyclic endoperoxides and thromboxane A_2 from platelet membrane arachidonic acid. The effect of aspirin is primarily due to its acetylation of the platelet enzyme cyclo-oxygenase which is saturated at low aspirin concentrations (30 uM) (Roth and Majerus, 1975). Although aspirin is rapidly cleared from the circulation, it has an irreversible effect on all of the platelets that are circulating at the time of its administration. This is because the platelet, being a non-nucleated cell, cannot regenerate cyclo-oxygenase that has been irreversibly acetylated by aspirin. Aspirin administered to subjects in doses of as low as 160 mg. inhibits platelet cyclo-oxygenase activity by greater than 80% (Burch et al, 1978a; Burch et al, 1978b). Larger doses have little additional effect. Aspirin also acetylates platelet membrane proteins and inhibits the platelet membrane enzyme collagen-glucosal transferase and these effects may also contribute to its inhibitory effect on platelet function (Rosenberg et al, 1971).

Aspirin also effects enzyme systems within the blood vessel wall. In 1976, Moncada et al demonstrated that the prostaglandin precursors PGG_2 and PGH_2 were converted by vessel wall microsomes into a prostaglandin with potent platelet anti-aggregating activity. This was subsequently identified as prostaglandin I_2 (PGI_2) or prostacyclin. PGI_2 can be produced by cultured vessel wall cells when the cells are incubated directly with arachidonic acid but they are incapable of producing thromboxane A_2 (Needleman et al, 1979; Kelton and Blajchman, 1980). PGI_2 is synthesized by endothelial cells as well as smooth muscle cells and is a potent vasodilator as well as a powerful inhibitor of platelet aggregation (Moncada et al, 1977; Baenziger et al, 1977). Both these effects on hemostasis are directly opposed to those of thromboxane A_2. The synthesis of PGI_2 by vessel walls, like the synthesis of thromboxane by platelets, is inhibited by aspirin but studies of the effects of aspirin on platelets and vessel wall cells have yielded conflicting results. Baezinger et al (1977) reported that platelet cyclo-oxygenase was much more sensitive to inhibition by aspirin than vessel wall cyclo-oxygenase. In contrast, Jaffe and Weksler (1979) reported that concentrations of aspirin that inhibited platelet cyclo-oxygenase also inhibited prostacyclin production by cultured endothelial cells but the report by Kelton et al (1978) indicated that much higher concentrations of aspirin in vivo were required to inhibit PGI_2 production than thromboxane A_2. The effect of aspirin on PGI_2 production is short lived, presumably because synthesis of cyclo-oxygenase can occur in vessel wall cells while the effect of aspirin on platelet cyclo-oxygenase is permanent (Kelton et al, 1978; Jaffe and Weksler, 1979).

DIPYRIDAMOLE

Dipyridamole is a pyrimido-pyrimidine compound that was introduced as a vasodilator. Dipyridamole inhibits platelet adhesion to glass beads (Rajah et al, 1977) and has been reported to inhibit adhesion to collagen in a non-flowing system (Cazenave et al, 1974). Dipyridamole also inhibits ADP-induced platelet aggregation and inhibits adrenalin and collagen-induced platelet release (Zucker and Peterson, 1970; Cucuiano et al, 1971). The precise mechanism of the antiplatelet effect of dipyridamole is uncertain. Dipyridamole inhibits the uptake of adenosine and glucose by platelets and the binding of ADP to platelet membranes (Philp and Lemieux, 1968). However, it is likely that the major mechanism of action is the inhibition of platelet phosphodiesterase (Mills and Smith, 1971). Phosphodiesterase enzymatically degrades cyclic AMP. Cyclic AMP activates phosphokinases which phosphorylate protein substrates. This results in a decrease in levels of cytoplasmic calcium which in turn inhibits platelet function. The inhibitory effect of dipyridamole is proportional to its blood concentration with blood levels of 3.5 $\mu M/L$ being required to produce a measureable effect of platelet aggregation ex-vivo (Rajah et al, 1977). Dipyridamole is bound to acid glycoprotein in plasma and when it is bound, it is not available to act on platelets (Subbarao et al, 1977). Some investigators have reported that dipyridamole has a greater effect ex vivo than in vitro. The difference is related to the anticoagulant used for the in vitro measurement of platelet function (Buchanan and Hirsh, 1978) and to differences in albumin concentrations when platelets are tested ex vivo and in vitro (Jørgensen and Stoffersen, 1978). Dipyridamole may have other effects on platelet function. Ally et al (1977) suggested that dipyridamole inhibited thromboxane A_2 synthesis, although Moncada and Vane (1978) presented evidence that did not support this hypothesis. Best et al (1979) demonstrated that dipyridamole inhibited platelet prostaglandin production. Moncada and Korbut (1978) suggested that circulating PGI_2 elevated levels of platelet cyclic AMP potentiating the effect of dipyridamole. This hypothesis is supported by the observation that dipyridamole and PGI_2 were synergistic in their inhibitory effect on platelet aggregation in vitro (DiMinno et al, 1978). There is also evidence that dipyridamole may stimulate prostacyclin release (Massoti et al, 1979).

Dipyridamole has been tested alone and in combination with aspirin in a number of thromboembolic disorders associated with reduced platelet survival time. Harker and Slichter (1970) reported that dipyridamole normalized shortened platelet survival in a dose dependent manner. Dipyridamole in doses of 400 mg. per day normalized platelet survival in patients with prosthetic heart valves, whereas 100 mg. per day had no effect. However, if aspirin, 1 gm. per day was added to dipyridamole in low doses, platelet survival was normalized. Harker and Slichter (1972) reported that while dipyridamole had no effect on the decreased platelet survival in venous thrombosis, it significantly prolonged decreased platelet survival in patients with arterial thrombosis, prosthetic heart valves and arteriovenous shunts. Ritchie and Harker (1977) reported that platelet survival was decreased in patients with coronary artery disease and this was normalized by a combination of dipyridamole and aspirin.

HYDROXYCHLOROQUINE

Hydroxychloroquine is an antimalarial drug which inhibits platelet function. Madow (1960) demonstrated that red cell sludging was reduced by several antimalarial agents, particularly hydroxychloroquine and subsequently Carter et al (1971) demonstrated that hydroxychloroquine inhibited ADP-induced platelet aggregation in vitro, although Pilcher (1975) did not demonstrate any ex vivo change in ADP or adrenalin-induced platelet aggregation after hydroxychloroquine ingestion.

The mechanism of the inhibitory effect of hydroxychloroquine on platelet function is not known. Hydroxychloroquine has no effect on the bleeding time and does not prolong reduced platelet survival.

SULPHINPYRAZONE

Sulphinpyrazone is a non-steroidal anti-inflammatory drug that inhibits platelet adhesion to collagen. At physiological hematocrit levels, however, inhibition of platelet adhesion requires concentrations of sulphinpyrazone that are unlikely to be achieved in humans (Davies et al, 1979). Sulphinpyrazone also inhibits platelet aggregation by adrenalin and collagen but there are conflicting reports on its effects on ADP-induced platelet aggregation (Wylie et al, 1979). Zucker and Peterson (1970) reported that sulphinpyrazone inhibited the release of platelet granule contents. Early investigators failed to demonstrate any antiplatelet activity of sulphinpyrazone ex-vivo, but Packham et al (1967) reported that sulphinpyrazone inhibited platelet function after administration to rabbits.

The reports on the effects of sulphinpyrazone on the bleeding time are conflicting. Weston et al (1977) reported that sulphinpyrazone prolonged the bleeding time using a standardized template method on humans but other investigators have been unable to detect any change (Winchester et al, 1977).

The original report that suggested that sulphinpyrazone might have antiplatelet activity was that of Smythe et al (1965) who performed platelet survival studies in patients with hyperuricemia associated with gout and myeloproliferative disorders and reported that platelet survival was prolonged following sulphinpyrazone treatment in doses of 400 mg. per day. Weily and Genton (1970) reported that the decreased platelet survival in patients with prosthetic heart valves was normalized by sulphinpyrazone in doses of 800 mg. per day whereas 400 mg. per day had no effect. Steele et al (1973) subsequently reported that platelet survival time was reduced in 18 of 28 patients with recurrent venous thrombosis and that in 7 of these patients, sulphinpyrazone normalized reduced platelet survival. Steele et al (1975) reported that patients with ischemic heart disease had decreased platelet survival which was normalized with sulphinpyrazone. Steele et al (1977) reported reduced platelet survival in 25 patients with transient cerebral attacks and that sulphinpyrazone increased the platelet survival time in 9 of these patients. Sulphinpyrazone has no effect on coagulation or on fibrinolysis (Smythe et al, 1965).

The mode of action of sulphinpyrazone is incompletely understood, but there is evidence that sulphinpyrazone is a competitive inhibitor of cyclo-oxygenase (Ali and MacDonald, 1977). There is also evidence that a metabolite of sulphinpyrazone may be a more potent inhibitor of platelet function than the parent compound (Buchanan and Hirsh, 1978). It was also demonstrated that the inhibition produced by the presumed metabolite was irreversible in contrast to the reversible nature of sulphinpyrazone platelet inhibition. While it is likely that the predominant inhibitory effect of sulphinpyrazone on platelet function is mediated through its effect on the prostaglandin pathway, Wong et al (1978) reported that sulphinpyrazone had a synergistic effect on the inhibition of collagen-induced platelet aggregation by aspirin suggesting that sulphinpyrazone may have other modes of action or that mutual competition for protein binding may result in higher concentrations of free aspirin or sulphinpyrazone.

CLINICAL EVALUATION OF ANTIPLATELET DRUGS

Antiplatelet drugs have been evaluated in prospective clinical trials in cerebral vascular disease, coronary artery disease, peripheral vascular disease, post-aortocoronary bypass graft surgery, rheumatic heart disease and prosthetic heart valves, arterio-venous shunts and fistula and for the prevention of venous thromboembolism.

CORONARY ARTERY DISEASE

There is evidence that platelets are involved in the genesis of ischemic heart disease by their incorporation into occlusive coronary artery thrombosis, by embolization of aggregates into the microcirculation of the heart and by their contribution to the production of atherosclerosis (Haerem, 1972; Ross and Glomset, 1976). There is also increasing evidence that coronary artery spasm may be caused by the release of vasoconstrictors from platelets such as thromboxane A_2 (Moncada and Vane, 1979) and which may result in myocardial infarction. The relationship between abnormal platelet function and ischemic heart disease is supported by the reports that some patients with angiographically demonstrated lesions in the coronary arteries have reduced platelet survival (Genton and Steele, 1977).

ASPIRIN

Evidence for an association between the regular use of aspirin and increased risk of myocardial infarction was provided by the results of the retrospective study reported by the Boston Collaborative Drug Surveillance Group (1974) but a more extensive study including over 1,000,000 patients found no association between regular aspirin ingestion and the incidence of myocardial infarction (Hammond and Garfinkel, 1975). Six prospective studies have tested the hypothesis that aspirin protects patients with myocardial infarction from recurrent infarction or death. The results are summarized in Table 1.

Table 1. RANDOMIZED TRIALS OF ASPIRIN FOR PREVENTION OF DEATH AFTER MYOCARDIAL INFARCTION

Author	Dose (mg/day)	No. of Patients	Placebo	Aspirin	% Reduction in Mortality
Elwood et al (1974)	300	1239	10.9	8.3	25
Coronary Drug Project Research Group (1976)	975	1529	8.3	5.8	30
Breddin et al (1979)	1500	626	10.4	8.5	18
Elwood and Sweetnam (1979)	900	1682	14.8	12.3	17
AMIS (1980)	1000	4524	9.7	10.8	-11
PARIS (1980)	975	1216	12.8	10.5	18

The first of these was reported by Elwood et al in 1974. This study was a multicentred, randomized, double-blind trial of aspirin in doses of 300 mg. per day and the results showed a non-significant trend for a decrease in mortality in the aspirin-treated patients. However, of interest, the greatest benefit was in the patients who were admitted to the study within five weeks of the infarction. This suggested that the patients who were treated within a short time of the infarction were more responsive to the antiplatelet agents. This important finding was not considered in a number of more important recent clinical trials and may be responsible for the apparent discrepant results between these studies and the study with sulphinpyrazone. The Coronary Drug Project Research Group (1976) reported the results of a prospective double-blind trial of 1529 patients who had a myocardial infarction at lease two years before they were randomized and in some cases, five years before randomization. The mortality in the aspirin-treated group was 5.8% and in the placebo was 8.3%. This difference was not statistically significant. Breddin et al (1979) reported the results of a German-Austrian study which showed a strong trend in favour of aspirin compared with placebo or an anticoagulant treated group but again the results were not statistically significant. Elwood and Sweetnam (1979) reported the results of a double-blind trial of aspirin in doses of 900 mg. per day in patients who were admitted to the study within 7 days of infarction. The mortality was 12.3% in the aspirin-treated patients and 14.8% in the placebo-treated group. This difference was not significant. In a subgroup analysis, aspirin significantly reduced the total mortality and secondary occurrence of myocardial infarction in males. However, there was a 25% non-compliance and high withdrawal rate in this trial which may negate the results. Two multicentered trials have recently been reported in the United States. The Aspirin Myocardial Infarction Study (AMIS) reported in 1980 was a prospective trial in which patients were randomized into aspirin or placebo 8 weeks to 5 years following myocardial infarction. There were 2067 patients in the aspirin treated group and 2257 in the placebo-treated group. Over a three year follow-up period, there was no significant difference in the overall mortality, cardiac-related mortality or recurrence of myocardial infarction. There was, however, a non-significant trend towards a reduction in the occurrence of stroke and transient cerebral ischemic attacks in the aspirin-treated group. However, the mean interval between the qualifying myocardial infarction and entry into the trial for each group was 25 months so that this trial may have omitted the patients who were most likely to respond to the antiplatelet agents. Thus, the negative findings of the AMIS trial may have been due in part to the inclusion of a large percentage of low risk and low responsive patients.

The Persantine Aspirin Reinfarction Study Research Group (1980) have reported the results of the PARIS trial. In this study 2026 persons who had recovered from myocardial infarction were randomized; aspirin plus dipyridamole (810), aspirin alone (810) or placebo (406) with an average follow-up of 41 months. The results showed a non-significant trend in favour of aspirin for the prevention of total mortality and sudden death.

Thus, none of the studies evaluating aspirin in patients with myocardial infarction have shown a significant reduction in mortality or coronary events but most have shown a trend in favour of aspirin. When the results of the studies are analyzed together, the trend in favour of aspirin is consistent and there is some evidence that there may be significant reduction in patients who began treatment within 6 months of infarction.

DIPYRIDAMOLE

There are numerous reports on the use of dipyridamole as a coronary vasodilator and several of those were prospective and randomized. However, most were uncontrolled and the results regarding angina were inconclusive. In none of these studies was there satisfactory evaluation of the frequency of myocardial infarction and death (Wirecki, 1967; Sbar and Schlant, 1967). There is one study of dipyridamole compared with placebo in patients with recent myocardial infarction but no difference between the treated and control groups was found (Gent et al, 1968).

SULPHINPYRAZONE

The Anturane Reinfarction Trial (1978) was a multicentered double-blind randomized trial comparing sulphinpyrazone (200 mg q.i.d.) with placebo in 1628 patients who had a myocardial infarction 25 to 35 days before entry. In the first report (1978) calculated annual death rates were compared. When cardiac deaths were compared, the placebo group had an annual death rate of 9.5% and the sulphinpyrazone group 4.9%. Sudden death occurred in 6.3% of patients in the placebo-treated patients compared with 2.7% in the sulphinpyrazone-treated group. These differences are statistically significant. A follow-up publication (1980) indicated that the reduction was entirely due to the lower incidence of sudden death in the 2nd to 7th month after myocardial infarction. There have

been some questions raised regarding the validity of the
report of this trial, particularly in regard to the definition of sudden death and the inclusion and exclusion criteria and the results of the trial are currently being
reassessed.

VALVULAR HEART DISEASE AND PROSTHETIC HEART VALVES

Systemic arterial embolism is a major complication in
patients with valvular heart disease and prosthetic heart
valves. Reduced platelet survival has been reported in
patients with rheumatic heart disease and prosthetic heart
valves and there is some evidence that systemic embolism
occurs more frequently in these patients with shortened
platelet survival (Steele et al, 1974; Weily and Genton, 1970).
Two antiplatelet drugs have been evaluated in the treatment
of patients with valvular disease or patients with prosthetic heart valves.

There have been two prospective trials reported that
indicate that the addition of dipyridamole or aspirin to
conventional oral anticoagulant therapy is more effective
than oral anticoagulants alone for the prevention of
thromboembolic complications in patients with prosthetic
heart valves. The effect of adding dipyridamole, 400 mg.
daily, or placebo to conventional anticoagulant therapy on
the frequency of arterial thromboembolism in patients with
aortic and mitral valve prostheses has been studied and over
one year's observation, arterial thromboembolism occurred
in 14% of placebo-treated patients compared to 1.3% of the
dipyridamole treated patients (Sullivan et al, 1971). The
frequency of death was similar in the two groups but none
of the dipyridamole-treated patients died of embolic events
whereas two patients in the placebo-treated group died of
cerebral embolism. The results of a prospective trial of
aspirin 1000 mg. daily, combined with warfarin compared
with warfarin alone in patients with aortic ball valve
prostheses has been reported in which a significantly
lower incidence of arterial embolism occurred in the combined treatment group compared with the warfarin-treated
group (Dale, 1977). The incidence of thromboembolic complications per 100 patients per year was 1.8 in the combined
treatment group compared with 9.3 in the group treated with
warfarin alone. Aspirin alone was found to be ineffective
in preventing arterial thromboembolism in this study. A
non-randomized trial of dipyridamole 450 mg. daily combined

with aspirin 3000 mg. daily has been reported in which the
incidence of systemic embolism in patients with prosthetic
heart valves was reduced without concurrent anticoagulant
therapy (Taguchi et al, 1975) but the validity of this
conclusion is limited by the study design. Brott et al
(1981) reported that aspirin 1300 mg. combined with dipyridamole 200 mg. daily was inadequate for thromboembolic
prophylaxis following Starr-Edwards aortic valve replacement. A number of non-randomized trials have been performed
that support the conclusions of the prospective controlled
trials that drugs that inhibit platelet function, when added
to conventional anticoagulant therapy, provide additional
benefit in the prevention of systemic arterial thromboembolism in patients with prosthetic heart valves (Bjork and
Henze, 1975; Arrants and Hairston, 1972).

CEREBRAL VASCULAR DISEASE

Cerebral vascular ischemia, including stroke and transient ischemic attacks (TIA) is frequently caused by thromboemboli. Autopsy studies have demonstrated platelet/fibrin
emboli in the cerebral and retinal arteries in patients
with cerebral vascular ischemic syndromes, and platelet
thrombi have been identified on ulcerated lesions in the
intracranial arteries in the patients with transient cerebral ischemic attacks and amaurosis fugax (Russell, 1961;
Gunning et al, 1964).

The antiplatelet drugs that have been evaluated in
cerebral vascular disease are aspirin, dipyridamole and
sulphinpyrazone.

ASPIRIN

Dyken et al (1973) reported the results of a retrospective study which suggested that aspirin was associated with
a reduced incidence of transient cerebral ischemic attacks.
Two large cooperative studies evaluating the effects of aspirin have been reported and both are highly suggestive
that aspirin is beneficial in preventing recurrent TIA,
stroke and death in patients with a previous history of TIA.
The study by Fields and colleagues in the USA was reported
in two parts (Fields et al, 1977, 1978). The objective of
the study was to assess the benefit of aspirin in doses of
1200 mg per day among patients who had carotid TIA. The 1st
report (1977) involved patients treated medically only; the

second (1978) reported the results in patients who had reconstructive operations of the carotid artery before entry into the study. One hundred and seventy-eight patients with carotid TIA were randomly allocated to aspirin 1200 mg. or placebo and followed to determine the incidence of subsequent TIA, death, cerebral infarction or retinal infarction. After the first six months of follow-up when death, cerebral or retinal infarction and recurrent TIA were grouped together, there was a statistically significant reduction in the aspirin-treated group compared with the placebo-treated group. The difference was most marked in patients with a history of multiple TIA and in patients who had lesions appropriate with the TIA symptoms. However, when the results were analyzed in terms of reduction in death and cerebral or retinal infarction alone, there was no significant difference between aspirin and placebo. In the second report, 125 patients with carotid TIA who had reconstructive operations were randomly assigned to aspirin 1200 mg. per day or placebo and followed using the same endpoints as the medical study. Aspirin had no significant effect on overall mortality nor in the frequency of retinal or cerebral infarctions. However, when deaths which were not stroke related were eliminated from the analysis, there was significant reduction in the frequency of fatal and non-fatal cerebral or retinal infarctions in the aspirin-treated group compared with the placebo group. However, because of the small number of patients and the short period of follow-up, the results from these studies did not prove conclusively that aspirin was effective in preventing cerebral infarction or death in patients with TIA.

The Canadian Cooperative Study Group (1978) reported the results of a prospective study in 585 patients with TIA which compared four treatments. The study was a double-blind trial comparing aspirin 1200 mg., aspirin 1200 mg. plus sulphinpyrazone 800 mg., sulphinpyrazone 800 mg. or placebo. Seventy-five percent of the patients in the study had symptoms of carotid TIA and 25% had vertebrobasilar TIA. The endpoints were continuation of TIA, stroke or death and the results showed a statistically significant risk reduction of 19% in the aspirin-treated group. A second analysis omitting TIA as an endpoint was carried out and showed a 31% risk reduction in stroke and death from aspirin. In the study, sulphinpyrazone did not produce a statistically significant reduction for any endpoint and did not exhibit any synergism or antagonism to aspirin. Of interest in this study was a striking difference in the response between males and females. There was a 48% risk reduction of stroke or death in men treated with aspirin, but there was no significant reduction in females. This sex difference has been found not to be unique to clinical trials in patients with cerebral vascular disease. In the study by Kaegi et al (1974) on arterio-venous shunts and in patients undergoing hip replacement by Harris et al (1977) the benefit was greater in male patients compared with female patients. In addition, when the studies by Fields et al (1977, 1978) were re-analyzed, a similar trend in favour of males was noted.

DIPYRIDAMOLE

Acheson et al (1969) reported on a double-blind study evaluating dipyridamole in patients with established cerebral vascular disease. One hundred and sixty-nine patients who had previous history of cerebral vascular ischemia were randomized into dipyridamole 400 mg. per day or placebo. After 14 months, the dose of dipyridamole was increased to 800 mg. and the patients treated for a further 11 months. There was no significant difference in the frequency of transient cerebral ischemic attacks, stroke or death in the dipyridamole and control groups.

SULPHINPYRAZONE

Evans (1972) reported the results of a double-blind crossover study of 20 patients with amaurosis fugax who received sulphinpyrazone in doses of 800 mg. per day or placebo. The results of the study showed a significant reduction in the episodes of TIA during the period of sulphinpyrazone treatment. However, the conclusions from this study were limited because the drug was tested for a six week period only and the more important endpoints of stroke and death were not considered. Blakely and Gent (1975) reported the results of a study in 291 elderly males randomized to receive either sulphinpyrazone 600 mg. or placebo and a subgroup analysis of the results showed that the patients with a history of stroke who were treated with sulphinpyrazone showed a reduction in the mortality compared with placebo. In a subsequent study, Blakely (1979) was unable to confirm the protective effect of sulphinpyrazone. The Canadian Cooperative Study Group (1978) reported in the multicentred study that sulphinpyrazone had no effect on TIA stroke and death in patients with previous TIA.

VENOUS THROMBOEMBOLISM

Venous thrombi consist mainly of fibrin and trapped red cells but there is evidence that some venous thrombi originate in valve pockets as aggregates of platelets which propagate with the formation of fibrin (Paterson, 1969; Sevitt, 1973). In addition, platelet thrombi may form at sites of direct injury to veins, such as after trauma to the legs and to the femoral veins during hip surgery (Clagett et al, 1975). Because of the role of platelets in the genesis of venous thrombosis, antiplatelet drugs may be of value in preventing the formation of thrombi that arise principally as platelet aggregates. Four antiplatelet drugs, aspirin, dipyridamole, hydroxychloroquine and sulphinpyrazone have been tested in prospective clinical trials for deep vein thrombosis prophylaxis. Two studies of the effect of aspirin on the frequency of fatal pulmonary embolism have been reported.

ASPIRIN

Aspirin has been evaluated as an antithrombotic drug in venous thromboembolism in humans in doses of 300 to 3500 mg. per day alone and in combination with other antiplatelet drugs. The results are summarized in Table 2. In general, aspirin has not been effective in reducing the frequency of venous thrombosis in general surgical patients and the results have been mixed in patients undergoing orthopaedic procedures. In two studies in general abdominothoracic surgery, aspirin in doses of 600 mg. and 2400 mg. (O'Brien et al, 1971) and 600 mg. (Medical Research Council, 1972) for the prevention of leg scan detected venous thrombosis showed no benefit from aspirin. Clagett et al (1974) found no benefit with 1300 mg. of aspirin in general surgical patients, but on re-analysis of the study, a significant difference was found between the control and treated group after four patients in the aspirin-treated group who developed venous thrombosis were excluded because they did not receive the drug. A combination of aspirin and dipyridamole has been reported to produce a significant reduction in the incidence of postoperative venous thrombosis in general surgical patients by Renney et al (1976).

TABLE 2. RANDOMIZED TRIALS OF ASPIRIN FOR THE PREVENTION OF DEEP VEIN THROMBOSIS (DVT)

Author	Dose (mg/day)	Surgery	No. of Patients	DVT % Control	DVT % Treated
Medical Research Council (1972)	600	General	303	22	28
Clagett et al (1974)	1300	General	105	20	12
Schöndorf & Hey (1978)	450	Ortho Hip	45	60	53
Harris et al (1977)	1200	Ortho Hip	95	43	25
Hume et al (1977)	1200	Ortho Hip	71	35	30
McKenna et al (1980)	900	Ortho Knee	43	75	78
	3500	Ortho Knee	--	75	8

There have been several trials of aspirin in patients undergoing orthopaedic procedures. Salzman and Harris (1971) reported the results of a comparison of aspirin, dipyridamole, warfarin and low-molecular weight Dextran and reported that aspirin was as effective as warfarin but that the dipyridamole was ineffective in preventing venous thrombosis. In a second study, Harris et al (1974) found that the rate was high in all groups but in the aspirin-treated group there were more large and multiple thrombi. Harris et al (1977) reported the results of a prospective controlled double-blind study of aspirin 1200 mg. per day in patients undergoing total hip replacement using venography as a diagnostic endpoint. There was a significant reduction in venous thrombosis in patients receiving aspirin which was limited to men. Hume et al (1977) reported the results of a non-randomized trial of aspirin and found a decreased incidence of venous thrombosis in the treated group.

Schöndorf and Hey (1978) reported that intravenous aspirin in doses of 900 mg. had no effect on objectively diagnosed venous thrombosis in patients undergoing elective hip surgery. A study reported by McKenna et al (1980) in patients undergoing elective knee surgery is of particular interest as aspirin was ineffective when used in doses of 900 mg. per day but markedly reduced the frequency of objectively detected venous thrombosis in doses of 3500 mg. per day. The reason for this dose dependent effect is not certain.

Zekert et al (1974) reported the results of a randomized double-blind trial of aspirin 1500 mg. per day for the prevention of postoperative pulmonary embolism and found a statistically significant reduction in fatal pulmonary embolism diagnosed at autopsy. Jennings et al (1976) reported on the use of 1200 mg. of aspirin daily in patients undergoing total hip replacement and found no pulmonary embolism in patients in whom the expected incidence of death from pulmonary embolism was 1 - 2%. Dechavanne et al (1975) and Silvergleid et al (1977) found that a combination of aspirin and dipyridamole does not reduce the incidence of venous thrombosis after elective hip replacement. Similar results were reported by Morris and Mitchell (1977).

DIPYRIDAMOLE

Browse and Hall (1969) using a clinical endpoint to detect venous thrombosis failed to demonstrate any benefit with dipyridamole and Morris and Mitchell (1977) found no benefit with dipyridamole. In contrast, however, both Renney et al (1976) and Plante et al (1979) reported aspirin plus dipyridamole was effective in decreasing leg scan detected venous thrombosis following elective general surgery.

HYDROXYCHLOROQUINE

There have been a number of clinical trials on the use of hydroxychloroquine for the prevention of venous thrombosis in general surgical and orthopaedic patients. The results are summarized in Table 3. Three trials have involved general surgical patients where the risk of thrombosis would be considered moderate. In these studies a significant reduction in deep vein thrombosis was detected in the hydroxychloroquine treated patients (Carter and Eban, 1971; Carter and Eban, 1974; Wu et al, 1977). Conflicting results have been reported in studies in patients undergoing orthopaedic procedures. Chrisman et al (1976) in a comparison of hydroxychloroquine and placebo for preventing venous thrombosis in orthopaedic patients reported that venous thrombosis diagnosed by impedance plethysmography and confirmed by venography was significantly reduced in the hydroxychloroquine treated group. On the other hand, studies by Hansen et al (1976) Hume et al (1977) and Cooke et al (1977) failed to demonstrate any benefit in patients undergoing orthopaedic procedures.

SULPHINPYRAZONE

There have been two prospective trials on the use of sulphinpyrazone for the prevention of postoperative deep vein thrombosis. In neither of these studies, was any benefit reported (Gruber et al, 1977; Rogers et al, 1978).

Table 3. RANDOMIZED TRIALS OF HYDROXYCHLOROQUINE FOR THE PREVENTION OF DEEP VEIN THROMBOSIS (DVT)

Author	Dose (mg/day)	Surgery	No. of Patients	DVT % Control	DVT % Treated
Carter et al (1971)	600	General	52	23	0
Carter & Eban (1974)	800	General	214	18	5
Wu et al (1977)	600	General	90	14	1
Chrisman et al (1976)	600	Ortho-Hip	100	16	2
Hansen et al (1976)	600	Ortho-Hip	98	66	50
Hume et al (1977)	600	Ortho-Hip	40	60	50
Cooke et al (1977)	1000	Ortho-Hip	50	48	40

ARTERIAL THROMBOSIS AND ARTERIOVENOUS THROMBOSIS

There have been only a small number of studies on the use of antiplatelet drugs in patients with arterial thrombosis and following arterial bypass surgery. These studies have been carried out in patients undergoing diagnostic or therapeutic arterial catheterization, after peripheral arterial surgery and after aortocoronary bypass graft.

There have been two prospective randomized studies evaluating aspirin in the prevention of thrombosis in patients undergoing arterial catheterization and both were negative. In one study, 150 consecutive patients having brachial artery catheterization were randomly assigned into aspirin or placebo-treated groups and there was no reduction in the frequency of thrombi detected by balloon catheter or by a decrease in the peripheral artery pulsations (Hynes et al, 1973). In the other study, the effect of aspirin on the reduction of blood flow, determined by oscillometry, in children undergoing cardiac catheterization was evaluated and there was no difference in the aspirin-treated group compared with the placebo-treated group (Freed et al, 1974).

A study comparing the effects of aspirin, warfarin or placebo on aortocoronary bypass graft patency rates has been reported in abstract form (McEnany et al, 1976) in which the patency rate of the grafts was 84% in the warfarin-treated group, 72% in the placebo-treated group and 80% in the aspirin-treated group. The difference between the warfarin and placebo-treated group was statistically significant but the difference between the aspirin and placebo-treated groups was not significant. In peripheral vascular surgery, the results of the study comparing the effect of sulphinpyrazone with placebo showed no benefit from treatment. (Blakely and Gent, 1975).

Harter et al (1979) carried out a double-blind trial of aspirin in doses of 160 mg. per day for the prevention of arteriovenous shunt thrombosis in 44 patients and reported a significant reduction in the aspirin-treated group compared with placebo. The effect of sulphinpyrazone on the frequency of shunt thrombosis in patients with arteriovenous shunts inserted to facilitate chronic hemodialysis has been reported. In these patients, thrombus formation occurred in the veins distal to the shunt and the effectiveness of sulphinpyrazone 600 mg. daily with or without concurrent anticoagulant treatment for the prevention of thrombosis at that site was shown in a prospective randomized crossover trial. The beneficial effect of sulphinpyrazone was maintained after the crossover phase of the study (Kaegi et al, 1974). In the second study (Kaegi et al, 1975) it was noted that the beneficial effect of sulphinpyrazone was present within the first week of treatment and disappeared rapidly when treatment was withdrawn. The reduction in shunt thrombosis by sulphinpyrazone has recently been confirmed (Michie and Wombolt, 1977).

There have been a number of observations made on the use of platelet suppressant therapy in patients with peripheral arterial ischemia and thrombocytosis. Aspirin, but not dipyridamole has been shown to produce prolonged control of the symptoms of a syndrome of recurrent, painful cyanotic digits associated with thrombocytosis and spontaneous platelet aggregation (Vreeken and van Aken, 1971; Bierme et al, 1972; Preston et al, 1974).

SUMMARY

Four drugs that inhibit platelet function have been evaluated for their antithrombotic effects in humans. These are aspirin, dipyridamole, hydroxychloroquine and sulphinpyrazone. Aspirin has been shown to reduce the number of transient ischemic attacks (TIA), stroke and death in patients with multiple TIA. The reduction in TIA was greatest in males who were normotensive and when there was an angiographically demonstrated lesion in the carotid artery that accounted for the symptoms. Aspirin reduced venous thrombosis and non-fatal and fatal pulmonary embolism in patients after surgery for fractured hip and after elective hip replacement. There is evidence that the prophylactic effect of aspirin may be greater in male patients. Aspirin reduced the frequency of arteriovenous shunt thrombosis. Aspirin abolished symptoms in patients with peripheral ischemia associated with thrombocytosis and spontaneous platelet aggregation. There is no conclusive evidence at the present time that aspirin is effective in patients with coronary artery disease. Dipyridamole in combination with oral anticoagulants is effective in reducing the frequency of systemic embolism in patients with prosthetic heart valve replacement but is ineffective in patients with transient cerebral ischemic attacks or for the prevention of venous thromboembolism. Hydroxychloroquine was effective in reducing postoperative venous thrombosis in patients undergoing general abdominothoracic surgery but the evidence that it was effective in patients undergoing orthopaedic surgery is inconclusive. Sulphinpyrazone may be effective in reducing the frequency of sudden cardiac deaths in patients in the first year after myocardial infarction when it is started within 25 to 35 days after the infarction. Sulphinpyrazone reduced the incidence of arteriovenous shunt thrombosis in patients undergoing chronic hemodialysis and in combination with anticoagulants, it reduced the frequency of recurrent venous thrombosis.

There have been no large scale trials of platelet suppressant drugs in clinical cancer and successful treatment of thromboembolic disorders cannot be used to predict success in the treatment of malignant disease.

REFERENCES

Acheson J, Danta G, Hutchinson EC (1969). Controlled trial of dipyridamole in cerebral vascular disease. Br Med J 1: 614.
Ali M, McDonald JWD (1977). Effects of sulphinpyrazone on platelet prostaglandin synthesis and platelet release of serotonin. J Lab Clin Med 89:868.
Ally AI, Manku MS, Horrobin DF, Morgan RO, Karmazin M, Karmali RA (1977). Dipyridamole: A possible potent inhibitor of thromboxane A_2 synthetase in vascular smooth muscle. Prostaglandins 14:607.
Anturane Reinfarction Trial Research Group (1978). Sulphinpyrazone in the prevention of cardiac death after myocardial infarction: The Anturane Reinfarction Trial. N Engl J Med 298:289.
Anturane Reinfarction Trial Research Group (1980). Sulphinpyrazone in the prevention of sudden death after myocardial infarction. N Engl J Med 302:250.
Arrants JE, Hairston P (1972). Use of persantine in preventing thromboembolism following valve replacement. Am Surg 38:432.
Aspirin Myocardial Infarction Study Research Group (1980). A randomized controlled trial of aspirin in persons recovered from myocardial infarction. JAMA 243:661.
Baenziger NL, Dillender MJ, Majerus PW (1977). Cultured human

skin fibroblasts and arterial cells produce a labile platelet-inhibitory prostaglandin. Biochem Biophys Res Comm 78:294.

Baumgartner HR, Muggli R, Tschopp TB, Turitto VT (1976). Platelet adhesion, release and aggregation in flowing blood: effects of surface properties and platelet function. Thromb Haemost 35:124.

Bell RL, Kennerly DA, Stanford N, Majerus PW (1979). Diglyceride lipase: a pathway for arachidonate release from human platelets. Proc Natl Acad of Sci USA 76:238.

Best LC, McGuier MB, Jones PBB, Holland TK, Martin TJ, Preston FE, Segal DS, Russell RGG (1979). Mode of action of dipyridamole on human platelets. Thromb Res 16:367.

Bierme R, Boneu B, Guirard B, Pris J (1972). Aspirin and recurrent painful toes and fingers in thrombocythaemia. Lancet 1:432.

Bjork V, Henze A (1975). Management of thromboembolism after aortic valve replacement with the Bjork-Shiley tilting disc valve. Scand J Thorac Cardiovasc Surg 3:183.

Blakely JA (1979). A prospective trial of sulphinpyrazone and survival after thrombotic stroke. VIth International Cong on Thromb and Haemost. Abstract #0382.

Blakely JA, Gent M (1975). Platelets, drugs and longevity in a geriatric population, In Hirsh J, Cade JF, Gallus AS (eds): "Platelets, Drugs and Thrombosis," Basel, Karger p 284.

Boston Collaborative Drug Surveillance Group (1974). Regular aspirin intake and acute myocardial infarction. Br Med J 1:440.

Breddin K, Loew D, Lechner K, Uberla K, Walter E (1979). Secondary prevention of myocardial infarction. Comparison of acetylsalicylic acid, pnenprocoumon and placebo. A multicentre two-year prospective study. Thromb Haemost 40:225.

Brott, WH, Zajtchuk R, Bowen TE, Davis J, Green DC (1981). Dipyridamole-aspirin as thromboembolic prophylaxis in patients with aortic valve prosthesis. J Thorac Cardiovasc Surg 81:632.

Browse NL, Hall JH (1969). Effect of dipyridamole on the incidence of clinically detectable deep vein thrombosis. Lancet ii:718.

Buchanan MR, Hirsh J (1978). A comparison of the effects of aspirin and dipyridamole on platelet aggregation in vivo and ex vivo. Thromb Res 13:517.

Buchanan GR, Martin V, Levine PH, Scoon K, Handin RI (1977). The effects of antiplatelet drugs on bleeding time and platelet aggregation in normal human subjects. Am J Clin Path 68:355.

Burch JW, Baenziger NL, Stanford N, Majerus PW (1978a). Sensitivity of fatty acid cyclo-oxygenase from human aorta to acetylation by aspirin. Proc of Natl Acad Sci USA 75: 5181.

Burch JW, Stanford N, Majerus PW (1978b). Inhibition of platelet prostaglandin synthetase by oral aspirin. J Clin Invest 61:314.

Carter AE, Eban R, Perrett RD (1971). Prevention of postoperative deep venous thrombosis and pulmonary embolism. Br Med J 1:312.

Carter AE, Eban R (1974). Prevention of postoperative deep venous thrombosis in the legs by orally administered hydroxychloroquine. Br Med J 3:94.

Cazenave JP, Packham MA, Guccione MA, Mustard JF (1974). Inhibition of platelet adherence to a collagen-coated surface by nonsteroidal anti-inflammatory drugs, pyrimido-pyrimidine and tricyclic compounds and lidocaine. J Lab Clin Med 83:979.

Cazenave JP, Kinlough-Rathbone RL, Packham MA, Mustard JF. (1978). The effect of acetylsalicylic acid and indomethacin on rabbit platelet adherence to collagen and the subendothelium in the presence of a low or high hematocrit. Thromb Res 13:971.

Chrisman OD, Snook GA, Wilson TC, Short JY (1976). Prevention of venous thromboembolism by administration of hydroxychloroquine. J Bone Joint Surg 58A:918.

Clagett GP, Brier DF, Rosoff CB, Schneider PB, Salzman EW (1974). Effect of aspirin on postoperative platelet kinetics and venous thrombosis. Surg Forum 25:473.

Clagett GP, Schneider P, Rosoff CB, Salzman EW (1975). The influence of aspirin on postoperative platelet kinetics and venous thrombosis. Surg 77:61

Cooke ED, Dawson MHO, Ibbotson RM, Bowcock SA, Ainsworth ME, Pilcher MF (1977). Failure of orally administered hydroxychloroquine sulphate to prevent venous thromboembolism following elective hip operations. J Bone Joint Surg 59A: 496.

Coronary Drug Project Research Group (1976). Aspirin in coronary heart disease. J Chron Dis 29:625.

Cucuianu MP, Nishizawa EE, Mustard JF (1971). Effect of pyridimo-pyrimidine compounds on platelet function. J Lab Clin Med 77:958.

Dale J (1977). Prevention of arterial thrombosis with acetylsalicylic acids in patients with prosthetic heart valves.

Thromb Haemost 38:66.
Davies JA, Essien E, Cazenave JP, Kinlough-Rathbone RL, Gent M, Mustard JF (1979). The influence of red blood cells on the effects of aspirin or sulphinpyrazone on platelet adherence to damaged rabbit aorta. Br J Haematol 42:283.
Dechavanne M, Ville D, Biala JJ, Kher A, Faivre J, Poussset MB, Dejour H (1975). Controlled trial of platelet antiaggregating agents and subcutaneous heparin in the prevention of postoperative deep vein thrombosis in high risk patients. Haemostasis 4:94.
Dejana E, Cazenave JP, Groves HM, Kinlough-Rathbone RL, Richardson M, Packham MA, Mustard JF (1980). The effect of aspirin inhibition of PGI_2 production on platelet adherence to normal and damaged rabbit aortae. Thromb Res 17:453.
DiMinno G, DeGaetano G, Garattini S (1978). Dipyridamole and platelet function. Lancet II:1258
Dyken ML, Kolar OJ, Jones FH (1973). Differences in the occurrence of carotid transient ischemic attacks associated with antiplatelet aggregation therapy. Stroke 4:732.
Elwood PC, Cochrane AL, Burr ML, Sweetnam PM, Williams G, Welsby E, Hughes SJ, Renton R (1974). A randomized controlled trial of acetylsalicylic acid in the secondary prevention of mortality from myocardial infarction. Br Med J 1:436.
Elwood PC, Sweetnam PM (1979). Aspirin and secondary mortality after myocardial infarction. Lancet II:1313.
Evans G (1972). Effect of drugs that suppress platelet surface interaction on incidence of amaurosis fugax and transient cerebral ischemia. Surg Forum 23:239.
Fields WS, Lemak NA, Frankowski RF, Hardy RJ (1977). Controlled trial of aspirin in cerebral ischemia. Stroke 8:301.
Fields WS, Lemak NA, Frankowski RF, Hardy RJ (1978). Controlled trial of aspirin in cerebral ischemia (Part II). Surg Group: Stroke 9:309.
Freed MD, Rosenthal A, Fyler D (1974). Attempts to reduce arterial thrombosis after cardiac catheterization in children. Use of percutaneous technique and aspirin. Am Heart J 87:283.
Gent AE, Brook CGD, Foley TH, Miller TN (1968). Dipyridamole: a controlled trial of its effect in acute myocardial infarction. Br Med J 4:366
Genton E, Steele P (1977). Platelet survival time - alteration with disease and drug treatment. In Mitchell JRA, Domenet JG (eds): "Thromboembolism, a New Approach to Therapy", New York, Acad Press p 104.
Goldsweig HG, Kapusta M, Schwartz J (1976). Bleeding,

salicylates and prolonged prothrombin time. Three case reports and a review of the literature. J Rheumatol 3:37.
Gruber UF, Fuser P, Frick J, Loosli J, Matt E, Segesser D, (1977). Sulphinpyrazone and postoperative deep vein thrombosis. Europ Surg Res 9:303.
Gunning AJ, Pickering GW, Robb-Smith AHT, Russell R. Ross (1964). Mural thrombosis of the internal carotid and subsequent embolism. Am J Med 33:155.
Haerem JW (1972). Platelet aggregates in intramyocardial vessels of patients dying suddenly and unexpectedly of coronary artery disease. Atherosclerosis 15:199.
Hammond EC, Garfinkel L (1975). Aspirin and coronary heart disease: Findings of a prospective study. Br Med J 2:269.
Hansen EH, Jessing P, Lindewald H, Ostergaard P, Olesen T, Malver EI (1976). Hydroxychloroquine sulphate in prevention of deep venous thrombosis following fracture of the hip, pelvis or thoracolumbar spine. J Bone Joint Surg 58A:1089.
Harker LA, Slichter SJ (1972). Platelet and fibrinogen consumption in man. N Engl J Med 287:999.
Harker LA, Slichter SJ (1970). Studies of platelet and fibrinogen kinetics in patients with prosthetic heart valves. N Engl J Med 283:1302.
Harris WH, Salzman EW, Athanasoulis C, Waltman AC, Baum S, deSanctis RW (1974). Comparison of warfarin, low-molecular weight dextran, aspirin and subcutaneous heparin in prevention of venous thromboembolism following total hip replacement. J Bone Joint Surg 56:1552.
Harris WH, Salzman EW, Athanasoulis CA, Waltman AC, DeSanctis RW (1977). Aspirin prophylaxis of venous thromboembolism after total hip replacement. N Engl J Med 297:1246.
Harter HR, Burch JW, Majerus PW, Stanford N, Delmez JA, Anderson CB, Weerts CA (1979). Prevention of thrombosis in patients on hemodialysis by low-dose heparin. N Engl J Med 301 (11):577.
Hirsh J, Street D, Cade JF, Amy H (1973). Relation between bleeding time and platelet connective tissue reaction after aspirin. Blood 41:369.
Hovig T (1963). Release of platelet aggregating substance (adenosine diphosphate) from rabbit blood platelets induced by saline "extract" of tendons. Thromb Diathes Hemorr 9:264.
Hume M, Bierbaum V, Kuriakose TX, Surprenant J (1977). Prevention of postoperative thrombosis by aspirin. Am J Surg 133:420.
Hynes KM, Gau GT, Rutherford BD, Kazmier FJ, Frye RL (1973). Effect of aspirin on brachial artery occlusion following brachial arteriotomy for coronary arteriography. Circulation

47:554.
Jaffe EA, Weksler BB (1979). Recovery of endothelial cell prostacyclin production after inhibition by low-doses of aspirin. J Clin Invest 63:532.
Jennings JJ, Harris WH, Sarmiento A (1976). A clinical evaluation of aspirin prophylaxis of thromboembolic disease after total hip arthroplasty. J Bone Joint Surg 58A:926.
Jørgensen KA, Stoffersen E (1978). Dipyridamole and platelet function. Lancet II:1258.
Kaegi A, Pineo GF, Shimizu A, Trivedi H, Hirsh J, Gent M (1974). Arterio-venous shunt thrombosis: prevention by sulphinpyrazone. N Engl J Med 290:304.
Kaegi A, Pineo GF, Shimizu A, Trivedi H, Hirsh J, Gent M (1975). The role of sulphinpyrazone in the prevention of arterio-venous shunt thrombosis. Circulation 52:497.
Kelton JG, Blajchman MA (1980). Prostaglandin I_2 (prostacyclin). Can Med J 122:175.
Kelton JG, Hirsh J, Carter CJ, Buchanan MR (1978). Thrombogenic effect of high-dose aspirin in rabbits. J Clin Invest 62:892.
Loew D, Vinazzer H (1976). Dose-dependent influence of acetylsalicylic acid on platelet functions and plasmatic coagulation factors. Haemostasis 5:239.
Madow BM (1960). Use of antimalarial drugs as "desludging" agents in vascular disease processes. JAMA 172:1630.
Majno G, Joris I (1978). Endothelium 1977: a review. Adv Exp Med Biol 104:169.
Malmgren R, Olsson P, Tornling G (1979). Uptake and release of serotonin in adhering platelets. Relationship to time and effect of acetylsalicylic acid. Thromb Res 15:803.
Masotti G, Poggesi L, Galanti G, Neri Serneri GG (1979). Stimulation of prostacyclin by dipyridamole. Lancet I:1412.
McEnany MT, DeSanctis RW, Hawthorne JW, Mundth ED, Weintraub RM, Austen WG, Salzman EW (1976). Effect of anti-thrombotic therapy on aortocoronary vein graft patency rates. Circulation Supp. 2,54:124
McKenna R, Galante J, Bachman F, Wallace DL, Kaushal SP, Meredith P (1980). Prevention of venous thromboembolism after total knee replacement by high-dose aspirin or intermittent calf and thigh compression. Br Med J 23:154.
Medical Research Council (Report of the Steering Committee) (1972). Effect of aspirin on postoperative venous thrombosis. Lancet II:441.
Michie DD, Wombolt DG (1977). Use of sulphinpyrazone to prevent thrombus formation in arteriovenous fistulas and bovine grafts of patients on chronic hemodialysis. Curr

Ther Res 22:196.
Mielke CH, Britten AFH (1970). Aspirin as an antithrombotic agent: Template bleeding time - test of antithrombotic effect. Blood 36:855,A91.
Mielke CH, Ramos JC, Britten AFH (1973). Aspirin as an antiplatelet agent: template bleeding time as a monitor of therapy. Am J of Clin Path 59:236.
Mills DC, Smith JB (1971). The influence of platelet aggregation of drugs that effect accumulation of adenosine 3':5' cyclic monophosphate in platelets. Biochem J 121:185.
Moncada S, Gryglewski R, Bunting S, Vane JR (1976). An enzyme isolated from arteries transforms prostaglandin endoperoxides to an unstable substance that inhibits platelet aggregation. Nature 263:663.
Moncada S, Higgs EA, Vane JR (1977). Human arterial and venous tissues generate prostacyclin (prostaglandin X), a potent inhibitor of platelet aggregation. Lancet I:18.
Moncada S, Korbut R (1978). Dipyridamole and other phosphodiesterase inhibitors act as antithrombotic agents in potentiating endogenous prostacyclin. Lancet I:1286.
Moncada S, Vane JR (1978). Pharmacology and endogenous roles of prostaglandin endoperoxides, thromboxane A_2 and prostacyclin. Pharm Rev 30:293.
Moncada S, Vane JR (1979). The role of prostacyclin in vascular tissues. Federation Proc 38:66.
Morris CK, Mitchell JRA (1977). Preventing venous thromboembolism in elderly patients with hip fractures, studies of low-dose heparin, dipyridamole, aspirin and flurbiprofen. Br Med J 1:535.
Mustard JR, Packham MA (1979). The reaction of the blood to injury. In Movat HZ (ed): "Inflammation, immunity and hypersensitivity" 2nd edition. New York, Harper & Row p557
Needleman P, Wyche A, Raz A (1979). Platelet and blood vessel arachidonate metabolism and interactions. J Clin Invest 63:345.
O'Brien JR, Tulevski V, Etherington M (1971). Two in vivo studies comparing high and low aspirin dosage. Lancet I:399.
Packham MA, Warrior ES, Glynn MF, Senyi AS, Mustard JF (1967). Alteration of the response of platelets to surface stimuli by pyrazole compounds. J Exp Med 126:171.
Paterson JS (1969). The pathology of venous thrombi. In Sherry S, Brinkhous KM, Genton E, Stengle JM (eds): "Thrombosis", Washington D.C: Natl Academy Sci p 321.
Philp RB, Lemieux JRV (1968). Comparison of some effects of dipyridamole and adenosine on thrombus formation, platelet adhesiveness and blood pressure in rabbits and rats.

Nature 218:1072.
Pilcher DB (1975). Hydroxychloroquine sulfate in prevention of thromboembolic phenomena in surgical patients. Am Surg 41:761.
Plante J, Boneu B, Vaysse C, Barret A, Bouzi M, Bierme R (1979). Dipyridamole-aspirin versus low doses of heparin in the prophylaxis of deep venous thrombosis in abdominal surgery. Thromb Res 14:399.
Preston FE, Emmanual IG, Winfield DA, Malia RG (1974). Essential thrombocythaemia and peripheral gangrene. Br Med J 3:548.
Rajah SM, Crow MJ, Penny AF, Ahmad R, Watson DA (1977). The effect of dipyridamole on platelet function: correlation with blood levels in man. Br J Clin Pharm 4:129.
Renney JTG, O'Sullivan EF, Burke PF (1976). Prevention of postoperative deep vein thrombosis with dipyridamole and aspirin. Br Med J 1:992.
Ritchie JL, Harker LA (1977). Platelet and fibrinogen survival in coronary atherosclerosis. Am J Card 39:595.
Rogers PH, Walsh PN, Marder VJ, Bosak GC, Lachman JW, Ritchie WGM, Oppenheimer L, Sherry S (1978). Controlled trial of low-dose heparin and sulphinpyrazone to prevent venous thromboembolism after operation on the hip. J Bone Joint Surg 60A (6):758.
Rosenberg FJ, Gimber-Phillips PE, Groblewski GE, Davison C, Phillips DK, Goralnick SJ, Cahill ED (1971). Acetylsalicylic Acid: Inhibition of platelet aggregation in the rabbit. J Pharmacol Exp Ther 179:410.
Ross R, Glomset JA (1976). The pathogenesis of atherosclerosis. N Engl J Med 295:369.
Roth GJ, Majerus PW (1975). The mechanism of the effect of aspirin on human platelets. J Clin Invest 56:624.
Russell RWR (1961). Observations on the retinal blood vessels in monocular blindness. Lancet II:1422.
Salzman EW, Harris WH, DeSanctis RW (1971). Reduction in venous thromboembolism by agents affecting platelet function. N Engl J Med 284:1287.
Sbar S, Schlant RC (1967). Dipyridamole in the treatment of angina pectoris: A double-blind evaluation. JAMA 201:865.
Schöndorf TH, Hey D (1978). Modified 'low-dose' heparin prophylaxis to reduce thrombosis after hip joint operation. Thromb Res 12:153.
Sevitt S (1973). Pathology and pathogenesis of deep vein thrombi. In Moser KM, Stein M (eds): "Pulmonary Thromboembolism". Chicago, Year Book Med Pub. p 93.
Silvergleid AJ, Bernstein R, Burton DS, Tanner JB, Silver-

man JF, Schrier SL (1977). Aspirin-Persantin prophylaxis in elective total hip replacement. Thromb Haemostas 38:166.

Smith JB, Ingerman C, Kocsis JJ, Silver MJ (1974). Formation of an intermediate in prostaglandin biosynthesis and its association with the platelet release reaction. J Clin Invest 53:1468.

Smythe HA, Ogryzlo MA, Murphy EA, Mustard JF (1965). The effect of sulphinpyrazone (Anturan) on platelet economy and blood coagulation in man. Can Med Assoc J 92:818.

Steele PP, Weily HS, Genton E (1973). Platelet survival and adhesiveness in recurrent venous thrombosis. N Engl J Med 288:1148.

Steele P, Weily HS, Davies H, Genton E (1974). Platelet survival in patients with rheumatic heart disease. N Engl J Med 290:537.

Steele P, Battock D, Genton E (1975). Effects of clofibrate and sulphinpyrazone on platelet survival time in coronary artery disease. Circulation 52:473.

Steele P, Carroll J, Overfield D, Genton E (1977). Effect of sulphinpyrazone on platelet survival time in patients with transient cerebral ischemic attacks. Stroke 8:396.

Subbarao K, Rucinski B, Rausch MA, Schmid K, Niewiarowski S, (1977). Binding of dipyridamole to human platelets and to a_1 acid glycoprotein and its significance for the inhibition to adenosine uptake. J Clin Invest 60:936.

Sullivan MM, Harken DE, Gorlin R (1971). Pharmacologic control of thromboembolic complications of cardiac-valve replacement. N Engl J Med 284:1391

Taguchi E, Matsumura H, Washizu T (1975). Effect of athromogen therapy, especially high dose therapy of dipyridamole after prosthetic valve replacement. J Cardiovasc Surg (Torino) 16:8.

The Canadian Cooperative Study Group (1978). A randomized trial of aspirin and sulphinpyrazone in threatened stroke. N Engl J Med 299:53.

The Persantine-Aspirin Reinfarction Study Research Group (1980). Persantine and aspirin in coronary heart disease. Circulation 62:449.

Vargaftig BB (1978). The inhibition of cyclo-oxygenase of rabbit platelets by aspirin is prevented by salicylic acid and by phenanthrolines. Eur J Pharmacol 50:231.

Vreeken J, van Aken WG (1971). Spontaneous aggregation of blood platelets as a cause of idiopathic thrombosis and recurrent painful toes and fingers. Lancet II:1394.

Weily HS, Genton E (1970). Altered platelet function in patients with prosthetic mitral valves. Effects of sulphinpyrazone therapy. Circulation 42:967.

Weiss HJ, Aledort LM (1967). Impaired platelet/connective tissue reaction in man after aspirin ingestion. Lancet II:495.

Weston MJ, Rubin MH, Langley PG, Westaby S, Williams R, (1977). Effects of sulphinpyrazone and dipyridamole on capillary bleeding time in man. Thromb Res 10:833.

Winchester JF, Forbes CD, Courtney JM, Reavey M, Prentice CRM (1977). Effect of sulphinpyrazone and aspirin on platelet adhesion to activated charcoal and dialysis membranes in vitro. Thromb Res 11:443.

Wirecki M (1967). Treatment of angina pectoris with dipyridamole: a long term double-blind study. J Chronic Dis 20:139.

Wong LT, Zawidzka Z, Thomas RB (1978). Effect of acetylsalicyclic acid, sulphinpyrazone and their combination on collagen-induced platelet aggregation in guinea pigs. Pharmacological Res Communications 10:939.

Wu TK, Tsapogas MJ, Jordan FR (1977). Prophylaxis of deep venous thrombosis by hydroxychloroquine sulfate and heparin. Surg Gynecol Obstet 145:714

Wylie JS, Chesterman CN, Morgan FJ, Castaldi PA (1979). The effect of sulphinpyrazone on the aggregation and release reactions of human platelets. Thromb Res 14:23.

Zekert F, Kohn P, Vormittag E, Poigenfurst J, Thien M (1974). Thromboembolic prophylaxe mit acetylsalicylsaure bei operationen wegen huftgelenknaher frakturen. Monatsschrift der unfallheilkunde 77:97.

Zucker MB, Peterson J (1970). Effect of acetylsalicylic acid other non-steroidal anti-inflammatory agents and dipyridamole on human blood platelets. J Lab Clin Med 76:66

Zucker MB, Rothwell KG (1978). Differential influence of salicylate compounds on platelet aggregation and serotonin release. Current Therapeutic Res 23:194.

PLATELET CANCER CELL INTERACTION IN METASTASIS FORMATION.
PLATELET AGGREGATION INHIBITORS: A POSSIBLE APPROACH TO
METASTASIS PREVENTION

H. Gastpar[1,2], J.L. Ambrus[2,3,4], C.M. Ambrus[2,5]

University of Munich Medical School, Munich,
West Germany[1]; Roswell Park Memorial Institute[2],
and the State University of New York at Buffalo,
Departments of Internal Medicine[3], Exper. Pathol.[4]
and Pediatrics[5], Buffalo, New York 14263

ABSTRACT

   Abnormal platelet aggregation on circulating and lodged cancer cells may play an important role in the early stages of metastasis formation. The immediate drop in the number of circulating platelets following intravenous injection of Walker-256 carcinosarcoma cells in rats represents the experimental counterpart of the morphologic finding of tumor cells associated with tumor clusters in the pulmonary arterioles and capillaries.

INTRODUCTION

   Following intravenous injection of Walker-256 carcinosarcoma cells into rats, several platelet aggregation inhibitors have been shown to inhibit, in a dose-dependent manner, cancer cell stickiness, reduction in platelet count and the mortality due to tumor cell embolism. Some of the substances are able to prolong significantly the circulation time of intravenously injected Ehrlich ascites tumor cells in mice. Two of the platelet aggregation inhibitors tested caused significant reduction of spontaneous lung metastases in the syngeneic Wilms' tumor of the rat and of the C-1300 neuroblastoma of the mouse.

   A clinical long-term pilot study of prophylaxis of metastasis in patients suffering from sarcomas and malignant lymphomas restricted to the head and neck region was started nine years ago employing the platelet aggregation inhibitor mopidamole (RA-233). Final statistical evaluation of the matched pairs show that the metastases in the mopidamole group occurred significantly later and to a lesser extent than in the untreated group.

COAGULATION PROCESSES IN TUMOR DISSEMINATION

The fate of disseminated tumor cells in the circulation is largely determined by physiochemical surface properties of the lining of the vasculature, which are foreign to the blood, and by the thromboplastic activities intrinsic to the cell (Gastpar, 1976a). The term "cancer cell stickiness" (Coman, 1961) describes a tumor-specific, population-variable property of the circulating tumor cells in terms of an increased tendency to adhere to foreign surfaces (Sträuli, 1966). Cell populations with a high degree of stickiness have a correspondingly high thromboplastic activity and transplantation rate (Koike, 1964; Kojima and Sakai, 1964). This specific stickiness may be induced by a transcellular diffusion of tumor-specific thromboplastic factors and ADP - and possibly also mucin in the case of certain adenocarcinomas. These factors lead to the formation of fibrin monomers and a pericellular fibrin film (Gastpar, 1979).

In previous investigations we were able to demonstrate that malignant tumors with a high frequency of metastasis correlate with a low incidence of circulating tumor cells in the peripheral blood. In contrast, in neoplastic diseases which metastasize only infrequently, the percentage of free-floating tumor cells in the peripheral venous blood was significantly higher (Figure 1). Obviously the disseminated tumor cells of metastasizing cancers are more sticky than the cells of non-metastasizing ones, resulting in a greater tendency to adhere to vascular endothelium (Gastpar, 1976c). Using intravital capillary microscopy and microcinematography, it has been demonstrated in animals that sticky tumor cells and tumor cell-platelet complexes are able to adhere to normal vascular endothelium. Within minutes parietal microthrombi occur and are immediately stabilized by a fibrin network upon which further platelets and leukocytes gather (Gastpar et al, 1961; Wood et al, 1961; Gastpar, 1970).

Many experimental data suggest that development of metastases from blood-borne cancer cells is, in some instances, closely related to disseminated intravascular coagulation (Sträuli, 1966). The immediate drop in the number of circulating platelets following intravenous injection of Walker-256 carcinosarcoma cells in rats (Gastpar,

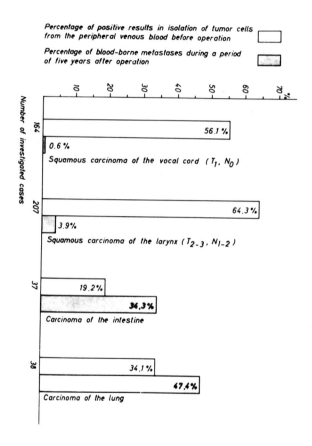

Figure 1. Correlation between primarily positive isolations of circulating tumor cells from peripheral venous blood and the frequency of blood-borne metastases during an observation period of five years after tumor resection (Gastpar, 1972.

1972; Hilgard, 1973) represents the hematologic counterpart of the morphologic findings of tumor cells associated with platelet clusters arrested in the pulmonary arterioles and capillaries (Jones et al, 1971; Warren and Vales, 1972; Hilgard and Gordon-Smith, 1974). The demonstration of fibrin monomers in the blood of the animals indicates that intravascular fibrin formation accompanied the consumption of platelets (Hilgard, 1973).

Recently Hara et al (1980) observed that human platelets contain a growth-promoting factor to four well-established malignant cell lines which is released during thrombin- or collagen-induced aggregation. We have demonstrated (Gastpar, 1970) that platelet aggregation may also occur on human cancer cells during their circulation in the blood, and Abrahamsen (1976) showed that patients with metastatic cancer have a permanent reduction of platelet survival and increased platelet consumption.

EFFECT OF PLATELET AGGREGATION INHIBITORS ON METASTASIS FORMATION IN ANIMALS

Gasic and Gasic (1962) observed that pretreatment of mice with neuraminidase resulted in a significant reduction in the frequency of pulmonary metastases after intravenous tumor cell transplantation. Further studies showed that neuraminidase induced marked thrombocytopenia in the animals and, conversely, that this antimetastatic effect was abolished by platelet transfusion (Gasic et al, 1968).

We were able to demonstrate (Gastpar, 1970) that the rate of immediately fatal tumor cell embolisms in the lungs after intravenous transplantation of 1×10^6 Walker-256 carcinosarcoma cells in rats was significantly reduced by pretreatment with the platelet aggregation inhibitors dipyridamole (RA 8) and several other pyrimido-pyrimidine derivatives, such as mopidamole (RA 233) and others (Figure 2). At the same time, these substances reduced the tendency of the circulating tumor cells to adhere to the endothelium of the mesenteric vessels in a dose-dependent manner as observed by intravital microscopy (Figure 3) and to lessen the fall in the platelet count occurring immediately after tumor cell transplantation (Figure 4) (Gastpar, 1972). Similar findings were obtained with the aggregation inhibitors bencyclane, pentoxifylline, sulfinpyrazone and the pyrimidopyrimidine derivative RX-RA 69 (Gastpar, 1973, 1974, 1981; Gastpar et al, 1978).

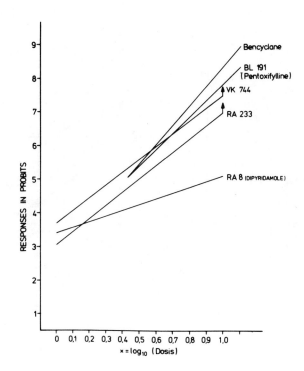

Figure 2. Probit regression lines (Finney, 1952) for the effect of bencyclane, pentoxifylline and three pyrimido-pyrimidine derivatives on prevention of cancer cell embolism mortality in rats after intravenous injection of 1 x 10^6 Walker-256 carcinosarcoma cells. The arrows at the top of the RA 233 and VK 744 lines indicate 100% rates (Gastpar, 1977).

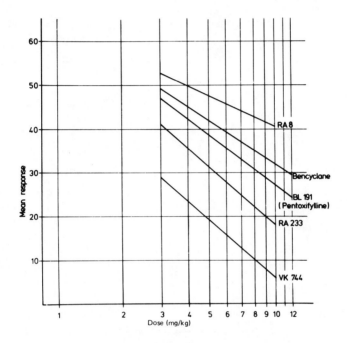

Figure 3. Dose-response relationship for the effect of bencyclane, pentoxifylline and three pyrimido-pyrimidine derivatives on cancer cell stickiness in vascular endothelium in the mesentery (cells/cm^2) of surviving rats after intravenous injection of 1 x 10^6 Walker-256 carcinosarcoma cells (Gastpar, 1977).

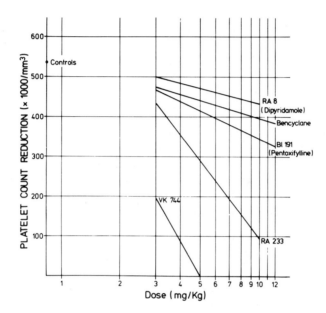

Figure 4. Dose-response relationship for the effect of bencyclane, pentoxifylline and three pyrimido-pyrimidine derivatives on platelet count reduction in the venous blood of surviving rats after intravenous injection of 1 x 10^6 Walker-256 carcinosarcoma cells (Gastpar, 1977).

Schumann (1976) also showed an inhibition of the adhesive tendency of tumor cells by bencyclane, both in vivo and in vitro. Gasic et al (1973) found a significant decrease in the frequency of metastasis formation in several, including spontaneously metastasizing mouse tumors, with the platelet aggregation inhibitor ASA (aspirin). We were able to show that the aggregation inhibitors mopidamole (RA 233) (Figure 5), bencyclane, and pentoxifylline, in therapeutic doses can significantly increase the circulation time of intravenously injected tumor cells in mice (Ambrus et al, 1978; Gastpar et al, 1977, 1978).

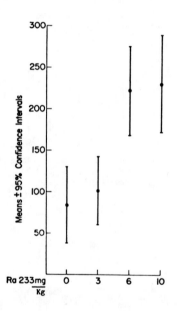

Figure 5. Effect of various intravenous doses of mopidamole (RA 233) on the circulation time (minutes) of intravenously transplanted polyploid Ehrlich ascites tumor cells in mice (Ambrus et al, 1978).

Wood and Hilgard (1972) did not find a significant effect of ASA on the frequency of metastasis formation after intravenous infusion of V_2-carcinosarcoma cells in rabbits. However, pulmonary metastases in mice induced by intravenous infusion of 5×10^5 Lewis lung carcinoma cells was reduced by intraperitoneal pretreatment of the animals with ASA, bencyclane or mopidamole although there was no effect on the rate of immediately fatal pulmonary tumor cell embolism. The number of pulmonary metastases in the animals affected showed a marked increase after pretreatment with all three substances (Hilgard et al, 1976).

The reason for these discrepancies is not at present readily explicable. Apart from the route of administration and the blood levels of the aggregation inhibitors with relatively short duration of action, a major part may be played by the transplantation route, the number of transplanted tumors cells and the nature of the transplanted tumor (Gastpar, 1976b). The studies by Gasic et al (1976), Kinjo et al (1979), and Donati et al (1980) showed that there were enormous differences between the platelet-aggregating, the procoagulant, and the fibrinolytic activities in various tumor strains. When using tumor cells with a high degree of platelet aggregation inducing activity, platelet aggregation can be inhibited in vitro by the apyrase. This suggests that tumor cells first induce the release of adenine nucleotides, with accelerated platelet aggregation as a secondary effect (Gasic et al, 1976).

A very important role in the regulation of energy metabolism of platelets is played by cyclic AMP (Schneider et al, 1974). The pool of cAMP is controlled by the two enzymes, adenylate cyclase and cAMP-phosphodiesterase (PDE). The fact that inhibitors of adenylate cyclase and activators of PDE affect platelet aggregation suggests a role for cAMP (Reuter and Gross, 1978). Pyrimido-pyrimidine derivatives, dipyridamole and mopidamole, as well as the methylxanthine derivative, pentoxifylline, are well-known inhibitors of PDE and thereby cause an increase in the cAMP level of platelets (Salzman and Weisenberger, 1972; Stefanovich et al, 1977), especially the membrane-related cAMP (Amer and Mayol, 1973). However, at therapeutic doses the above drugs have only a moderate and short lasting effect in vitro on cAMP levels on human platelets and on platelet aggregation.

This discrepancy between low in vitro and the high

in vivo activity may be explicable by the new observations that dipyridamole (Masotti et al, 1979; Blass et al, 1980) and pentoxifylline (Weithmann, 1980) at therapeutic doses stimulate the biosynthesis and/or the release of prostacyclin (PGI_2) from the vessel wall, which in turn stimulates platelet adenylate cyclase, resulting in further elevation of cAMP levels in platelets. A combination of pentoxifylline and prostacyclin (10^{-7} M) in vitro greatly stimulates cAMP levels as well as inhibition of aggregation over and beyond the action of prostacyclin itself. It may be speculated that the interaction between the intact vessel wall and platelets is required for the full activity of these derivatives in vivo (Weithmann, 1980). Masotti et al (1979) and Blass et al (1980) showed that the pyrimido-pyrimidine derivative dipyridamole (RA 8) also increases prostacyclin syntheses. The enhancement of prostacyclin release from the vessel wall is followed by elevated cAMP synthesis in the platelets. The cAMP level, stimulated in this manner, may be further increased by the inhibitory effect on platelet-PDE by dipyridamole, mopidamole or pentoxifylline.

Cyclic AMP, however, not only has a major influence on the integrity of the platelets but it also controls the arrest of mitosis by chalones (Iversen, 1960). Transformed cells show a lower concentration of cAMP than the corresponding normal cells (Sheppard, 1972) and the addition of cAMP derivatives inhibited cell division (Heidrick and Ryan, 1970; Smets, 1972; Wijk et al, 1972). In various tumor cell cultures it was found that cAMP derivatives induced an inhibition of proliferation (Ryan and Heidrick, 1968; Gericke and Chandra, 1969; Keller, 1972), sometimes causing differentiation of tumor cells to normal (Johnson et al, 1971; Sheppard, 1971). The PDE activity of leukemic lymphocytes is 5 to 10 times greater than that of normal lymphocytes (Hait and Weiss, 1977), but Chatterjee and Kim (1975) observed that the steady-state cAMP levels of spontaneously metastasizing rat mammary carcinomas were 1.3-2.0 times higher than in the non-metastasizing types, inversely proportional to their PDE-activities. Stefanovich et al (1981) showed that in human tumor biopsies certain types of PDE activities were higher than in corresponding normal tissues from the same patients.

There is some evidence to suggest that the aggregation inhibitors with a stimulating effect on cAMP synthesis in vivo, and an inhibiting action on PDE, (dipyridamole, mopidamole, pentoxifylline) may also have a direct effect on tumor cells by inhibiting their cAMP turnover. The xanthine derivative theophylline has a prophylactic effect against growth of CELO-virus transformed cells (Reddi and Constantinides, 1972). Unpublished studies by Young of the Sloan Kettering Institute also point in this direction, showing that ^{3}H-thymidine incorporation into human leukemic cells was inhibited by the pyrimidine derivative mopidamole. Gordon et al (1979) demonstrated that pentoxifylline is able to diminish spontaneous lung metastases significantly ($p < 0.01$) in the syngenetic Wilms' tumor (nephroblastoma) of the rat and the C 1300 neuroblastoma of the mouse, while it has no effect in the NIH-renal adenocarcinoma of the mouse. Similar experiments indicate that mopidamole works in the same way. The antimetastatic effect of PDE-inhibitors may be potentiated by simultaneous treatment with dibutyryl cAMP. Similar effects appear to be present in PGI_2-releasing agents (Ambrus and Gastpar, 1981).

CLINICAL RESULTS OF METASTASIS PROPHYLAXIS IN MALIGNANT HUMAN TUMORS USING THE PLATELET AGGREGATION INHIBITOR MOPIDAMOLE

On the basis of our in vivo findings in animal tumors using several aggregation inhibitors, and the good clinical tolerance of the pyrimido-pyrimidine derivative mopidamole, nine years ago we started a long-term clinical study to investigate the possibility of metastasis prophylaxis using mopidamole in a total of 38 patients with primary nodular sarcoma and malignant lymphoma in the head and neck region (Table 1).

Table 1. Histological tumor diagnoses of 38 matched pairs treated with pyrimido-pyrimidine derivative, mopidamole (RA 233)

Number of Patient Pairs	Diagnosis
19	Retothelial Sarcoma
6	Lymphosarcoma
5	Hodgkin's Sarcoma
2	Melanosarcoma
2	Anaplastic Sarcoma
2	Spinocellular Sarcoma
1	Angioplastic Sarcoma
1	Rhabdomyosarcoma
38	

We are referring to a very carefully selected group of patients, tumor stage classification $T_1N_1M_0$, in whom no indications of metastasis existed, either clinically or by x-rays, lymphographically or scintigraphically or by surgical exploration.

The patients received mopidamole in daily oral doses of 3 x 500 mg, or 3 x 250 mg when under 16 years of age. The preparation was administered until side effects occurred, recurrent tumor growth at the original site was observed or metastases were found. Otherwise, the above-mentioned dose schedule was continued for a period of five years. Mopidamole is very well tolerated. In spite of the extensive duration of treatment, only moderate side effects were thus far observed in only three cases which might have been caused by the drug.

We arranged for a comparable patient group five years ago. This group of 38 matched pairs were comparable for the following factors: (1) age ($\pm$ 5 years); (2) sex; (3) site of tumor; (4) histologic tumor type; (5) surgical procedures; (6) x-ray techniques and doses used; and (7) time of surgery ($\pm$ 6 months).

From a statistical point of view, the use of these criteria made our matching technique free of bias. The study is now completed. Sixteen patients were under prophylactic treatment with mopidamole over the full five-year time period. Up to now, that is 65-109 months, they are without metastases or tumor recurrence.

Another group of nine patients terminated the prophylactic treatment for various reasons after 20-46 months but remained under clinical surveillance. Of these, only one patient developed metastases, 22 months after termination of treatment. In the other eight patients, neither recurrent tumor growth nor development of metastases has occurred in a period of 62-109 months. In all 24 patients of the mopidamole group (63.2%) neither recurrent tumor growth nor metastases occurred over a five year period. Within a period of 3-59 months, the tumor recurred or metastases developed in 14 patients (36.8%), six of whom have died in the meantime (16.8%).

Six of the 38 patients in the control group (15.8%) exhibited no recurrence of tumor growth or development of metastases. In the remaining 32 patients, recurrent tumor growth or metastases were observed within a period of 7-39 months (84.2%). Seventeen of these 32 patients (44.7%) have died in the meantime.

Analysis by the "Life Table Technique" of Cutler and Ederer (1958) (Figure 6) showed a highly significant difference between the two groups ($\hat{z}38 = 4.8$; $p < 0.001$). The final statistical evaluation is not yet available, but it can be stated that metastases in the mopidamole group occurred significantly later and to a lesser extent than in the untreated group. The general significance of the antimetastatic effect of mopidamole, however, has to be substantiated in other human malignant tumors as well.

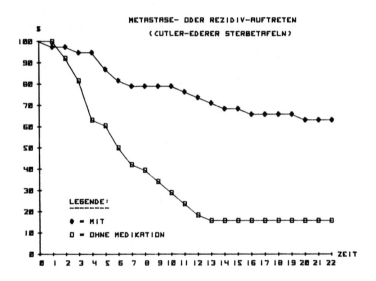

Figure 6. Cumulative relation between observation period (Quarters) after surgery/radiation and manifestation of metastases or recurrences of tumor growth on the basis of 38 matched pairs. +, treated with mopidamole; □, untreated controls.

REFERENCES

Abrahamsen AF (1976). Platelet turnover in metastatic cancer and the effect of platelet aggregation inhibitors. Z Krabsforsch 86:109.

Ambrus JL, Ambrus CM, Gastpar H (1978). Study of platelet aggregation in vivo. VI. Effect of pyrimido-pyrimidine derivative (RA 233) on tumor cell metastasis. J Med 9:183.

Ambrus JL, Gastpar H (1981). Unpublished data.

Amer MS, Mayol RF (1973). Studies with phosphodiesterase. III. Two forms of the enzyme from human blood platelets. Biochim Biophys Acta 309:149.

Blass KE, Block HU, Forster W, Ponike K (1980). Dipyridamole: a potent stimulator of prostacyclin (PGI_2) biosynthesis. Br J Pharmacol 68:71.

Chatterjee SK, Kim U (1975). Adenosine-3'5'-cyclic monophosphate levels and adenosine-3'5'-monophosphate phosphodiesterase activity in metastasizing and non-metastasizing rat mammary carcinomas. J Natl Cancer Inst 54:181.

Coman DR (1961). Adhesiveness and stickiness: two independent properties of cell surface. Cancer Res 21:1436.

Cutler SJ, Ederer F (1961). Maximum utilization of the life table method in analyzing survival. J Chron Dis 8:699 (1958).

Donati MB, Semeraro N, Poggi A (1980). Some interactions between cancer cells and components of the hemostatic system. In Grundmann E (ed); "Metastatic Tumor Growth, Cancer Campaign," Vol 4, Stuttgart: Fischer, p 101.

Finney DJ (1952). "Probit Analysis." 2nd Ed. Cambridge: University Press.

Gasic G, Gasic T (1962). Removal of sialic acid from the cell coat in tumor cells and vascular endothelium and its effect on metastasis. Proc Natl Acad Sci (USA) 48:1172.

Gasic G, Gasic T, Stewart CC (1968). Antimetastatic effects associated with platelet reduction. Proc Natl Acad Sci (USA) 61:46.

Gasic C, Gasic T, Galanti N, Johnson T, Murphy S (1973). Platelet tumor cell interaction in mice. The role of platelets in the spread of malignant disease. Int J Cancer 11:704.

Gasic G, Koch PAG, Hsu B, Gasic T, Niewiarowski N (1976). Thrombogenic activity of mouse and human tumors: Effect on platelets, coagulation and fibrinolysis and possible significance for metastasis. Z Krabsforsch 86:263.

Gastpar H, Graeber F, Herrmann A, Loebell E. (1961) Intravitalmikroskopische Beobachtungen von Tumorzellen in der Blutbahn (16 mm Film) Arch Ohr-Nas-Kehlk-Heilk 178:534.
Gastpar H (1967). Discussion. Thrombos Diathes haemorrh Suppl 24:52
Gastpar H (1970). Stickiness of platelets and tumor cells influenced by drugs. Thrombos Diathes haemorrh Suppl 42:291.
Gastpar H (1972). Inhibition of cancer cell stickiness by anticoagulants, fibrinolytic drugs and pyrimido-pyrimidine compounds. Hemat Rev 3:1.
Gastpar H (1973). Die Hemmung der "Cancer Cell Stickiness" durch Bencyclan-hydrogenfumarat (Fludilat). Fortschr Med 91:1322.
Gastpar H (1974). The inhibition of cancer cell stickiness by the methylxanthine derivative pentoxifylline (BL 191) Thrombos Res 5:277.
Gastpar H (1976a). Maligne Geschwülste und Thrombosen. Med Welt 27 (NF):1737.
Gastpar H (1976b). Discussion. In Gastpar H (ed): "Onkohaemostaseologie." Stuttgart: Schattauer (S. 149).
Gastpar H (1976c). Beziehungen zwischen positiven Tumorzellbefunden im peripheren Venenblut, Fernmetastasierung und thromboembolischen Ereignissen bei Patienten mit Karzinomen verschiedener Lokalisation. Laryng Rhinol 55:70.
Gastpar H (1977). Platelet cancer cell interaction in metastasis formation: A possible approach to metastasis prophylaxis. J Med 8:103.
Gastpar H, Ambrus JL, Thurber LE (1977). Study of platelet aggregation in vivo. II. Effect of bencyclane on circulating metastastic tumor cells. J Med 8:53.
Gastpar H, Ambrus JL, Ambrus CM (1978). Study of platelet aggregation in vivo. VII. Effect of pentoxifylline on circulating tumor cells. J Med 9:265.
Gastpar H (1978). Die Hemmung der "Cancer Cell Stickiness," ein Modell zur Prüfung von Thrombozytenaggregationshemmern in vivo. IV. Die Wirkung von Sulfinpyrazon. Fortschr Med 96:1823.
Gastpar H (1979). Der Einfluss von Antikoagulantien und Aggregationshemmern auf die Metastasierung. In Krokowski E (ed): "Neue Aspekte der Krebsbekämpfung," Stuttgart: Thieme, p 111.
Gastpar H (1981). Unpublished experimental data.
Gericke D, Chandra P. (1969). Inhibition of tumor growth by nucleotide cyclic 3'5'-monophosphates. Hoppe Seylers Z Physiol Chem 350:1469.

Gordon S, Witul M, Cohen G, Williams P, Gastpar H, Murphy GP, Ambrus JL (1979). Studies on platelet aggregation inhibitors in vivo. VIII. Effect of pentoxifylline on spontaneous tumor metastasis. J Med 10:435.

Hait WN, Weiss B (1977). Characteristics of the cycle nucleotide phosphodiesterases of normal and leukemic lymphocytes. Biochim Biophys Acta 497:86.

Hara Y, Steiner M, Baldini MG (1980). Platelets as a source of growth-promoting factor(s) of tumor cells. Cancer Res 40:1212.

Heidrick ML, Ryan WL (1970). Cyclic nucleotides on cell growth in vitro. Cancer Res 30:376.

Hilgard P (1973). The role of blood platelets in experimental metastases. Br J Cancer 28:429.

Hilgard P, Gordon-Smith EC (1974). Microangiopathic haemolytic anaemia and experimental tumor-cell emboli. Br J Haematol 26:651.

Hilgard P, Heller H, Schmidt CG (1976). The influence of platelet aggregation inhibitors on metastasis formation in mice (3LL). Z Krebsforsch 86:243.

Iversen OH (1969). Chalones of the skin. In Wolstenhome GEW, Knight J (eds): "Homeostatic regulators." London: Churchill, p 29.

Johnson GS, Freidman RM, Paston I (1971). Restoration of several morphological characteristics of normal fibroblasts in sarcoma cells treated with adenosine-3'5'-cyclic monophosphate and derivatives. Proc Natl Acad Sci (USA) 68:425.

Jones DS, Wallace AC, Frazer EE (1971). Sequence of events in experimental metastases of Walker-256 tumor: light, immunofluorescent and electron microscopic observations. J Natl Cancer Inst 46:493.

Keller R (1972). Suppression of normal and enhanced tumor growth in rats by agents interacting with intracellular cyclic nucleotides. Life Sci 11:485.

Kinjo M, Oka K, Naito S, Kohga S, Tanaka K, Oboshi S, Hayata Y, Yasumoto K (1979). Thromboplastic and fibrinolytic activities of cultured human cancer cell lines. Br J Cancer 39:15.

Koike A (1964). Mechanisms of blood-borne metastases. I. Some factors affecting lodgement and growth of tumor cells in the lungs. Cancer 17:450.

Kojima K, Sakai I (1964). On the role of stickiness of tumor cells in the formation of metastasis. Cancer Res 24:1887.

Masotti G, Poggesi L, Galanti G, Neri Seneri G (1979). Stimulation of prostacyclin by dipyridamole. Lancet I:1412.

Reddi PK, Constantinides SM (1972). Partial suppression of tumor production by dibutyryl cyclic AMP and theophylline. Nature 238:286.

Reuter H, Gross R (1978). Platelet metabolism. In Gastpar H (ed): "Collagen Platelet Interaction," Stuttgart: Schattauer, p. 87.

Ryan WL, Heidrick ML (1968). Inhibition of cell growth by 3'5'-monophosphate. Science 162:1484.

Salzman EW, Weisenberger H (1972). Role of cyclic AMP in platelet function. Adv Cyclic Nucleotide Res 1:231.

Schneider W (1974). Regulation of energy metabolism in human blood platelets by cyclic AMP. In Baldini MG, Ebbe S (eds): "Platelets: Production, Function, Transfusion, and Storage." New York: Grune & Stratton, p 177.

Schumann J (1976). Der Einfluss von Bencyclan-hydrogenfumarat (Fludilat) auf die Haftfähigkeit von Tumorzellen in vivo und in vitro. In Gastpar H (ed): "Onkohämostaseologie." Stuttgart: Schattauer, p 143.

Sheppard JR (1971). Restoration of contact-inhibited growth to transformed cells by dibutyryl adenosine-3'5'-monophosphate. Proc Nat Acad Sci (USA) 68:1316.

Sheppard JR (1972). Difference in cyclic adenosine-3'5'-monophosphate levels in normal and transformed cells. Nature 236:12.

Smets LA (1972). Contact inhibition of transformed cells incompletely restored by dibutyryl cyclic AMP. Nature 239:128.

Stefanovich V, Jarvis P, Grigoleit HG (1977). The effect of pentoxifylline on the 3'5'-cyclic-AMP-system in bovine platelets. Int J Biochem 8:359.

Stefanovich V, Ambrus JL, Ambrus CM, Karakousis C, Takita H (1981) Cyclic nucleotides in normal and malignant human tissues. J Med (in press).

Sträuli P (1966). Intravascular clotting and cancer localization. Thrombos Diathes haemorrh Suppl 20:147.

Warren BA, Vales O (1972). The adhesion of thromboplastic tumour emboli to vessel walls in vivo. Br J Exp Pathol 53:301.

Weithmann KU (1980). The influence of pentoxifylline on interaction between blood vessel wall and platelets. IRCS Med Sci 8:293.

Wijk R van, Wicks WD, Clay K (1972). Effects of derivatives of cyclic-3'5'-adenosine monophosphate on the growth, morphology and gene expression of hepatoma cells in culture. Cancer Res 32:1905.

Wood S Jr, Holyoke ED, Yardley JH (1961). Mechanism of metastasis production by blood-borne cells. Proc 4th Cancer Conf. New York: Academic Press, p 167.

Wood S Jr, Hilgard P (1972). Aspirin and tumor metastasis. Lancet II:1416.

DISCUSSION

Dr. H. Gastpar (Universitat-Hals-Nasen-Ohrenklinik, Munich). One of my concerns in terms of designing antiplatelet therapy in tumors is the kind of preliminary studies from in vitro and animal models in vivo. As an example, aspirin in vitro had no effect on the adherence in the Baumgartner chamber. We have a paper in press using collagen-linked Sepharose which indicates platelet adherence and release after aspirin treatment so that obviously the technique used can give different results in terms of whether a particular in vitro phenomenon has been affected by a particular drug. Clearly, the utility of various in vitro studies as an indicator of what we can expect in vitro, particularly with humans, is a major question. Perhaps we should ignore all the in vitro results and consider only clinical trials.

Dr. H. J. Day (Temple University Medical Center, Philadelphia). I would like to ask Dr. Gastpar what was the rationale for choosing this unusual group of tumors for the use of actual therapy. There is a variety of sarcomas with lesions with tremendous differences in terms of their natural progression. It must have been an extreme challenge to find age controls matched with respect to sex and histology. It makes one worry about the dramatic results.

Dr. Gastpar. Yes, of course. In fact, it was not a prospective study.

Dr. L. R. Zacharski (VA Medical Center, White River Junction). You could absolutely match the histology?

Dr. Gastpar. Yes. At that time, we had many of these tumors in our hospital and we had no difficulty in finding matching patients.

Dr. S. Karpatkin (New York University Medical Center, New York). I would like to raise the question as to the rationale for the use of this type of treatment in patients.

In all the animal studies which have been done the antiplatelet agents have, to my knowledge, been given either before the tumor has been introduced into the animal or very soon after. When the patient comes to you he already has the tumor and probably has silent or micrometastases all over his body. Now, if that is so, why are antiplatelet agents going to be successful in this kind of patient and, if they are successful, what is the theoretical rationale for this success?

Dr. Gastpar. Yes, of course, some of these patients have micrometastases. On the other hand, some patients come to us at a very early stage and, normally, we may run tests or operate at this time.

Dr. Zacharski. I think one main rationale for running these tests in humans is that there is no real alternative treatment. The question then becomes, is the treatment worse than the disease? The answer for all of these drugs that we are interest in is "no."

Dr. J. L. Ambrus (Roswell Park Memorial Institute, Buffalo). I also want to address the question brought up earlier as to whether in vitro experiments are pertinent. Of course, they are never quite as pertinent as in vivo experiments; however, there are effects we can see in in vitro experiments. Some of the agents we have been examining are obviously platelet aggregation inhibitors, but some of these also increase membrane fluidity of red cells. Now, there is a great deal of difference as to whether we add these agents to a suspension of red cells, a suspension of platelets, or to whole blood. The reason for that probably is that whole blood also has leukocytes and apparently some of these agents induce prostacyclin which can then inhibit aggregation.

I want to comment on the very interesting fact that certain of these therapeutic trials appear to be more effective in males than in females. In a series of clinical studies we have found that estrogen, particularly as oral contraceptives rich in estrogen or in treatment of prostatic carcinoma in the male, have a very significant effect on the blood coagulation; they can increase levels of blood coagulation factors and they can also increase platelet aggregability and these may be responsible for the very interesting differences demonstrated in male and female patients.

PLATELET AGGREGATION INHIBITORS AND METASTATIC SPREAD OF NEOPLASTIC CELLS

Julian L. Ambrus, Clara M. Ambrus, and Helmuth Gastpar

Roswell Park Memorial Institute, Buffalo, NY and the University of Munich Medical School, Munich West Germany

In previous studies we have investigated the causes of death in cancer patients in our hospital (Ambrus et al 1975a, 1975b, 1976). The most important cause of death was superinfection primarily with gram negative antibiotic resistant bacteria. The second major cause of death was a category designated as hemorrhagic and/or thromboembolic complications including disseminated intravascular coagulation (DIC) (which includes features from both of these phenomena). Significant correlation was found between the demonstration of circulating tumor cells in blood samples from patients and the incidence of thromboembolism. In a series of studies at the Roswell Park Memorial Institute Hospital and at the University of Munich, circulating tumor cells appeared or their concentration increased during radical cancer surgery (Gastpar 1972). Many of the circulating tumor cells were found to be surrounded by or tailed by a cap of aggregating platelets. It was thought that platelet aggregation induced by tumor cells may be, in part, responsible for disseminated intravascular coagulation and for thromboembolic complications of cancer. This phenomenon may also be involved in the arrest of cancer cells in the microcirculation. Masses of tumor cells, aggregating platelets and fibrin fibers forming around these cells may be arrested in the microcirculation. In the resulting clot, tumor cells are protected from attack by natural killer cells, humoral and cellular immunity. High levels of plasminogen activators in tumor cells (Ossowski et al 1973; Reich, 1973) will generate plasmin which in turn will continuously digest fibrin fibers and the resulting fibrin decomposition products (FDP) may be used by tumor cells as nutrients.

High levels of FDP may also be demonstrated in the blood of certain groups of cancer patients (Ambrus et al 1976). Tumor cells grow along fibrin fibers and eventually break through the capillaries forming metastases which become clinically detectable.

In a previous study (Ambrus et al 1955) we developed methods to inject radioisotope labelled ascitic tumor cells intravenously into the jugular vein of mice while continuously recording circulating radioactivity from the tail. We found that intravenously injected tumor cells disappeared rapidly from the circulation. On the other hand, if animals were anticoagulated with heparin circulating tumor cells could be maintained in the circulation throughout the experiment (approximately 8 hours). We have explored several groups of platelet aggregation inhibitors for their ability to prolong circulating life span of intravenously injected cancer cells. The technique has been described previously (Ambrus et al 1976; Gastpar et al 1977). Figure 1 shows the structural formula of a series of pyrimido-pyrimidine derivatives which we have studied. Of these agents, RA-233 appears to be most active. Figure 2 shows the effect of various doses of RA-233 on the circulation time of intravenously injected polyploid Ehrlich ascites tumor cells in mice. Doses above 6 mg/kg injected 10 minutes before tumor cell injection approximately tripled circulation times.

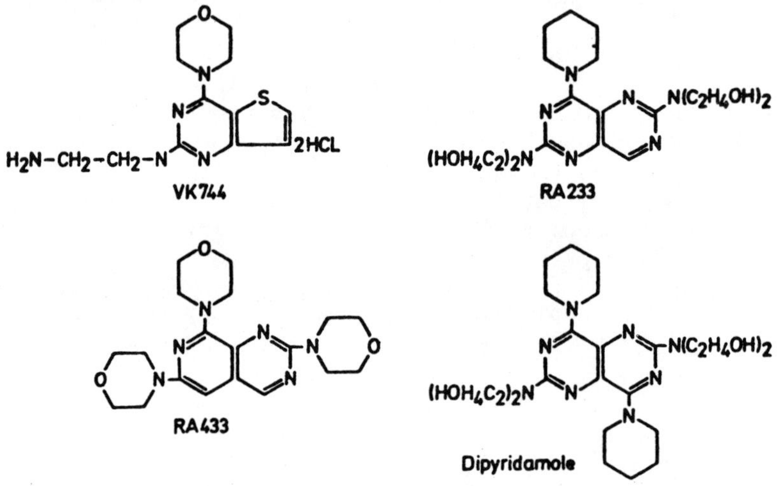

Figure 1. Structural formula of pyrimido-pyrimidine derivatives.

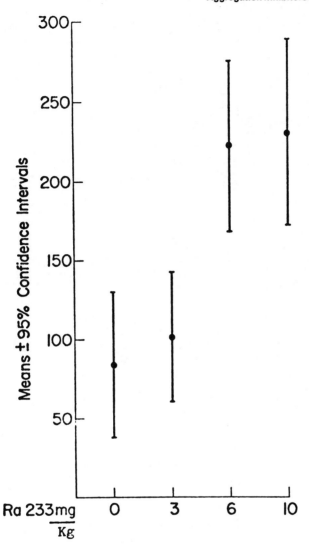

Figure 2. Effect of various doses of RA 233 given i.v. 10 minutes before i.v. tumor cell injection on circulation time (minutes) of polyploid Ehrlich ascites tumor cells.

Figure 3 shows the structural formula of pentoxifylline, the compound we found most effective among the methylxanthine derivatives examined.

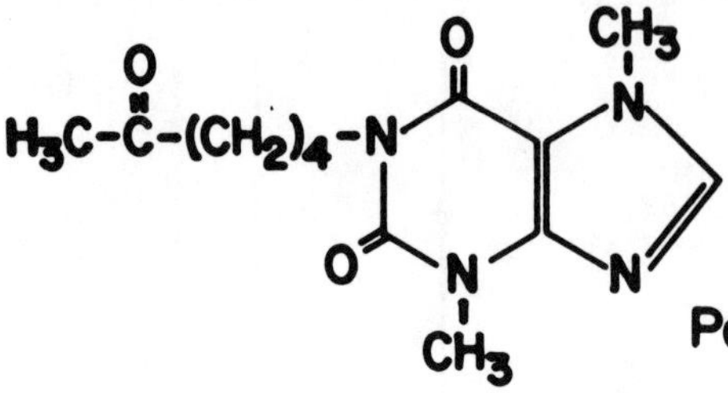

$$H_3C-\overset{O}{C}-(CH_2)_4-N\cdots\text{(xanthine ring with two } CH_3 \text{ groups)}$$

Pentoxifylline (Trental)

Figure 3. Structural formula of pentoxifylline

Figure 4 shows the effect of pentoxifylline on circulation time of intravenously injected polyploid Ehrlich ascites cells. Doses above 18 mg/kg given i.v. 10 minutes before injection of the tumor cells approximately tripled circulation time.

Figure 5 shows the structural formula of some imidazoquinazolinone compounds. Of these agents, BL-4162 was found to be most effective. Figure 6 shows the doses of 5 mg/kg i.p. or 10 mg/kg orally approximately tripled circulation times of intravenously injected polyploid Ehrlich ascites cells.

The question arose, however, whether prolongation of circulating life spans of tumor cells means increased propensity for metastasis formation of spontaneously metastasizing tumor cells. (Gordon et al, 1979) Figure 7 shows a study in which Furth-Columbia Wilms' tumor (nephroblastoma) cells were implanted subcutaneously into Furth-Wistar rats as described by Tomashefsky et al (1971), Saroff et al (1975) and Muntzing et al (1976). Control animals were untreated and treated animals received 12 mg/kg pentoxifylline intraperitionially twice daily for three weeks. At the end of three weeks, all animals were sacrificed and pulmonary metastases counted macroscopically. Figure 7 shows that all control animals had pulmonary metastases, but 77.7% of the treated animals were without metastases.

In a similar study, C-1300 neuroblastoma was implanted subcutaneously into A/J mice as described by Fioriani (1978),

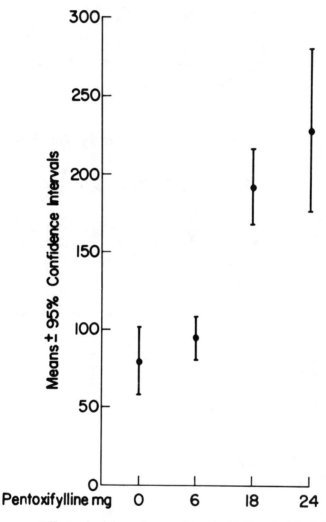

Figure 4. Effect of various doses of pentoxifylline (mg/kg) i.v. 10 minutes before i.v. injection of 2.5×10^7 polyploid Ehrlich ascites tumor cells on circulation time of the latter. Mean circulation times ± 95% confidence intervals are shown in minutes.

BL 4162

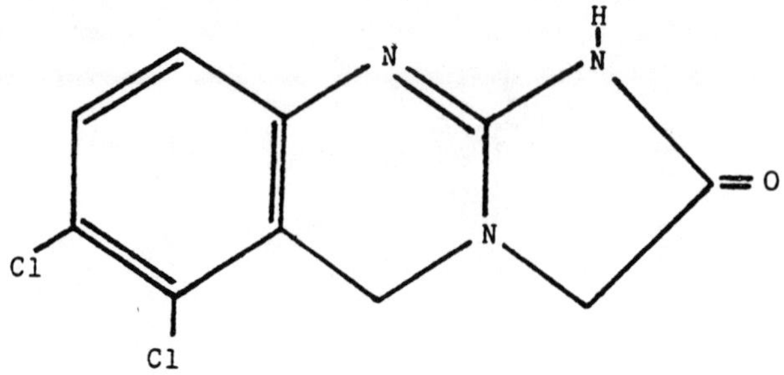

BL 3459

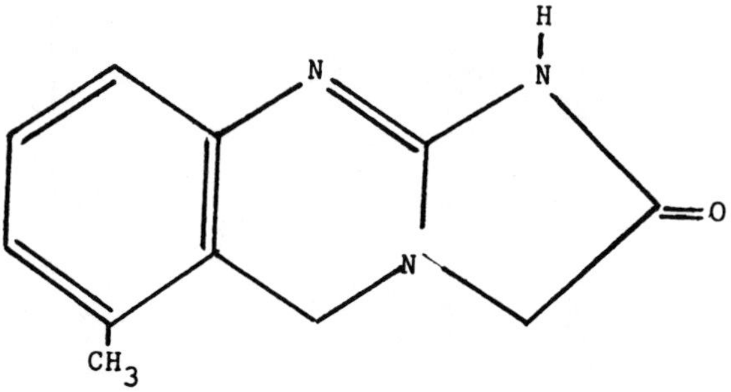

Figure 5. Structural formula of BL-4162 and BL-3459.

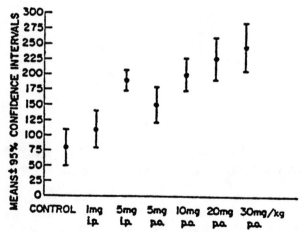

Figure 6. BL-4162, effect of various doses (mg/kg) i.v. or p.os. 10 min. before i.v. injection 2.5 x 10^7 polploid Ehrlich ascites tumor cells on circulation time of the latter.

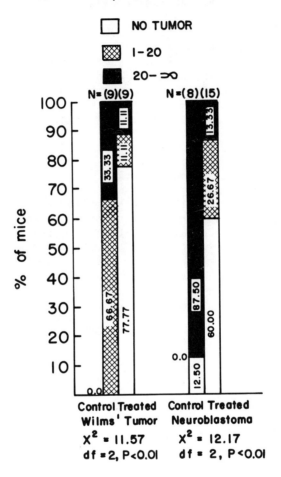

Figure 7

Finklestein et al (1972), Murphy et al (1975) and Sufrin et al (1976). One month after inoculation the animals were sacrificed and pulmonary metastases counted. Figure 7 shows that 87.5% of the controls had pulmonary metastases but only 26.6% of the treated animals.

In a third transplantable spontaneously metastasizing tumor, the results were different. NIH renal adenocarcinoma was implanted subcutaneously into BALB/cCr mice and the animals sacrificed in one month as described previously by Murphy and Hrushesky (1973), Murphy and Williams (1974), Hrushesky and Murphy (1974), Sufrin and Murphy (1977), Murphy et al (1975), Muntzing et al (1976), and Williams et al (1973).

Figure 8 shows that there were no significant differences between treated and control groups. It appears that factors governing metastasis formation are different for different neoplastic cell lines. Similar results were obtained in preliminary studies with other phosphodiesterase inhibitors and prostaglandin synthesis increasing agents including Bay 6575 (Tables 1-4), including pyrimido-pyrimidine derivatives. Recent studies have suggested that agents which we have selected for their cyclic AMP (cAMP) phosphodiesterase inhibitory activity also work through other mechanisms of action, for example, RA-233 and pentoxifylline increase the synthesis and/or release of prostacyclin (prostaglandin I_2).

It appears that platelet aggregation on the abnormal surfaces of tumor cells plays an important role in causing these cells to settle in the microcirculation and to initiate metastases. Platelet aggregation inhibitors can prolong circulating time of tumor cells and prevent metastasis formation probably by exposing circulating tumor cells for longer periods to normal killer cells and cellular immune phenomena and by preventing the development of favorable nutritional conditions of being enclosed in clots and platelet aggregates. Growth stimulating factors originating from aggregating platelets may also play an important role in this phenomenon. Platelet aggregation inhibitors used in these studies were cAMP phosphodiesterase inhibitors and/or prostacyclin synthesis increasing and/or releasing agents. We have not used cyclo-oxygenase inhibitors since in preliminary studies these have inhibited metabolic pathways leading to the synthesis of Thromboxane A_2 but also meta-

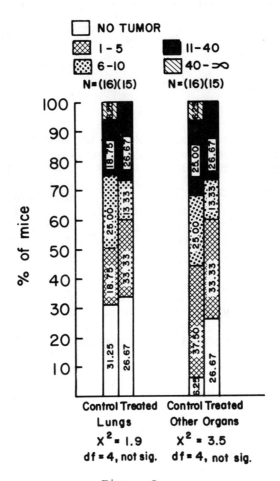

Figure 8

TABLE 1

FURTH-COLUMBIA WILMS' TUMOR IN FURTH-WISTAR MALE RATS
(100-125 gm body weight, 20 animals in each group)
% OF ANIMALS WITH INDICATED NUMBER OF PULMONARY METASTASES.

No. Metastases/ Rat	Controls	Bay 6575 1 mg/kg i. p. 2 x daily
0	0	30
1-20	60	45
20+	40	25

TABLE 2

C-1300 NEUROBLASTOMA IN A/J MALE MICE
(20-25 gm body weight, 20 animals in each group)
% OF ANIMALS WITH INDICATED NUMBER OF PULMONARY METASTASES.

No. Metastases/ Mouse	Controls	Bay 6575 1 mg/kg i.p. 2 x daily
0	0	25
1-20	15	70
20+	85	.5

TABLE 3

HM-KIM BREAST CANCER IN WISTAR FEMALE RATS
(100-125 gm body weight, 20 animals in each group)
% OF ANIMALS WITH INDICATED NUMBER OF PULMONARY METASTASES.

No. Metastases/ Rat	Controls	Bay 6575 1 mg/Kg i.p. 2 x daily
0	10	50
1-20	85	50
20+	5	0

TABLE 4

NIH ADENOCARCINOMA (RENAL) IN BALB/cCr MALE MICE
(20-25 gm body weight, 20 animals in each group)
% OF ANIMALS WITH INDICATED NUMBER OF PULMONARY METASTASES.

No. Metastases/ mouse	Controls	Bay 6575 1 mg/Kg i.p. 2 x daily
0	30	40
1-20	60	50
20+	10	10

bolic pathways leading to the synthesis of prostacyclin, thus one effect tended to offset the other. Thromboxane A_2 is one of the most powerful physiologic platelet aggregation inducing agents. Therapeutic implications of these findings are currently under study.

SUMMARY

Cytologic studies revealed that on the abnormal membranes of circulating tumor cells platelets aggregate and may promote the settling of tumor cells in capillary beds.

A method was developed to measure platelet aggregation in vivo. An arterio-venous anastomosis was established in stumptailed monkeys (<u>Macaca</u> <u>arctoides</u>) and into this anastomosis a 20 micron pore diameter metal screen was inserted. Blood pressure was recorded before and after the screen. Intravenous injection of 0.1 - 5 micrograms of ADP or serotonin resulted in rapid platelet aggregation on the screen, increase in pre-screen pressure and decrease in post-screen pressure. From these data an aggregation index could be calculated. A series of chemicals were selected in this model which prevented in vivo platelet aggregation in this system. This included, pyrimido-pyrimidine derivatives, methylxanthine derivatives, and imidazoquinazolinones.

When isotope labelled ascites tumor cells were injected intravenously into experimental animals, these same agents in the doses in which they proved to be effective platelet aggregation inhibitors in vivo, prolonged the circulation time of tumor cells.

In a series of experiments, spontaneously metastasizing tumors were implanted into animals and metastatic dissemination determined. The agents found effective in prolonging circulation time of labelled tumor cells also decreased incidence of spontaneous metastases.

Further studies suggested, however, that this effect is not based entirely on platelet aggregation inhibtory effect. In biopsies of human tumors and surrounding normal tissue, we found that tumor tissue has an invariably higher level of low affinity cAMP phosphodiesterase than the corresponding normal tissue. In tissue cultures of human tumors, phosphodiesterase inhibitors exhibited significant growth inhibitory effect. This mechanism in addition to the platelet aggregation inhibitory effect might have contributed to the decrease incidence of spontaneous metastases in the above experiments.

REFERENCES

Ambrus JL, Ambrus CM, Byron JW, Goldberg ME, Harrisson JWE (1955). Study of metastasis with the aid of labeled ascites tumor cells. Ann NY Acad Sci 63:938.
Ambrus JL, Ambrus CM, Mink IB, Pickren JW (1975a). Causes of death in cancer patients. J Med 6:61.
Ambrus JL, Ambrus CM, Pickren JW, Soldes S, Bross I (1975b). Hematologic changes and thromboembolic complications in neoplastic disease and their relationship to metastasis. J Med 6:433.
Ambrus JL, Ambrus CM (1976). Blood coagulation in neoplastic disease. In Gastpar H (ed): "Onkohamostaseologie (Hematologic Problems in Cancer)," Stuttgart-New York: F K Schattauer Verlag, p 167.
Ambrus JL, Ambrus CM, Gastpar H, Sapvento PJ, Weber FJ, Thurber LE (1976). Study of platelet aggregation in vivo I. Effect of bencyclan. J Med 6:439.
Finklestein JZ and Gilchrist GS (1972). Recent advances in neuroblastoma. Calif Med 27:116.
Fioriani M, Butler R, Bertolini L, Revoltella R (1978). Early events during C1300 neuroblastoma cell interaction with syngeneic lymphocytes. Eur J Cancer 14:217.
Gastpar H (1972). Inhibition of cancer cell stickiness by

anticoagulants. Fibrinolytic drugs, and pyrimido-
pyrimidine derivatives. In Ambrus JL (ed): "Hematologic Reviews," New York:Marcel Dekker, Inc, Volume 3, p 1.
Gastpar H, Ambrus J, Thurber LE (1977). Study of platelet
aggregation in vivo II. Effect of bencyclan on circulating metastatic tumor cells. J Med 8:53.
Gordon S, Witul M, Cohen H, Sciandra J, Williams P, Gastpar
H, Murphy GP, Ambrus JL (1979). Studies on platelet
aggregation inhibitors in vivo. VIII. Effect of
pentoxifylline on spontaneous tumor metastasis. J Med
10:435.
Hrushesky WJ, Murphy GP (1974). Evaluation of chrmothera-
peutic agents in a new murine renal carcinoma model.
J Natl Cancer Inst 52:1117.
Muntzing J, Williams PD, Murphy GP (1976). The growth
characteristics of metastases from experimental renal
tumors. Res Commun Chem Pathol Pharmacol 13:541.
Murphy GP, Hrushesky WJ (1973). A murine renal cell
carcinoma. J Natl Cancer Inst 50:1013.
Murphy GP, Williams PD (1974). Testing of chemotherapeutic
agents in murine renal cell adenocarcinoma. Res Comm
Chem Pathol Pharmacol 9:265.
Murphy GP, Williams P, Keogh B (1975). Chemotherapy of a
murine neuroblastoma model. J Surg Oncol 7:521.
Ossowski L, Unkeless JC, Tobia A, Quigley JP, Rifkin DP,
Reich E (1973). An enzymatic function associated with
transformation of fibroblasts by oncogenic viruses.
II. Mammalian fibroblast cultures transformed by DNA
and RNA tumor viruses. J Exp Med 137:112.
Reich E (1973). Tumor associated fibrolysis. Fed Proc
32:2174.
Saroff J, Chu TM, Gaeta JF, Williams P, Murphy GP (1975).
Characterization of a Wilms' tumor model. Invest
Urol 12:320.
Sufrin G, Murphy GP (1976). Pharmacokinetic studies in the
chemotherapy of neuroblastoma using the C1300 murine
system. Oncology 33:173.
Sufrin G, Murphy GP (1977). Pharmacokinetic studies of a
treantplantable murine renal adenocarcinoma. Invest
Urol 15:9.
Tomashefsky PO, Furth J, Lattimer JK, Tannerbaum M, Priest-
ley J (1971). The Furth-Columbia rat Wilms' tumor.
Trans Amer Assoc Genitourin Surg 63:28.
Williams PD, Bhanalaph T and Murphy GP (1973). Unilateral
nephrectomy. Its effect on primary murine renal
adenocarcinoma. Urology 11:619.

ANTIMETASTATIC AND ANTITUMOR EFFECT OF PLATELET AGGREGATION INHIBITORS

J.L. Ambrus, C.M. Ambrus, H. Gastpar, E. Huberman, R. Montagna, W. Biddle, S. Leong, J. Horoszewicz

Roswell Park Memorial Institute, Buffalo, NY; Associated Biomedic Systems, Buffalo, NY; Oakridge National Laboratory, Oakridge, TN; University of Munich Medical School, West Germany

In the previous papers presented at this conference (Ambrus et al, 1981; Gastpar et al, 1981) we described experiments and clinical trials suggesting that agents which inhibit platelet aggregation also decrease metastatic potential of certain clinical and experimental tumors. These agents probably act by increasing intracellular cAMP by inhibiting cAMP phosphodiesterases and by stimulation of production and/or release of prostacyclin which in turn stimulates adenylate cyclase for increased synthesis of cAMP. We have found that these agents have additional activities which may contribute to the clinical and experimental antimetastatic effect. In this presentation we will deal only with one of the agents under study by our group: the pyrimido-pyrimidine derivative RA-233 (Figure 1). Figures 2-5 show experiments in which four transplantable tumor lines were implanted into rats or mice and metastatic patterns determined 3-4 weeks later. RA-233 was given i.v. in doses of 6 or 12 mg/kg/12 hours or in oral doses of about 100 mg/kg/24 hours in the drinking water. In all instances there was decreased incidence of metastasis in Columbia-Furth Wilms' tumor, C1300 neuroblastoma and HM(SMT-2A) Kim mammary carcinomas, but not in NIH renal adenocarcinoma. This suggests that from this point of view different tumors have different characteristics.

In a series of experiments we demonstrated the effect of RA-233 in inhibiting platelet aggregation and in prolonging the circulating life span of i.v. injected labeled tumor cells (Ambrus et al, 1977, 1978).

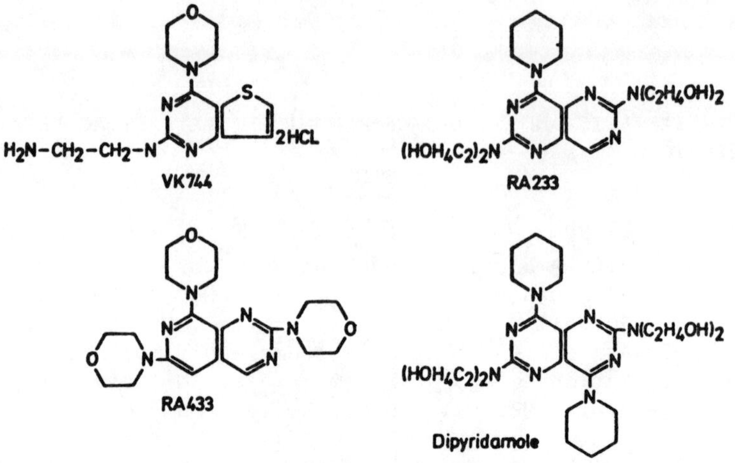

Figure 1. Structural formula of pyrimido-pyrimidine derivatives.

The clinical results (Gastpar et al, 1981) suggested, however, that other mechanisms may also be involved. For this reason this question was further investigated.

Table I shows that in a number of biopsies from tumors and surrounding normal tissues from the same organs of the same patients, cAMP phosphodiesterase levels (particularly of the low affinity types) were higher in the tumor tissue than in the normal tissue. It can be expected that inhibition of phosphodiesterases may have unfavorable effects on tumor cells.

Figure 7 shows inhibitory activity of RA-233 on a number of human tumors. The characteristics of these cell lines and those related to Table IV are summarized in Table II.

Table III summarizes a recent experiment in which it was found that RA-233 induces differentiation in a cell culture of a human promyleocytic leukemia cell line.

Table IV shows that a number of human neoplastic cell lines are inhibited by RA-233 and also by human fibroblastic interferon (HFIF) produced in our Institute. When the two

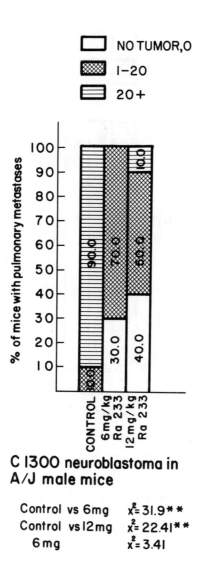

Figure 2

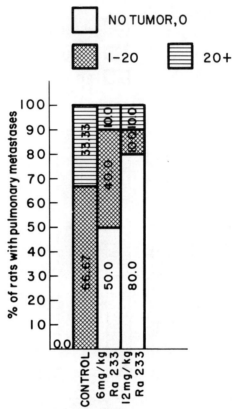

Figure 3

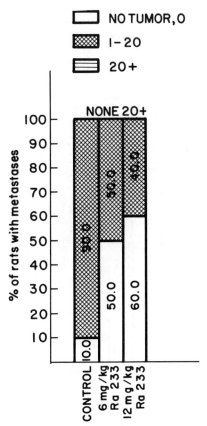

Figure 4

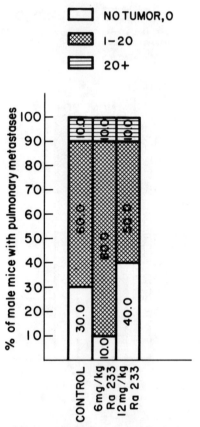

Figure 5.

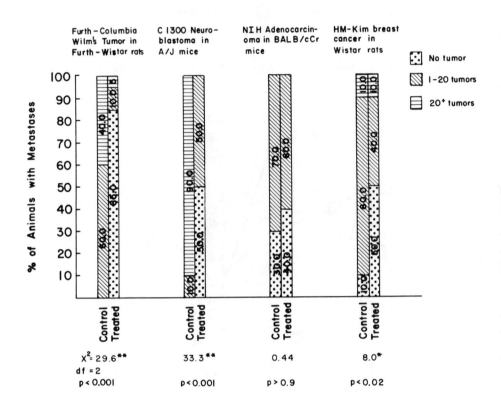

Figure 6. Effect of oral RA-233 100 mg/kg/24 hours on 4 lines of metastatic tumors, number of pulmonary metastases 3-4 weeks after transplantation.

are combined, potentiation of the inhibitory effect occurs. This may be related to the finding that HFIF stimulates adenylate cyclase for increased cAMP production, while RA-233 prevents cAMP decomposition by inhibiting cAMP phosphodiesterases. RA-233 also further stimulates adenylate cyclase through stimulation of PGI_2 synthesis and/or release. If low level interferon production is part of normal antitumor defense mechanism, potentiation of this effect by pyrimido--pyrimidine derivatives may be an additional mechanism of action of the antitumor-antimetastasis effect of RA-233.

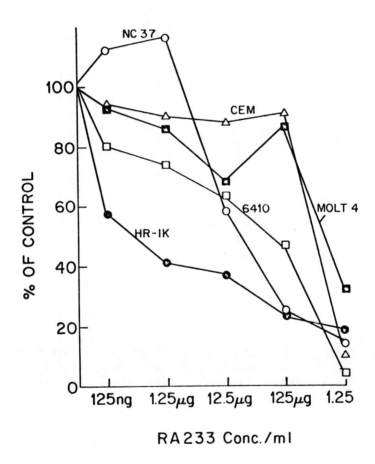

Figure 7. Inhibitory activity of RA-233 on human tumors. ^{3}H thymidine added at 48 hours, incorporation measured at 72 hours.

Table I

Patient No.	Diagnosis	cAMP Phosphodiesterase X				cGMP High Affinity phosphordiesterase XX	
		High Affinity		Low Affinity			
		Normal Tissue	Tumor Tissue	Normal Tissue	Tumor Tissue	Normal Tissue	Tumor Tissue
3456 M	Malignant Melanoma	0.224 0.948	0.743 1.026	5.93 21.52	23.64 52.36	0.025 0.067	0.068 0.100
151617 M	Malignant Melanoma	1.418	0.716	33.39	13.56	0.155	0.065
1819 M	Malignant Melanoma	0.595	0.582	14.67	12.33	0.070	~0.045
2021 M	Malignant Melanoma	0.010	0.736	3.53	9.30	0.014	0.048
2223 M	Malignant Melanoma	0.088	0.619	5.86	26.11	0.015	0.069
2425 M	Malignant Melanoma	0.346	0.756	6.62	72.30	0.039	0.071
2829 M	Malignant Melanoma	0.000	0.229	4.01	9.63	0.010	0.014
3031 M	Malignant Melanoma	0.720	1.005	14.63	23.09	0.034	0.081
3233 M	Malignant Melanoma	0.429	0.276	10.56	16.02	0.055	0.030
3637 M	Malignant Melanoma	0.238	0.268	12.93	12.79	0.046	0.030
4243 M	Malignant Melanoma	0.042	0.887	12.88	17.67	0.021	0.099
12 S	Rhabdomyo-sarcoma	0.899	0.426	15.85	6.94	0.045	0.066
910 S	Leio-myosarcoma	0.000	0.130	9.67	3.72	0.016	0.018
1314 S	Fibro-sarcoma	0.201	0.698	6.23	17.01	0.024	0.054
2627 S	Synovial cell sarcoma	0.775	0.471	14.63	27.06	0.088	0.082
11 12 X	Adenocarci-noma	0.015	0.031	3.21	0.41	0.012	0.000
34 35 X	Ependy-moma	0.424	1.364	9.37	27.96	0.052	0.090

X nMol cAMP hydrolyzed/mg protein/minute
XX nMol cGMP hydrolyzed/mg protein/minute

Table II

Cell Line	Origin	Media, Monolayer (M) or in Suspension (S)
ES-1	Malignant melanoma	RPMI 1640 M
DAUDI	Burkitt's lymphoma (B cell) (carries EB virus)	RPMI 1640 S
HT-29	Colon adenocarcinoma	MEM M
RT-4	Transitional bladder cell carcinoma	MEM M
LMCaP	Prostatic carcinoma	RPMI 1640 M
BG-27	Diploid foreskin fibroblastic	MEM M
HR-IK	Burkitt's lymphoma (B cell) (carries EB virus)	RPMI 1640 S
CCRF-CEM	Acute lymphatic leukemia (T cell)	RPMI 1640 S
6410	Acute myeloid leukemia (B cell)	RPMI 1640 S
MOLT-4F	Acute lymphoid leukemia (T cell)	RPMI 1640 S
NC-37	Normal lymphocyte (B cell) (carries EB virus)	RPMI 1640 S

Table III

Myeloid Differentiation in Human Promyelocytic Leukemia (HL-60) Cells after 6 Days Treatment with RA-233.

RA-233 mg/ml	# Cells/100mm Petri dish x 10^7	% Morphologically mature cells	Lysozyme Activity (µg equiv./10ml/ 10^7 cells)
0.000	3.4	10	3
0.025	3.5	13	3
0.250	0.5	73	17

Table IV

Effect of HFIF and RA-233 on in vitro cell growth.

Cells	Drug Concentration	% Control Cells Population Increase ± S.D.	% Control Expected if Additive
DAUDI	0.001 mg/ml RA-233	91 ± 3	
	50 ref. U/ml HFIF	53 ± 4	48
	RA-233 + HFIF	36 ± 2	
ES-1	0.01 mg/ml RA-233	66 ± 4	
	50 ref. U/ml HFIF	48 ± 3	32
	RA-233 + HFIF	13 ± 5	
LNCaP	0.1 mg/ml RA-233	80 ± 2	
	100 ref. U.ml HFIF	76 ± 2	61
	RA-233 + HFIF	32 ± 3	
RT-4	0.1 mg/ml RA-233	86 ± 2	
	100 ref. U/ml HFIF	58 ± 4	50
	RA-233 + HFIF	37 ± 5	
HT-29	0.1 mg/ml RA-233	16 ± 2	
	100 ref. U/ml HFIF	75 ± 4	12
	RA-233 HFIF	2 ± 1	
BG-27	0.1 mg/ml RA-233	65 ± 3	
	100 ref. U/ml HFIF	86 ± 3	56
	RA-233 + HFIF	33 ± 5	

Drugs administered during the lag phase of cell growth. Results expressed as a percentage of control cell population increases.

References

Ambrus JL, Ambrus CM, Gastpar H, Thurber L, Miller R, Fretwell B, Lane KP (1977). Study of platelet aggregation in vivo IV. Effect of pyrimido-pyrimidine derivatives. J Med 8:287.

Ambrus JL, Ambrus CM, Gastpar H (1978). Studies on platelet aggregation in vivo VI. Effect of a pyrimido-pyrimidine derivative (RA-233) on tumor cell metastasis. J Med 9:183.

Ambrus JL, Ambrus CM, Gastpar H (1981). Platelet aggregation inhibitors and metastatic spread of neoplastic cells. Chapter in this conference.

Gastpar H, Ambrus JL, Ambrus CM (1981). Platelet cancer cell interaction in metastasis formation. Platelet aggregation inhibitors: a possible approach to metastasis prevention. Chapter in this conference.

DISCUSSION

Dr. N. Hanna (Frederick Cancer Institute, Frederick). Do you find any differences in the primary tumor growth in the RA233-treated mice?

Dr. J. L. Ambrus (Roswell Park Memorial Institute, Buffalo). Essentially, if we measure tumor volume or tumor weight, we find little or no difference, and we were very much surprised at that. On the other hand, we do find histologic changes both in the original and in subsequent transplanted tumors.

Dr. Hanna. Inhibition of aggregation could let the tumor cells circulate for a long time so that the defense mechanism could take care of these cells.

Dr. Ambrus. Exactly. Only a few concepts can be presented in 29 minutes. For example, if you use serum from whole blood in the culture medium you have platelet material which results in adherence. On the other hand, if you have serum from platelet-poor plasma, that same tumor growth is not supported. Tentatively we have formulated a hypothesis, that continuous delivery of platelet growth factor to tumors may be an important aspect of growth, but it is a very early kind of speculation.

Dr. Hanna. I am afraid that under certain conditions the tumor cells which are resting somewhere can disseminate and grow, supported by the platelets.

Dr. Ambrus. Exactly. It is a very complex system. Platelet aggregation inhibitors can make things worse in certain situations For example, a tumor may be normally arrested by the lung and give only pulmonary metastasis. Certain treatments might help pass the pulmonary circulation and you could have metastasis all over the body. One single concept will not solve the cancer problem.

Dr. R. Greig (Smith, Kline, and French, Inc.,Philadelphia). Is there any role for fibronectin?

Dr. Ambrus. Fibronectin is present in plasma and on the surface of normal cells but is lost when they are transformed with a virus or with a carcinogen. On the other hand, when fibronectin is added to the culture medium with one of these transformed cell lines, it immediately starts showing contact inhibition and morphologic changes so that it starts looking much like a normal cell. However, when these normally differentiated cell lines are injected into nude mice, they grow out as malignant tumors.

Dr. J. C. Hoak (University of Iowa, Iowa City). What is the evidence that they undergo morphological differentiation?

Dr. Ambrus. We find the most important morphological criterion to be contact inhibition. Tumor cells in this type of system pile up, but once we get morphological differentiation contact inhibition is reestablished.

Dr. Hoak. No other histological changes? In other words, how could you distinguish contact inhibition from cells that were just inhibited in growth but still had all of their malignancy?

Dr. Ambrus. Contact inhibition is the major criterion but there are some minor ones, and a major biochemical criterion is the lysomal enzymes.

Dr. E. Pearlstein (New York University Medical Center, New York). What is the basis of the inherited change that you apparently observe after you treat the transformed cells with a combination of RA-233 and interferon?

Dr. Ambrus. I do not even know whether this conclusion is true, but if there is a phenotypic recurrence of tumor it appears to be slower than it would be without this treatment.

Dr. P. Hilgard (Bristol-Meyers International Corporation, Brussels). Would you consider the decreasing numbers of cells after incubation with the various agents as being cytoxic activity, or is it reversible?

Dr. Ambrus. I do not think that it is cytotoxic activity because, both by counting cells and by measuring thymidine incorporation, we find a decrease in proliferative activity. However, once we remove this culture from the inhibition and just continue culture, or inject them into new nude mice, they grow according to a normal slope as you would expect. So, I suspect that we have done something to the cells which, however, at these levels, is probably reversible.

Dr. M. B. Donati (Instituto di Recherche Farmacologische, Milan). I wonder whether you have any data on possible effects on the microcirculation of agents which may ultimately be vasodilatory.

Dr. Ambrus. This is a very important consideration. These agents inhibit platelet aggregation. They decrease stiffness and increase deformability of red cells, they inhibit phosphodiesterases and cell growth and probably decrease prostaglandin E_2. Obviously these mechanisms may cause improvement of the microcirculation. Similar agents which can increase red cell flexibility can increase the effect of various chemotherapeutic agents suggesting that, by increasing the microcirculation in the tumor, you may have more drug get into the cancer cell. Rheology is, indeed, a very important component of these types of chemotherapy.

Dr. Hanna. If, say, 5% of the cells are differentiated and resistant to the drug and the other 95% not differentiated, can you eliminate the possibility that when you inject them into the nude mice you select those that are resistant?

Dr. Ambrus. That is a very good possibility. So far we have no tumors growing in these nude mice at all. If tumors develop, we will subculture them in order to answer your point. On the other hand, if tumors do not develop we must assume that we have destroyed their malignant potential one way or another.

Dr. Greig. Did you check the effect of your metal screen?

Dr. Ambrus. Yes, your point is well taken. With Nucleopore membranes we found differences in the two surfaces in the in vitro filtration apparatus. On the other hand, the metal screens which are inserted in vivo into animals have a pore size of 20 microns, which is huge as compared to the 3-5 or 8 micron pore size in the Nucleopore membrane used in vitro. In these screens we get occlusion, but we do not get significant differences in scanning electron microscopy by looking at the two sides.

THE BIOLOGIC BASIS FOR ANTICOAGULANT TREATMENT OF CANCER

Leo R. Zacharski

Dartmouth Medical School, Hanover, NH and the
VA Hospital, White River Jct., VT 05001

INTRODUCTION

The interest in anticoagulants as an approach to the management of malignancy is based upon observations which suggest that a cause and effect relationship exists between activation of coagulation reactions and the growth and spread of cancer (Zacharski et al, 1979). Although work on the interaction between blood coagulation and cancer has been underway for several decades, little information is available on the therapeutic value of coagulation- and platelet-inhibitory drugs in human tumors. The purpose of this paper is to survey data, appearing since this subject was last reviewed (Zacharski, 1981), that are relevant to our understanding of the possible role of coagulation reactions in human malignancy. In addition, an interpretation of the existing literature will be offered that is intended to provide a conceptual framework for understanding the biologic basis for the anticoagulant approach to the treatment of cancer.

INDUCTION OF COAGULATION IN HUMAN MALIGNANCY

The clotting mechanism is known to be activated in patients with malignancy (Zacharski et al, 1979). Such activation may be manifested by localized thromboembolic events which occur with increased infrequency, especially when studied at autopsy (Ambrus et al, 1975). Alternatively, coagulation may be activated throughout the circulation (disseminated intravascular coagulation, or DIC) and manifested by changes in a variety of coagulation tests

performed on peripheral blood. There is little doubt that
the accelerated turnover of fibrinogen and platelets, which
is characteristic of the malignant state and which worsens
with approaching death, is due to activation of the clotting mechanism with resulting conversion of prothrombin to
thrombin (Slichter and Harker, 1974). Such thrombin may be
generated within the circulation. However, thrombin generation in malignancy probably more characteristically occurs
in the extravascular compartment; that is, in the vicinity
of the tumor (Rickles et al, 1981). This undoubtedly explains why the extent of platelet and fibrinogen consumption parallels the extent of the disease and is ameliorated
by successful therapy (Slichter and Harker, 1974).

Thrombin acts upon the fibrinogen molecule to cleave
two pairs of peptides, termed fibrinopeptide A (FPA) and
fibrinopeptide B (FPB), from the parent molecule (Rickles
et al, 1981). These fibrinopeptides can be detected in
plasma by radioimmunoassay and, when elevated, provide a
sensitive measure of thrombin activity.

DIC also occurs in acute non-lymphocytic leukemia.
Systematic studies of FPA in this condition have shown that
levels are regularly elevated, that levels decline (but not
always to normal) with induction of remission, and that
levels increase with impending relapse (Myers et al, 1981).
Thus, FPA levels appear to parallel disease activity in acute
leukemia.

Elevated FPA levels are also observed in patients with
various solid tumors (Rickles et al, 1981). Results of
serial studies of FPA have suggested that persistent elevation of FPA levels was associated with treatment failure.
As was found in acute leukemia, an upward trend in FPA
values was observed with disease progression and impending
death. Of special interest was the finding in these studies
that long-term anticoagulation with warfarin was associated
with a significant reduction in FPA values. By contrast,
short-term anticoagulation by means of heparin infusion
failed to reduce FPA levels; thus, the reaction responsible
for FPA production appeared to be taking place outside the
vascular compartment.

Elements of clots are known to occur in association
with local tumor deposits in many different tumor types
(Dvorak et al, 1981). This fibrin, demonstrable by a

variety of techniques, may assume at least three different patterns when associated with tumor tissue. In the first and most commonly observed pattern, the fibrin, evident by standard microscopic techniques, is present in thrombi adjacent to intravascular tumor deposits. In the second and third patterns, the fibrin is extravascular and therefore clearly demonstrable only by special techniques (such as electronmicroscopy or immunofluorescence). In the second pattern, the fibrin is laced diffusely between the cells of the tumor. This pattern is observed in lymphomas (Dvorak, 1981). In the third pattern, the fibrin is a component of the stroma which surrounds microscopic tumor nodules. The latter pattern may be seen in carcinoma of the breast (Dvorak et al, 1981).

ROLE OF THE COAGULATION MECHANISM IN THE GROWTH AND SPREAD OF HUMAN MALIGNANCY

Compelling evidence for a cause-and-effect relationship between the clotting mechanism and the growth and spread of neoplasia has come from studies of experimental tumor models (Zacharski et al, 1979, 1981b). In a substantial number of studies in several animal species it has been shown that various anticoagulant, fibrinolytic and anti-platelet drugs ameliorate the course of a number of kinds of tumors. Two recent studies serve to illustrate the coagulation-cancer interaction. In one study, a strong, positive correlation was found between the platelet-aggregating properties of a series of renal tumor cell lines cultured *in vitro* and the metastatic potential of these cells upon injection into rats (Pearlstein et al, 1980). Since the coagulant and fibrinolytic properties of these cells were not reported, it is not known whether these features of the cell lines also correlated with metastatic potential. In the other study, administration of prostacyclin (PGI_2, a potent endogenous platelet-inhibitory prostaglandin, to mice bearing the B16 melanoma resulted in marked inhibition of metastatic spread (Honn et al, 1981). The effect of PGI_2 was enhanced by theophylline which potentiates the PGI_2-induced elevation in platelet cyclic AMP. Conversely, inhibition of endogenous PGI_2 synthesis enhanced metastasis formation. PGI_2 plus theophylline had no effect on sequestration of isotopically tagged tumor cells by various organs and had no effect when incubated with the tumor cells prior to injection. These results suggest that these agents favorably influenced metastasis formation through an effect on the host rather than on the tumor.

Several questions are usually posed in discussions of the possible involvement of blood coagulation reactions in human malignancy. Does malignancy occur in individuals with hereditary hemorrhagic disorders? Is the incidence of malignancy reduced in patients treated previously with anticoagulants? Is the course of malignancy modified by anticoagulant administration?

The first of these questions was addressed in a study by Forman (1979). Through a postal survey, 61 instances of malignancy were identified in a total of 10,500 individuals with congenital factor deficiency states. There did not appear to be either a reduced incidence or death rate from cancer. Interestingly, there seemed to be an excess of soft tissue sarcomas in this population. Unfortunately, this particular approach is a weak test of the relationship in question because most such individuals have factor deficiencies which exist within the intrinsic pathway of coagulation. It is most likely that coagulation induction in cancer is by way of the extrinsic (tissue factor-initiated) pathway which requires the presence of factors II (prothrombin), V, VII, and X. Individuals with deficiencies of these factors are so uncommon that the opportunity to determine whether deficiency of these factors blocks the development of cancer is unlikely to be forthcoming.

Michaels (1974) attempted to address the second question. In a study of patients previously anticoagulated for cardiovascular diseases, he discovered the expected incidence of malignancy but an unexpectedly reduced death rate from malignancy. However, patients with malignancy might not have been originally included in the anticoagulant treated group or anticoagulants might have been withdrawn from some individuals in the study group upon making the diagnosis of malignancy. Either of these possibilities might contribute to the apparent reduction in deaths from malignancy by eliminating cancer patients from the anticoagulated group.

In an effort to verify these findings, a study was undertaken by Annegers and Zacharski (1980) of the incidence and death rate from malignancy in 378 anticoagulated patients followed for a total of 3,541 person-years. The overall incidence of malignancy, and the number of new malignancies observed in this cohort while on anticoagulation and 3 years or more after initial anticoagulant treatment were similar to that expected in the general population. The number of deaths from malignancy in the anticoagulated group was less, but not significantly less, than that expected. However, even this modest difference in deaths disappeared at three years or more following initiation of anticoagulation suggesting that anticoagulant therapy was undertaken in patients who were initially free of malignancy.

This epidemiologic approach is also a weak test of the relationship between the clotting mechanism and the growth and spread of malignancy. Thus, the possibility exists that sample sizes were too small to permit real differences that might be present for anticoagulated patients to be seen. The problem of sample size is pointed up when anticoagulated patients who acquired cancer were sorted by organ type. In the study of Annegers and Zacharski (1980), the mean number of patients in each category of malignancy was less that 15. Furthermore, such studies provide no direct information on the effect of the anticoagulant on the natural history of the malignancy. Finally, observations are limited to the effects of a single anticoagulant, namely warfarin.

Other kinds of epidemiologic studies might, however, provide more useful insights. For example, it is customary in certain institutions to use agents such as heparin or dextran for prevention of thrombosis. It would be of interest to determine whether such treatment, when given to patients with malignancy, modified the course of their tumor. Thus, the time to recurrence or death could be determined in patients given mini-dose heparin during resection of lung or bowel tumors and comparison made with patients receiving no heparin. Likewise, large scale trials of platelet-inhibitory drugs are being undertaken for prevention of arterial thrombotic disease. Perhaps the occurrence of malignancy or death from malignancy could be added as end-point parameters in such studies.

Administration of anticoagulants to patients with cancer is obviously the best test of the significance of the involvement of coagulation reactions in malignant diseases. Reports of the effect of anticoagulants given to patients with a number of kinds of cancer have been reviewed (Zacharski, 1981). Types of tumors that have been so treated include both Hodgkin's and non-Hodgkin's lymphomas;

chronic and acute leukemias, osteogenic sarcoma; gliomas; and carcinomas of the breast, ovary, uterine cervix, colon, pancreas and lung. Since that review, several relevant reports have appeared in the literature. Powles and associates (1980) performed a double blind trial of a conjugate of paracetamol and aspirin in 160 women with carcinoma of the breast. Patients randomized to receive the drug had lower thromboxane B_2 levels, indicative of platelet inhibition, than placebo-treated controls. Observation over an 18-month period failed to reveal a difference between treatment groups for relapse rate, development of bony metastasis or survival.

Al-Saleem and co-workers (1980) administered indomethacin together with corticosteroids in xeroderma pigmentosum complicated by squamous cell carcinoma of the skin. Regression of the skin cancer was observed which was attributed to the ability of indomethacin to reduce prostaglandin synthesis.

A preliminary report has appeared of an ongoing study of a variety of tumors resected from the region of the head and neck (Ambrus et al , 1980). Patients received RA-233, a dipyridamole derivative with potent platelet-inhibitory properties. A highly significant reduction in recurrence was observed in RA-233-treated patients in comparison to historical controls.

No attempt will be made at this time to critically evaluate these previous reports of anticoagulant therapy for cancer. It should be noted, however, that promising findings are sometimes clouded by inadequacies in experimental design or conflicting results. Problems that appear in some of these studies include the lack of adequate controls, failure to define cell types of malignancy under investigation, and the use of combinations of drugs that have quite different mechanisms of action and possibly even antagonistic effects (Zacharski et al , 1981a). It has been difficult, if not impossible, therefore, to identify the effects of a given agent under specified conditions. This problem is critical since there is reason to believe that different agents may modify certain, but not all, tumors under some, but not all, circumstances (Zacharski et al , 1981a).

With these considerations in mind, a VA Cooperative Study Group was formed that was committed to systematic evaluation of anticoagulants in the treatment of cancer by means of prospective, controlled, randomized trials (Zacharski et al , 1979). The first trial undertaken by this group began in April of 1976 and ended in May of 1981. Warfarin was the agent chosen for this initial investigation. This choice was based on the fact that warfarin had demonstrable efficacy in certain experimental tumors and possible efficacy in certain human malignancies, compliance with treatment could be objectively assessed, that most physicians have had experience with its use and that it is reasonably well understood pharmacologically (in terms of its effect on the coagulation mechanism), that its toxicity can be recognized and treated, and that it can be given on an outpatient basis for an extended period of time.

From the outset, this group recognized that there was no way to predict whether the tumor types available to this study would be responsive to the agent under investigation. Therefore, caution was exercised in experimental design such that maximum information could be obtained that would be of use in planning other studies. The time consuming process of analysis of data from this study is in progress at the time of this writing. However, results are available from one tumor stratum within this study (Zacharski et al, 1981a). Of 50 patients admitted to this study with small cell carcinoma of the lung, 25 were randomized to receive conventional chemotherapy and 25 were randomized to receive conventional chemotherapy plus warfarin. At the time these results were reported, 20 of the warfarin-treated and 21 control patients had died. The median survival for warfarin-treated patients was 50 weeks and for control patients 24 weeks ($p = 0.026$). This difference in survival could not be accounted for by differences between treatment groups for age, sex, performance status or extent of disease at the time of randomization. In addition, there was no difference between groups for the amount of conventional chemotherapy given; the kinds of chemotherapeutic agents used after patients were taken off study; mean values for leukocytes, platelets or hemoglobin; or the average number of physician visits per month. A significant survival advantage was observed for warfarin in patients with extensive disease at the time of randomization and in patients who failed to achieve a complete or partial remission. Survival in the control group was similar to that observed for patients with SCCL seen simultaneously at participating VA Medical Centers who were comparable to study patients, but not admitted to the study for various reasons. Survival for the control group was also similar to that observed by a different group of investigators using the same chemotherapy regimen. While the incidence of remission was not significantly improved with warfarin, the median duration of remission and the median length of time to first evidence of disease progression beyond that at randomization were increased.

The incidence of bleeding episodes was increased in warfarin-treated patients. However, most of these episodes were minor and the warfarin was either continued or discontinued briefly during these. Two warfarin-treated patients had their warfarin permanently discontinued because of severe bleeding. One warfarin-treated and one control patient had bleeding associated with their terminal event. Of interest was the fact that nine patients randomized to receive warfarin had objective evidence of intracerebral metastasis at some time while receiving warfarin. Intracerebral bleeding did not occur in these or any other patient in this group.

In June, 1981, a second study was initiated by this VA Cooperative Study group (Zacharski, 1981b). A platelet antagonist (RA-233) was selected for this trial because data existed indicating efficacy of this agent in experimental tumors. In addition, preliminary evidence for an effect in human malignancy has also been reported (Ambrus et al, 1980). This drug has a low degree of toxicity and compliance can be monitored by means of blood levels. It therefore lends itself to a double-blind experimental design.

DISCUSSION

It is perhaps surprising that so much significance should be attached to coagulation activation in cancer. Cause and effect relationships for coagulation reactions certainly pertain to the achievement of hemostasis. Might not the occurrence of such reactions in malignancy merely represent a manifestation of, for example, tumor necrosis and, therefore, be nonspecific? To hold to such a view would, however, require that a wealth of evidence from experimental tumors as well as the earliest glimmer of evidence from human malignancy be ignored.

It is appropriate to discuss the mechanism(s) by which anticoagulants might exert their effect on tumors in order to more fully appreciate the potential value of such drugs in tumor therapy. While much remains to be learned about such mechanisms, several pieces to this perplexing puzzle have already been fitted together. Admittedly, anticoagulants may exert their antineoplastic effect by mechanisms unrelated to their effect on blood coagulation. However, the fact that efficacy has been shown for a variety of anticoagulants that differ considerably in terms of the mechanism by which they inhibit coagulation, provides strong circumstantial evidence that their effect is indeed mediated by modification of blood coagulation (Zacharski et al, 1979).

Evidence relating the antineoplastic effect of warfarin to its anticoagulant properties is somewhat more complete. Tumor inhibition can be achieved by inducing a deficiency of vitamin K, as well as by warfarin treatment, that is of sufficient severity to render the animal hypocoagulable (Hilgard, 1977). Warfarin enantiomers that lack anticoagulant properties are incapable of tumor inhibition (Poggi, et al , 1978). In warfarin-treated animals, tumor inhibition can be negated by simultaneous administration of vitamin K (Brown, 1973), but not by simultaneously replenishing circulating levels of vitamin-K-dependent clotting factors (Hilgard and Maat, 1979). This apparent paradox may be resolved by the observation that tumor cells recovered from warfarin-treated animals have reduced coagulant activity (Poggi et al , 1981). Since warfarin appears to have the capacity to inhibit expression of tissue factor coagulant activity by cells (Edwards and Rickles, 1978; Zacharski and Rosenstein, 1979), it may be postulated that warfarin exerts its beneficial effect not by inhibiting the production of active vitamin K-dependent plasma clotting factors, but by inhibiting the production of active tumor cell tissue factor. Should the reduced cellular coagulant activity be directly related to inhibition of tumor spreading it would be reasonable to postulate that expression of tumor cell coagulant activity is related to the well-being of the tumor, but is not in the best interests of the host. The previously cited study of the effects of PGI_2 and theophylline on tumor metastasis (Honn et al , 1981) also serves to illustrate the importance of host responses to tumor cells in the control of tumor metastasis.

Granted that coagulation-inhibitory drugs work as they were intended -- by inhibiting coagulation rather than by some other mechanism, how might coagulation reactions support tumor proliferation? Again, our knowledge is limited but there are at least two general possibilities. The first possibility is that tumor cells possess coagulants that initiate a sequence of coagulation reactions which lead to fibrin formation. Thrombin, which must be generated in this sequence, has been shown to have mitogenic properties (reviewed by Black, 1980). Thus, a positive feedback loop may exist in which the most coagulant-laden cells are presented with the strongest stimulus for proliferation. Alternatively (or perhaps complementarily), the fibrin that forms may afford the tumor a scaffold upon which it can grow or may serve as a barrier to host cells that might otherwise be able to invade or destroy the tumor cells (Dvorak et al , 1981a).

The second possibility is that tumor cells possess surface constituents capable of attracting platelets (Pearlstein et al , 1980). These platelets may attach to tumor cells and liberate growth factor(s) that enhance tumor proliferation (reviewed by Zacharski et al , 1981b). Once again a positive feedback loop is possible in which the tumor cells that are most capable of aggregating platelets are presented with the greatest amount of platelet-derived growth factor. Alternatively (or perhaps complementarily), platelets that cluster about tumor cells within the circulation may serve as a bridge between the tumor cells and sites of damaged endothelium allowing tumor cells that might otherwise not attach to do so (Marcum et al , 1979). These two general possibilities are not mutually exclusive nor are they necessarily the only possibilities.

The importance of mechanisms of anticoagulant effect on tumors can be illustrated by the difficulties that might arise in attempting to interpret a negative therapeutic trial. The trial may, of course, have been unrewarding because the hypothesis was incorrect in general or incorrect for the particular tumor under investigation. On the other hand, the trial may have been unsuccessful because of the experimental design used. For example, under certain circumstances a negative interaction may occur between anticoagulants and chemotherapeutic agents given simultaneously resulting in negation of the beneficial effect of the anticoagulant. A negative interaction of this kind has been demonstrated in an experimental tumor system (Carmel and Brown, 1977). Such an interaction could conceivably occur in human tumors. Examples can also be cited from the experimental tumor literature in which one anticoagulant drug exerted a beneficial effect, but another did not (Hagmar, 1972). How might this be explained? Imagine a type of tumor for which formation of an intravascular coagulum is necessary in order to facilitate attachment of tumor cells to the vessel wall. In this hypothetical instance, heparin may block such intravascular coagulation and thus prevent metastatic seeding. However, it might have no effect whatever on established, fibrin-laden extravascular tumor deposits because the heparin fails to leave the intravascular compartment. On the other hand, warfarin, which is a weaker and indirect inhibitor of intravascular thrombin generation, might not be expected to have an effect on metastasis formation by circulating tumor cells. However, it might indeed influence extravascular tumor growth because it can gain access to these cells, where it inhibits the metabolic pathways involved in cellular coagulant production. A rational experimental design for heparin, then, would be to administer this agent at times of maximal seeding of tumor cells, such as might occur with manipulation of the tumor in the course of its removal at surgery.

There is obviously a danger of oversimplification at any time that hypotheses are formulated in an area of investigation that is in its infancy. Certainly the properties of tumor cells that allow them to interact with coagulation reactions represents but one very narrow aspect of the complex pathobiology of cancer. But it is, nevertheless, an intriguing aspect and it is all too tempting to speculate on the possible ramifications of the concepts under consideration. For example, it is conceivable that a hypothesis related to the clotting mechanism may explain the long period of dormancy that sometimes exists, for example, between the time a tumor is resected and recurrence months to years later. The clonalselection from a larger population of slowly proliferating cells of a cell having as its singular phenotype the capacity to, for example, aggregate platelets or induce fibrin formation could trigger the positive feedback mechanisms described above causing rapid cell proliferation.

If current best estimates are correct, the anticoagulant approach to the containment of malignancy is neither a direct toxic assault on the tumor (as in chemotherapy) nor a technique for bolstering host defenses that are presumed to be faulty (as in immunotherapy). Rather, it is a means for inhibition of certain reactions within the host that are required by the tumor in order for the tumor to compete successfully against the host.

An analogy has been drawn between the coagulation response to a tumor and the initial coagulation response to a wound (Dvorak et al , 1981a). However, unlike the wound that eventually heals by invasion of the fibrin clot by fibrous tissue, malignancy has been viewed as "a wound that cannot be healed" (Dvorak et al , 1981a). But, even this may be an understatement. It may be that malignant cells would cease to be "a wound" in the absence of this particular host response. Stated otherwise, without the response of the host, the assault of the tumor would fail. The hypothesis is that the "success" of the tumor depends upon how skillfully it has learned to manipulate coagulation-related reactions in its local environment.

It is unsettling, to say the least, to contemplate the possibility that the coagulation-related responses which serve so well in tissue repair are the very processes which lead to the downfall of the host when usurped by the invading neoplasm.

It remains for future experimentation to determine the validity of this hypothesis. Necessary information can only come from carefully controlled experiments planned on the basis of insight into presumed mechanisms. Attention must also be given to further elucidation of mechanisms responsible for observations made thus far as well as to the practical aspects of anticoagulant administration in cancer, for instance, definition of drug toxicity. It should be pointed out, however, that adequate rationale appears to exist for testing of the anticoagulant hypothesis of cancer treatment in humans, especially in tumor types for which current best therapy is inadequate, and for which future planning is stalemated. In addition, no special new expertise appears to be required before clinical trials of a variety of coagulation and platelet-inhibitory drugs are undertaken. Such trials are particularly attractive because the drugs in question are available in quantity and at low cost.

In summary, the anticoagulant approach to the management of malignancy is based on a novel concept of treatment that emphasizes the importance of the ability of tumor cells to provide for themselves an environment that is conducive to their growth and spread. This concept is attractive because a number of testable hypotheses may be derived from it and because it points to ways of modifying the biologic response of patients to malignancy by means that are readily available and cost-effective.

REFERENCES

Al-Saleem T, Ali ZS, Qassab M (1980). Skin cancers in xeroderma pigmentosum: response to indomethicin and steroids. Lancet 2:264.

Ambrus JL, Ambrus CM, Pickern J, Solder S, Boss I (1975). Hematologic changes and thromboembolic complications in neoplastic disease and their relationship to metastasis. J Med 6:433.

Ambrus JL, Ambrus CM, Gastpar H (1980). Studies on platelet aggregation and platelet interaction with tumor cells. In DeGaetano G, Garattini S (eds): "Platelets: A Multidisciplinary Approach", New York: Raven Press, p 467.

Annegers JF, Zacharski LR (1980). Cancer morbidity and mortality in previously anticoagulated patients. Thrombos Res 18:399.

Black PH (1980). Shedding from the cell surface of normal and cancer cells. In Klein G, Weinhouse S (eds): "Advances in Cancer Research" vol 32, New York: Academic Press, p 75.

Brown, JM (1973). A study of the mechanisms by which anticoagulation with warfarin inhibits blood borne metastasis. Cancer Res 33:1217.

Carmel RJ, Brown JM (1977). The effect of cyclophosphamide and other drugs on the incidence of pulmonary metastasis in mice. Cancer Res 37:145.

Dvorak HF (1981). Personal communication to the author.

Dvorak HF, Orenstein NS, Dvorak Am (1981). Tumor-secreted mediators and the tumor microenvironment: relationship to immunological surveillance. Lymphokines 2:203.

Edwards RL, Rickles FR (1978). Delayed hypersensitivity: effect of systemic anticoagulation. Science 200:541.

Forman WB (1979). Cancer in persons with inherited blood coagulation disorders. Cancer 44:1059.

Hagmar B (1972). Cell surface charge and metastasis formation. Acta Path Microbiol Scand 80:357.

Hilgard P (1977). Experimental vitamin K deficiency and spontaneous metastases. Br J Cancer 35:891.

Hilgard P, Maat B (1979). Mechanism of lung tumor colony reduction caused by coumarin anticoagulation. Europ J

Cancer 15:183.
Honn KV, Cicone B, Skoff A (1981). Prostacyclin: a potent antimetastatic agent. Science 212:1270.
Marcum JM, McGill M, Jamieson GA (1979). Platelet-tumor cell-vessel wall interaction *in vitro*: aggregation and perfusion studies. Blood 54 (Suppl 1): 251a.
Michaels L (1974). The incidence and course of cancer in patients receiving anticoagulant therapy. Retrospective and prospective studies. J Med 5:98.
Myers TJ, Rickles FR, Barb C, Cronlund M (1981). Fibrinopeptide A in acute leukemia: relationship of activation of blood coagulation to disease activity. Blood 57:518.
Pearlstein E, Salk PL, Yogeeswaran G, Karpatkin S (1980). Correlation between spontaneous metastatic potential, platelet aggregating activity of cell surface extracts and cell surface sialylation in 10 metastatic-variant derivatives of a rat renal sarcoma cell line. Proc Natl Acad Sci USA 77:4336.
Poggi A, Mussoni L, Kornblihtt L, Ballabio E, deGaetano G, Donati MB (1978). Warfarin enantiomers, anticoagulation and experimental tumor metastasis. Lancet I:163.
Poggi A, Donati MB, Garattini S (1981). Fibrin and cancer cell growth: problems in the evaluation of experimental models. In Donati MB, Davidson JF, Garattini S (eds): "Malignancy and the Hemostatic System", New York: Raven Press, p 89.
Powles TJ, Dady PJ, Williams J, Easty GC, Coombes RC (1980). Use of inhibitors of prostaglandin synthesis in patients with breast cancer. In Samuelsson B, Ramwell PW, Paoletti R (eds): "Advances in Prostaglandin and Thromboxane Research", vol 6, New York: Raven Press, p 511.
Rickles FR, Edwards RL, Barb C, Cronlund M (1981). Abnormalities of blood coagulation in patients with cancer: fibrinopeptide A generation and tumor growth. Submitted for publication.
Slichter SJ, Harker LA (1974). Hemostasis and malignancy. Ann NY Acad Sci 230:252
Zacharski LR, Rosenstein R (1979). Reduction of salivary tissue factor (thromboplastin) activity by warfarin therapy. Blood 53:366.
Zacharski LR, Henderson WG, Rickles FR, Forman WB, Cornell CJ Jr, Forcier RJ, Harrower HW, Johnson RO (1979). Rationale and experimental design for the VA Cooperative Study of anticoagulation (warfarin) in the treatment of cancer. Cancer 44:732.

Zacharski, LR (1981). Anticoagulation in the treatment of cancer in man. In Donati M, Davidson JF, Garattini S (eds): "Malignancy and the Hemostatic System", New York: Raven Press, p 113.

Zacharski, LR, Henderson WG, Rickles FR, Forman WB, Cornell CJ Jr, Forcier RJ, Headley E, Kim S-H, O'Donnell JF, O'Dell R, Tornyos K, Kwaan H (1981a). Effect of sodium warfarin on survival in small cell carcinoma of the lung: VA Cooperative Study #75. J Am Med Assoc 245:831.

Zacharski, LR, Rickles FR, Henderson WG, Martin JF, Forman WB, van Eeckhout JP, Cornell CJ Jr, Forcier RJ (1981b). Platelets and malignancy: rationale and experimental design for the VA Cooperative Study of RA-233 in the treatment of cancer. Submitted for publication.

DISCUSSION

Dr. D. Cowan (Case Western Reserve University, Cleveland). How long were your patients treated; for one year, two years or how long?

Dr. L. R. Zacharski (VA Medical Center, White River Junction). We only have about six months where we can see these patients.

Dr. Cowan. So the patients were still receiving chemotherapy during the relapse you observed?

Dr. Zacharski. Yes.

Dr. Cowan. You found that the effect of warfarin was to prolong remission in patients, is that right? Is this now an unmaintained remission or is this a remission in patients who are still on therapy?

Dr. Zacharski. They are very definitely still on chemotherapy.

Dr. Cowan. So that you really have a subgroup of your total group; in other words, the patients who do not achieve remission do not have an effect from warfarin.

Dr. Zacharski. That is not so; in fact, it is precisely the opposite. I think these results, if confirmed, should really change our thinking on the role of anticoagulants or antiplatelet agents on the course of human malignancy. It looks to us as if the patients who got a benefit were patients whom you would expect would not get a benefit: patients who did not get a remission and patients who had extensive disease.

Dr. J.L. Ambrus (Roswell Park Memorial Institute, Buffalo). I should like to speculate about the possible mechanism of warfarin in your experiments, Dr. Zacharski. Of course, you are familiar with the ancient literature which claims that warfarin has a possible relation to tumor cells?

On the other hand, you were using oat cell type carcinoma which we know is exceedingly susceptible to chemotherapy. One therefore might expect that the chemotherapy which you gave your patients would produce a lot of tumor destruction and would release a lot of thromboplastic factors. This might result in occlusion of the microcirculation to the remaining tumor and prevent the penetration of additional chemotherapeutic agent. By using anticoagulants you are probably preventing this occlusion. The presence or absence of the anticoagulants may make the difference as to whether the microcirculation may close in and thus protect the tumor from any therapeutic agent you may have administered. I wonder whether you would like to speculate as to what you think might be taking place?

Dr. Zacharski. I would have to say I do not really know, but the answer could be either very long or very short. We do rationalize inhibition primarily in terms of in vitro data but in vivo the results may be different.

We have crude survival figures. We decided that, if we were going to do a trial of warfarin, we at least ought to give the patients a chance to become anticoagulated. We defined an appropriate patient as one who lives at least two weeks in either the control or the warfarin group. This allows the patient to achieve some degree of anticoagulation after two weeks and would also allow the patient to get at least one cycle of chemotherapy. Thus, the minimum criteria for inclusion are one cycle of chemotherapy, two weeks survival and warfarin administration for over a two-week period.

ANTIMETASTATIC ACTION OF RX-RA 69, A NEW POTENT PDE-INHIBITOR IN THE LEWIS LUNG CARCINOMA OF THE MOUSE

Rosemarie Lichtner and Walter Haarmann

Dr. Karl Thomae GmbH, Biologische Forschung
Birkendorfer Strasse 65
D-7950 Biberach, West Germany

SUMMARY

RX-RA 69, a pyrimido-pyrimidine derivative, is a new potent PDA-inhibitor which inhibits tumor cell induced platelet aggregation in vitro. Also ex vivo an inhibition was found after pretreating mice with 1 mg/kg RX-RA 69 orally one hour before collecting the blood. The same dosage is able to prevent the drop in platelet count induced by injecting a large number of tumor cells into mice. The same dosage schedule in the amputation experiment resulted in a reduction of incidence of matastases in all treated groups; the best results, however, were obtained by starting the treatment two days after the amputation. In the spontaneous model an inhibition of metastases formation by 10-20 mg/kg RX-RA 69 starting from the fourth day after tumor cell implantation was found in four independent experiments. The mechanism of the antimetastatic action of this drug remains to be evaluated.

INTRODUCTION

Since thrombus formation around intravascular tumor cells prior to their movement through the endothelium has been observed in a vessel of the rabbit ear chamber (Wood, 1971), anticoagulants have been used to reduce metastases spread under various experimental conditions. These and other observations have led to attempts to prevent metastasis by using agents that inhibit platelet aggregation. Some investigators reported a decrease in metastases formation after aspirin treatment (Gasic et al, 1973) while others were not able to confirm these results (Wood and Hilgard, 1972). The phosphodiesterase (PDE)-inhibitors dipyridamole and its derivative RA 233, which are able to diminish platelet adhesiveness to circulating tumor cells (Gastpar, 1971) and to prolong their circulation time (Ambrus et al, 1978) failed to inhibit spontaneous and artificial metastases in a spontaneous metastasizing animal tumor, the Lewis lung carcinoma of the mouse (Hilgard et al, 1976).

Recently some PDE-inhibitors were found to produce marked reduction of metastasis in spontaneous animal models. Isobutylmethylxanthine (IBX) was effective with the Lewis Lung carcinoma (Janik et al, 1980) and pentoxyfilline in two different tumor systems (Gordon et al, 1979). We therefore tested the new potent pyrimido-pyrimidine derivative RX-RA 69 in the Lewis lung carcinoma of the mouse.

MATERIALS

Tumor

The tumor was obtained from M.B. Donati, M.D. (Mario-Negri Institute, Milan) and is maintained intramuscularly in male C 57 Bl/6 mice, weighing 18-22 g, by passages every 2-3 weeks. Single cell suspensions are obtained by mild trypsin digestion under stirring for 30 min at $37^\circ C$ (0.3% trypsin in Dulbecco's modified Eagle Medium). After passing the suspension through a folded sterile gauze pad, the cells are washed in PBS-buffer (phosphate-buffered saline) at $4^\circ C$ (350 g) resuspended in PBS with 10% fetal calf serum, centrifuged and washed again in ice-cold PBS. Cell viability as determined by trypan blue exclusion is about 80-90%.

The mice were injected with 10^5 freshly prepared tumor cells intramuscularly in the left posterior leg. The primary tumor became palpable on day 7-8 after tumor transplantation and on day 24-26 all animals were sacrificed.

For the i.v. injection of tumor cells or the aggregation experiments the cells were washed twice, the last washing being carried out in saline (for the i.v. injection) or in saline buffered with 10 mM MOPS (morpholinopropane sulphonate). Cell viability was 50-60%.

Substance

RX-RA 69 is a derivative of pyrimido-pyrimidine (8-benzylthio-4-morpholino-2-piperazino-pyrimido [5,4-d] pyrimidine) (Figure 1) and is a very potent inhibitor of platelet phosphodiesterase (EC_{50}: 10^{-9}M).

RX-RA 69:

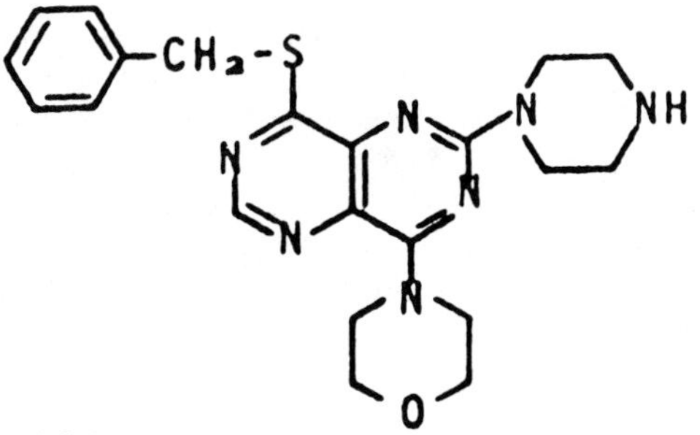

MG: 423,6

Figure 1: 8-Benzylthio-4-morpholino-2-piperazino-pyrimido (5,4-d) pyrimidine.

RX-RA 69 was developed as an antithrombotic agent. In vitro, platelet aggregation induced by ADP or collagen is inhibited with an EC_{50} between 10^{-8} and 10^{-9} M. In vivo, bleeding time in mice is prolonged by more than 50% in doses of about 0.15 mg/kg given orally one hour earlier. In rats, platelet aggregation and experimental thrombosis are influenced in doses as low as 0.1 mg/kg orally given one hour earlier. Like many other inhibitors of PDE this compound has blood pressure lowering effects at higher doses (results to be published).

In our experiments to prevent metastases formation the substance was supplied in the drinking water at a daily dose between 10-20 mg/kg. RX-RA 69 was stable in the drinking water for at least one week.

Many tumor cell lines are able to induce platelet aggregation in vitro and cause thrombocytopenia in vivo. The isolated cells are prepared from solid or ascites tumors or tissue cultures (Gasic et al, 1973). The platelet-aggregating activity is assumed to be associated with the tumor cell membrane (Hara et al, 1980; Pearlstein et al, 1980). Isolated Lewis lung carcinoma cells, prepared by mild enzymatic digestion from solid tumors, are also able to induce platelet aggregation in vitro in PRP (platelet-rich plasma) of mice (Figure 2) at room temperature. Blood was collected by heart puncture and pooled. Because tumor cell-platelet interaction is strongly dependent on divalent metal-ions in plasma (Gasic et al, 1973), all experiments were carried out with heparinized blood (5 units/ml blood) without citrate. Aggregation with $1-3 \times 10^6$ tumor cells/ml involves a short lag period followed by a single-phase aggregation curve. Inhibition by RX-RA 69 occurs in a dose-dependent manner (Figure 2) after incubating the PRP for 10 min.

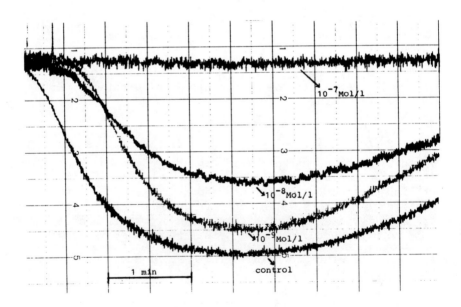

Figure 2. Aggregation of mouse PRP by tumor cells (1.35×10^6/ml). Dose-dependent inhibition by RX-RA 69 in vitro.

Pretreating mice with 1 mg/kg RX-RA 69 orally one hour before collecting the blood also resulted in an inhibition of tumor cell-induced platelet aggregation ex vivo (Figure 3).

RX-RA 69:

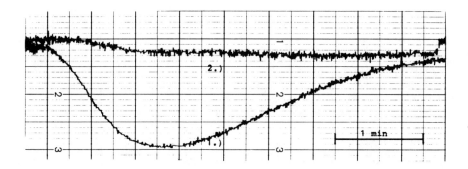

Figure 3. Aggregation of mice-PRP by tumor cells (2.36×10^6).
  1. Untreated control
  2. Inhibition by RX-RA 69 ex vivo (1 mg/kg p.o. 1 hour before testing)

The i.v. injection of cells from certain tumors into animals results in a rapid thrombocytopenia usually correlated with the number of cells injected. It has been shown that anticoagulation of the animals before tumor cell challenge effectively prevents the acute platelet drop. Platelet inhibitors, like Ditazole (Mussoni et al, 1977) or pyrimido-pyrimidine derivatives (Gastpar, 1971) are also able to prevent the platelet drop induced by injection of tumor cells.

RX-RA 69 is also able to prevent the drop in platelet count when given orally at 1 mg/kg one hour before i.v. injection of isolated Lewis lung cells in the tail vein (Table 1). Each animal was taken as its own control, the blood samples were taken by puncturing the retro-orbital complex under slight ether anaesthesia. We did not observe a fall in platelet count after injection of 200,000 cells/animal, reported by others (Mussoni et al, 1977), but under

Table 1. Effect of RX-RA 69 (1 mg/kg body weight, orally) on 3LL cell induced platelet aggregation in vivo (mean $\pm$ S.E.)

Treatment	Number of animals	Number of tumor cells injected	Platelets x 10^3 / mm^3 before injection (initial value)	Platelets x 10^3 / mm^3 5 min after injection	Reduction as percent of the initial value
Effect of manipulations	16	0	721 $\pm$ 35	605 $\pm$ 34	83.9
Control	18	$2 \cdot 10^5$	690 $\pm$ 25	592 $\pm$ 32	85.8
RX-RA 69	18	$2 \cdot 10^5$	690 $\pm$ 34	617 $\pm$ 35	89.5
Control	24	$1 \cdot 10^6$	708 $\pm$ 22	411 $\pm$ 25	58.0
RX-RA 69	26	$1 \cdot 10^6$	663 $\pm$ 21	522 $\pm$ 23	78.7

*$p < 0.05$ (Student's t-test)

our experimental conditions 10^6 tumor cells/animal were needed. Table 1 also shows that the manipulations alone (taking ca. 0.1 ml blood for the initial value of platelet count and i.v. injection of 0.1 ml volume) reduce the platelet number by 10-15%.

The effect of long-term treatment with RX-RA 69 on primary tumor growth and spontaneous lung metastases compared with the corresponding control is demonstrated in Table 2. In four independent experiments we found an inhibition of spontaneous metastases formation when we started the treatment on day 4 after tumor cell implantation without any effect on primary tumor weight. Starting the treatment on day 0 the results are not so clear. There is a tendency to decreased metastases number and weight but with a tendency to increased primary tumor weight.

We therefore studied the action of RX-RA 69 on Lewis lung cells lodging in the lungs as spontaneous metastases after removal of the primary tumor. Table 3 shows that pretreating the animals with 1 mg/kg RX-RA 69 one hour before amputation of the tumor-bearing leg results in reduction of the incidence of metastases without reducing the total number of metastases per animal. Starting the treatment two days after amputation, when the cells are arrested and no longer in the circulation, the best results were obtained, with reduced incidence and number of metastases.

DISCUSSION

In the present study we have shown that RX-RA 69, a new, very potent antiplatelet drug, is inhibiting tumor cell platelet interaction in vitro (Figure 2) and ex vivo (Figure 3). Injection of a large number of tumor cells into mice resulted in a drop in the platelet count. This thrombocytopenia was prevented by pretreating mice with 1 mg/kg RX-RA 69 orally one hour earlier (Table 1). The same treatment diminished the incidence of spontaneous metastases after removal of the primary tumor (Table 3). The best result was obtained when the treatment was started two days after amputation. This result is in accordance with the work of Janik et al (1980). They suggested that the PDE-inhibitor isobutylmethylxanthine acts directly on the tumor cell causing a reduction in growth rate of malignant cells by enhancing the degree of interaction between growth-arrested normal

Table 2. Effect of treatment with RX-RA 69 on primary and metastatic growth of 3LL (mean ± S.E.). The treatment periods are given within parentheses.

Number of Experiment	Treatment	Number of Animals	Tumor Weight (g)	Lung Weight (mg)	Metastases Number	Metastases Weight (mg)
1	Control	12	9.4 ± 0.4	312 ± 37	17 ± 3	134 ± 34
	RX-RA 69 (d.0-24)	10	10.8 ± 0.3*	218 ± 13*	12 ± 1	39 ± 11
	RX-RA 69 (d.4-24)	11	9.1 ± 0.5	185 ± 11*	5 ± 1*	17 ± 10*
2	Control	12	9.1 ± 0.4	289 ± 32	15 ± 2	93 ± 25
	RX-RA 69 (d.0-25)	12	9.3 ± 0.6	217 ± 19	12 ± 1	79 ± 25
	RX-RA 69 (d.4-25)	12	9.7 ± 0.4	210 ± 11*	8 ± 1*	59 ± 10
3	Control	12	9.6 ± 0.4	266 ± 34	11 ± 1	93 ± 43
	RX-RA 69 (d.0-24)	12	9.6 ± 0.4	226 ± 27	12 ± 2	66 ± 26
	RX-RA 69 (d.4-24)	12	9.4 ± 0.5	194 ± 17	6 ± 2*	36 ± 19
4	Control	12	10.5 ± 0.3	269 ± 26	12 ± 2	81 ± 24
	RX-RA 69 (d.4-24)	12	10.0 ± 0.3	185 ± 10*	6 ± 1*	21 ± 5*

*$p < 0.05$ (Student's t-test).

Table 3. Effect of treatment with RX-RA 69 combined with surgical removal of the tumor bearing leg (at day 8) on metastasis growth. Animals were sacrificed at day 24 (mean ± S.E.) The treatment periods are given within parentheses.

Treatment	Total numb. of animals	Animals with metast.	Animals with metast./total numb. of anim. (%)	Lung Weight (mg)	Metastases	
					Number	Weight (mg)
Control	14	9	64	248 ± 19	2.5 ± 0.7	56 ± 17
RX-RA 69 (1h before amp.)	13	5	38	241 ± 24	2.6 ± 1.3	59 ± 29
RX-RA 69 (1h before amp. and day 8-24)	11	4	36	247 ± 41	2.5 ± 0.5	43 ± 36
RX-RA 69 (day 10-24)	12	3	25	205 ± 13	0.5 ± 0.2*	18 ± 15

*$p < 0.05$ (Student's t-test)

cells and malignant cells. RX-RA 69 may exert its antimetastatic effect by the same mechanism. However, this experiment may indicate that the antimetastatic action of RX-RA 69 in this animal model is not mediated by inhibition of the initial arrest of tumor cells or tumor cell platelet emboli or that the antiplatelet effect during the initial phase of tumor cell lodgement is of minor importance. Also the dependence of this effect on timing of RX-RA 69 administration (Table 2) makes it unlikely that the antimetastatic effect is mediated by platelets. A time-dependent process seems to be involved, perhaps the development of antitumor immunity. Similar observations were made by Hengst et al (1980) who showed that the importance of timing in cyclophosphamide therapy in a murine tumor system has an immunological basis.

Further experiments are necessary to explain the antimetastatic action of RX-RA 69 in this animal tumor. The next steps of investigation will be to look for the importance of PDE-inhibition on the tumor cell itself and on the immune system of the host. However, further pharmacological actions cannot be excluded.

REFERENCES

Ambrus JL, Ambrus CM, Gastpar H (1978). Studies on platelet aggregation in vivo. VI. Effect of a pyrimido-pyrimidine derivative (RA 233) on tumor cell metastasis. J Med 9:183.
Gasic GJ, Gasic TB, Galatini N, Johnson T, Murphy S (1973). Platelet-tumor-cell interactions in mice. The role of platelets in the spread of malignant disease. Int J Cancer 11:704.
Gastpar H (1971). The inhibition of the cancer cell stickiness by pyrimido-pyrimidine derivatives induced by inhibition of platelet aggregation. Acta Med Scand 190:269.
Gordon S, Witul M, Cohen H, Sciandra J, Williams P, Gastpar H, Murphey GP, Ambrus JL (1979). Studies on platelet aggregation inhibitors in vivo. VIII. Effect of pentoxifylline on spontaneous tumor metastasis. J Med 10:435.
Hara Y, Steiner M, Baldini MG (1980). Characterization of the platelet-aggregating activity of tumor cells. Cancer Res 40:1217.
Hengst JCD, Mokyr MB, Dray S (1980). Importance of timing in cyclophosphamide therapy of MOPC-315 tumor-bearing mice. Cancer Res 40:2135.

Hilgard P, Heller H, Schmidt CG (1976). The influence of platelet aggregation inhibitors on metastasis formation in mice (3LL). Z Krebsforsch 86:243.

Janik P, Assaf A, Bertram JS (1980). Inhibition of growth of primary and metastatic Lewis Lung Carcinoma cells by the phosphodiesterase inhibitor isobutylmethylxanthine. Cancer Res 40:1950.

Mussoni L, Poggi A, Donati MB, de Gaetano G (1977). Ditazole and Platelets. III. Effect of Ditazole on tumor-cell induced thrombocytopenia and on bleeding time in mice. Hemostasis 6:260.

Pearlstein E, Salk PL, Yogeeswaran G, Karpatkin S (1980). Correlation between spontaneous metastatic potential, platelet-aggregating activity of cell surface extracts, and cell surface sialylation in 10 metastatic-variant derivatives of a rat renal sarcoma cell line. Proc Natl Acad Sci USA 77:4336.

Wood S Jr (1971). Mechanism of establishment of tumor metastases. Pathobiology Ann 1:281.

Wood S Jr, Hilgard P (1972). Aspirin and tumour metastasis. Lancet II:1416.

BLOOD PLATELETS AND TUMOUR DISSEMINATION

P. Hilgard

Bristol Myers International Corporation
Brussels, Belgium

According to the current concept of the pathogenesis of tumour dissemination, the release of cancer cells from the primary tumour, their transport in lymphatic and/or blood vessels and their subsequent arrest at a distant site represent the initial sequence of events. The distribution of cancer cells throughout the body is not a random phenomenon but follows a sequential pattern specific for each tumour type (Viadana et al., 1973). Numerous clinical and experimental observations indicate that only a minute fraction of circulating cancer cells will eventually lead to secondary tumours, whereas the bulk of cells do not survive in the circulation. A prerequisite for extravascular metastatic tumour growth is the arrest of the blood-borne cancer cell embolus at the vascular endothelium and its subsequent penetration of the vessel wall. The first morphological evidence for an association of intravascular cancer cells with blood platelets and fibrous material was found at the beginning of this century (Schmidt, 1903). It was, however, more than 50 years later that the possible pathogenic significance of this phenomenon was recognised. In the rabbit ear chamber, S. Wood, Jr. observed the fate of the intravascular tumour cells in vivo after their injection into the circulation: shortly after the initial attachment of the tumour cell embolus to the vascular endothelium, micro thrombi (consisting of blood platelets and fibrin) formed around the adherent cells, leading to their fixation at the site of their primary arrest in the capillary bed (Wood, 1958). Using time-lapse cinematography it was documented that these intravascular tumour cells, sheltered by the surrounding thrombotic material, penetrated the vessel wall and reached the perivascular tissue where they could proliferate into 'secondary' tumours.

Having gained access to the blood or lymph stream, the tumour cell or the tumour cell embolus is exposed to several mechanical and biochemical attacks frequently leading to its death. In addition, tumour cells will interact with other blood cells in the circulation and considerable interest has focused on the interaction between blood platelets and circulating cancer cells (Hilgard, 1978). Most of our present knowledge about the role of blood platelets in the establishment of tumour metastases is derived from two animal models: syngeneic or allogeneic tumour cell suspensions were injected into the blood stream of experimental animals and the subsequent development of 'tumour colonies' in various organs was evaluated by quantitative techniques. Alternatively, transplanted solid tumours, which gave rise to spontaneous metastases at distant sites, were used as models of disease. In either of these experimental models the host's platelet number or function was altered in vivo by various pharmacological manipulations.

The close association in lung capillaries of components of the coagulation system, in particular blood platelets, with intravascular Walker 256 cells is shown in figures 1 and 2.

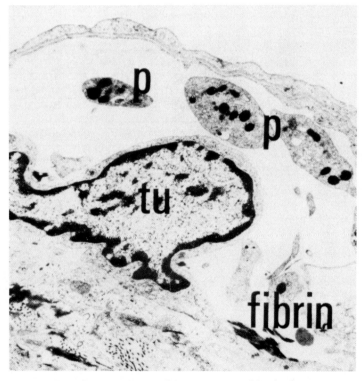

Figure 1: EM of a Walker 256 tumour cell (tu) in a pulmonary capillary 15 minutes after i.v. injection. (p = platelets). From Hilgard, 1980 (courtesy of G. Fischer Verlag).

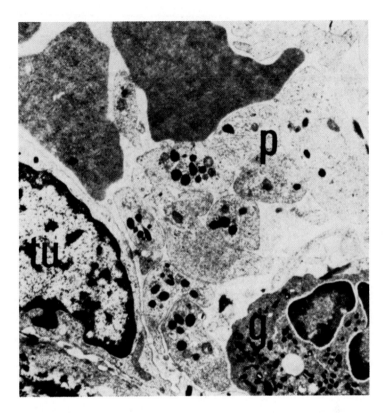

Figure 2: Walker 256 tumour cell (tu) in a lung capillary 15 minutes after intravenous injection (p = platelets g = granulocytes). From Hilgard, 1980 (courtesy of G. Fischer Verlag).

Shortly after the intravenous injection of tumour cells, small amounts of fibrin and platelets were found in the vicinity of the cancer cells adherent to the vascular endothelium. Many of the platelets had lost their normal ultrastructural appearance, indicating their activation.

The sequence of intravascular events following i.v. challenge of rats with Walker 256 tumour cells is shown in table 1.

Time after injection	Histological Findings (Light Microscopy + EM)
15 minutes	Tumour cells surrounded by platelets and fibrin-like material (occasionally polymerized fibrin)
6 hours	Tumour cells still intravascular, platelets and fibrin disappeared
9 - 24 hours	Destruction of vessels wall, tumour cell penetration
48 hours	First visible extravascular proliferation

Table 1: Intravascular events (lung) after i.v. injection of Walker 256 tumour cells into rats. From Hilgard 1980 (courtesy of G. Fischer Verlag).

This data was obtained by sequential microscopic analysis and supports the hypothesis that thrombosis and platelet aggregation are features of the early stages of secondary tumour growth. Similar findings were reported by other investigators (Sindelar et al., 1975, Chew and Wallace, 1976).

Furthermore, the intravenous injection of tumour cell suspensions into rodents resulted in a rapid decrease of circulating blood platelets in the animals (Hilgard, 1973, Poggi et al., 1977). The platelet decrease was strictly correlated to the number of tumour cells injected (figure 3).

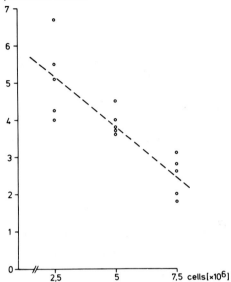

Figure 3: Number of circulating blood platelets five minutes after the i.v. injection of various amounts of Walker 256 tumour cells into rats. From Hilgard, 1980 (courtesy of Raven Press).

The distribution of 51-Cr labelled blood platelets in various organs following the i.v. injection of Walker 256 tumour cells, is shown in figure 4. The accumulation of platelets in the lungs - the site of primary arrest and 'metastatic tumour growth' in this experimental system - reflects the morphological association of platelets with tumour cells.

Gasic et al., (1968) have shown that the injection of a tumour cell suspension into thrombocytopenic rats and mice resulted in a decrease of 'metastatic' lung colonies in these animals; this 'antimetastatic' effect could be elicited either by neuraminidase pretreatment (Gasic et al., 1968) or by the injection of the antiplatelet serum (Gasic et al., 1968, Ivarsson 1976). These observations suggested that the platelet - tumour cell interaction was an important feature

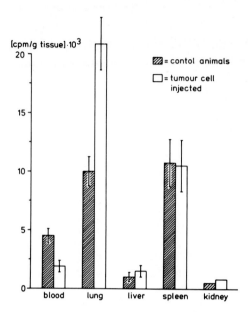

Figure 4: Distribution of 51-Cr labelled platelets in various organs 5 minutes after the i.v. injection of 10^6 Walker 256 tumour cells into rats. From Hilgard, 1980 (Courtesy of Raven Press).

in the haematogenous dissemination of malignant tumours and indicated that the pharmacological alteration of platelet function could alter the pattern of lung colony formation ('haematogenous metastases') in experimental animals. Further support for this concept was derived from Gastpar's data which showed that the incidence of lethal tumour emboli in the lungs of animals rapidly injected with a tumour cell suspension could be diminished by pretreatment with platelet aggregation inhibitors (Gastpar 1970).

In some experimental systems aspirin treatment was effective in reducing the number and incidence of tumour colonies resulting from the intravenous challenge with tumour cells (Gasic et al., 1973, Kolenich et al., 1972). This antimetastatic effect of aspirin was attributed by the authors to the drug-induced inhibition of platelet function. Powles et al. (1973) however, suggested that aspirin might excert its 'antimetastatic action' by inhibition of lysosomal enzymes or related mechanisms which are intrinsic to the tumour cell. The effects of various platelet aggregation inhibitors on the incidence and number of lung colonies were investigated in C57Bl mice, challenged intravenously with Lewis Lung Carcinoma cells. Aspirin, the dipyridamole derivative RA233 and bencyclan treatment was initiated prior to the tumour cell injection. As shown in table 2, none of the drugs

Pretreatment	Dose/kg	No. colonies	p-value
Control	-	0.9	-
Aspirin	200 mg	1.5	> 0.1
RA233	50 mg	1.2	> 0.1
Bencyclan	40 mg	2.4	= 0.01
Heparin	1000 USP-U	0.1	< 0.01

Table 2: Number of lung colonies after i.v. injection of 10^5 Lewis lung carcinoma cells into mice pretreated with various drugs.

administered to the animals was effective in reducing the number of tumour colonies. Similar experiments were carried out employing different experimental tumours in rabbits and rats, however the results obtained were identical (table 3).

Tumour	Pretreatment	Dose/kg	No. colonies
V2-carcinoma (rabbit)	control	-	22.7
	aspirin	50 mg	20.8
Walker 256 (rat)	control	-	1.5
	dipyridamole	50 mg	2.3
	heparin	750 USP-U	0.1*

* significant ($p < 0.01$)

Table 3: number of lung colonies after i.v. injection of tumour cells into rodents pretreated with various drugs.

Bencyclan treatment resulted in a signficant increase in tumour colonies in mice injected with the Lewis Lung tumour (table 2). This is an interesting finding because Gastpar et al. (1977) showed that this drug - given in the same dose as in the above experiments - prolonged the circulation time of intravenously injected tumour cells. In contrast to the 'antiplatelet drugs', heparin treatment prior to the intravenous tumour cell challenge was highly effective in reducing the number of tumour colonies in rats and mice (tables 2 and 3). Anticoagulation with heparin also prevented the thrombocytopenia which followed the intravenous injection of tumour cells (Hilgard, 1973). This finding indicated that the generation of thrombin at the site of tumour cell arrest at the vessel wall is responsible for the aggregation of blood platelets around adherent tumour cells.

Figure 5 shows the peripheral platelet count over a period of 15 minutes following tumour cell injection into rats pretreated with various anticoagulants. The number of lung colonies in the same experiment is given in table 4.

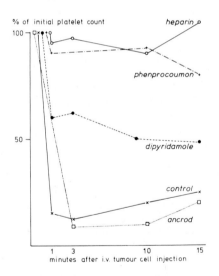

Figure 5: Peripheral platelet counts following the i.v. injection of Walker 256 tumour cells into rats pretreated with various anticoagulants. From Hilgard, 1980 (Courtesy of G. Fischer Verlag).

Pretreatment	no. of colonies	p. value
Control	25	-
Heparin	3	< 0.01
Phenprocoumon	7	< 0.01
Dipyridamole	26	> 0.1
Ancrod	8	< 0.01

Table 4: Number of lung colonies after i.v. injection of Lewis Lung Carcinoma cells into mice pretreated with various anticoagulants.

As mentioned before, a substantial decrease in the number of circulating platelets is found following the i.v. injection of tumour cell suspensions into control animals. Pretreatment with the anticoagulants heparin or phenprocoumon almost completely prevents the platelet decrease, whereas dipyridamole was less effective and ancrod treatment had no effect on the subsequent thrombocytopenia. If these data are compared with those shown in table 4, it becomes evident that the reduction of lung colonies as a result of pretreatment with different anticoagulants does not necessarily correlate with the degree of thrombocytopenia induced by the injection of tumour cells.

When constant numbers of tumour cells were injected into rats via different routes the subsequent platelet decrease was strictly dependent upon the route of injection (figure 6).

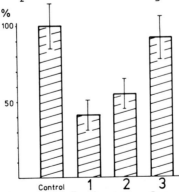

Figure 6: Platelet counts 5 minutes after injection of 1.5×10^6 Yoshida sarcoma cells by different routes (1 = femoral vein, 2 = portal vein, 3 = aorta): From Hilgard, 1980 (courtesy of Raven Press).

The intravenous injection into the femoral vein produced a more pronounced thrombocytopenia than the injection into the portal vein. In contrast, only a slight decrease of blood platelets was found if the tumour cells were injected into the adominal aorta. These findings suggest that the thrombocytopenia was not an expression of the interaction of blood platelets with tumour cells in the ciruclation but was rather caused by mechanisms which were related to the primary capillary bed in which the cells were arrested.

Since the injection of tumour cells directly into the circulation is a highly artificial situation, spontaneously metastasising transplantable tumours have been introduced as experimental approaches to tumour dissemination. In these models the initial stages of metastasis formation, i.e. the release of cancer cells from the primary tumour and their arrest and attachment to the bascular endothelium, are much more difficult to study. Metastasis formation is a dynamic process which takes place as long as a viable tumour mass is present in the host. Table 5 shows the results of typical

Treatment	No. of spontaneous metastases	
	control	treated
Aspirin	21.0	22.4
RA 233	19.0	23.7
Bencyclan	22.7	23.5
Cyproheptacline	23.3	29.3

Table 5: Effect of long-term treatment with various 'antiplatelet drugs' on spontaneous lung metastases of the Lewis Lung Carcinoma in mice. From Hilgard, 1980 (courtesy of Raven Press)

experiments using the subcutaneously transplanted and spontaneously metastasising Lewis Lung Carcinoma in C57B1 mice. Treatment with various 'antiplatelet drugs' was maintained throughout the entire period of the experiment. None of the treatments given had any significant influence on the development of spontaneous metastases in the animals. Whereas Gasic et al.

observed a reduction of lung metastases in an experimental mammary carcinoma after aspirin treatment (Gasic et al., 1973), Schmahl (1975) was equally unable to verify an effect of this drug on the frequency and distribution of spontaneous metastases in a carcinogen-induced autochthonous rat tumour. Two piperazine-derivatives, ICRF 158 and ICRF 159 are potent inhibitors of platelet function in vivo (both preventing ADP-induced thrombus formation), however, only ICRF 159 was shown to prevent spontaneous metastases (Atherton et al., 1975). This strongly suggests that 'antiplatelet' and 'antimetastatic' effects are not synonymous and that the 'antimetastatic' action of some platelet aggregation inhibitors might be related to other pharmacological properties. In addition, the mechanism by which tumour cells can aggregate platelets in vitro appears to be different from other known mechanisms of platelet aggregation (Pearlstein and Karpatkin, 1979).

Although some experimental evidence accumulated during the last decade indicating an inter-relationship between platelets and tumour dissemination, only one therapeutic trial has attempted to evaluate the relevance of the animal data in human malignant disease. One of the main pitfalls in designing adequate clinical trials for the use of antiplatelet drugs as antimetastatic agents, is the fundamental difference between the experimental and the clinical situation. Most malignant human tumours will already be disseminated at the time of diagnosis of the primary tumour, thus the efficacy of antimetastatic treatment is difficult, if not impossible, to assess. Nevertheless, a clinical trial with the pyrimido-pyrimidine compound RA 233 indicated that this drug might be an adequate adjuvant therapy to cancer surgery (Gastpar, 1977). As mentioned earlier, many of the inhibitors of platelet function exert their effects through non-platelet specific mechanisms such as inhibition of phosphodiesterase or prostaglandin biosynthesis. In the light of our present knowledge about the possible role of prostaglandins in tumour growth and in the evolution of tumour metastases the whole pharmacology of antiplatelet drugs has to be carefully considered (Tisdale, 1980). The delicate balance of diametrically opposing activities of the transient metabolites of arachidonic acid, many of which are also biosynthesised by tumour cells, should be taken into account when interpreting experimental and clinical data obtained with 'antiplatelet drugs'.

REFERENCES

Atherton A, Busfield D, Hellman K (1975). The effects of an antimetastatic agent (+/-)-1,2-bis (3,5-dioxopiperazin-1-yl) propane (ICRF 159) on platelet behaviour. Cancer Res 35:953.

Chew EC, Wallace AC (1976). Demonstration of fibrin in early stages of experimental metastases. Cancer Res 36:1904.

Gasic G, Gasic T, Stewart CC (1968). Antimetastatic effects associated with platelet production. Proc Nat Acad Sci (USA) 61:46.

Gasic G, Gasic T, Galanti N, Johnson T, Murphy S (1973). Platelet-tumor-cell interactions in mice. The role of platelets in the spread of malignant disease. Int J Cancer 11:704.

Gastpar H (1970). Stickiness of platelets and tumor cells influenced by drugs. Thrombos Diathes Haemorrh Suppl 42:291.

Gastpar H, Ambrus J, Thurber LE (1977). Study of platelet aggregation in vivo: II. Effect of bencyclan on circulatory metastatic tumor cells. J Med 8:53.

Hilgard P (1973). The role of blood platelets in experimental metastases. Brit J Cancer 28:429.

Hilgard P (1978). Blood platelets and experimental metastases. In de Gaetano G, Garrattini S (eds): "Platelets - A Multidisplinary Approach," New York: Raven Press, p 457.

Hilgard P (1980). Metastatic spread and altered blood coagulability. In Grundmann E (ed): "Metastatic Tumour Growth," Stuttgart: G. Fischer Verlag, p 107.

Ivarsson L (1976). Metastasis formation after intravenous tumor cell injection in thrombocytopenic rats. Eur Surg Res 8:51.

Kolenich JJ, Mansour EG, Flynn A (1972). Haematological effects of aspirin. Lancet ii:714.

Pearlstein E, Karpatkin S (1979). Requirement of a plasma factor for platelet aggregating material (PAM) extracted from tumor cells. Thrombos Haemostas 42:351.

Poggi A, Polentarutti N, Donati MB, de Gaetano G (1977). Blood coagulation changes in mice bearing Lewis lung carcinoma, a metastasizing tumor. Cancer Res 37:272.

Powles TJ, Easty GC, Easty DM, Neville AM (1973). Antimetastatic effect of aspirin. Lancet ii:100.

Schmal D (1975). Versuche zur beeinflussung der metastasierung autochthoner hepatome bei ratten mit antikoagulantien. Arzneim Forsch 25:1621.

Schmidt MB (1903). Die verbreitungswege der karzinome und die beziehung generalisierler sarkome zu den leukamischen neubildungen. In "Jena", Stuttgart: G. Fischer Verlag.

Sindelar WF, Tralka TS, Ketcham AS (1975). Electron microscopic observations on formation of pulmonary metastases. J Surg Res 18:137.

Tisdale MJ (1980). Prostaglandins and metastases. In Hellmann K, Hilgard P, Eccles S (eds): "Metastasis - Clinical and Experimental Aspects," The Hague: Nijhoff, p 136.

Viadana E, Bross IDJ, Pickren JW (1973). An autopsy study of some routes of dissemination of cancer of the breast. Brit J Cancer 27:336.

Wood, Jr S (1958). Pathogenesis of metastasis formation observed in vivo in the rabbit ear chamber. Arch Path (Chicago) 66:550.

DISCUSSION

Dr. M. B. Donati (Instituto di Recherche Farmacologische, Milan). You have said that this carcinoma is not the best model. Would that be due to the fact that this model induces thrombocytopenia?

Dr. P. Hilgard (Bristol-Meyers International Coporation, Brussels). I do not think I can answer that question. I know that lung tumors induce a thrombocytopenia during the course of tumor growth. I do not know if other tumors do the same. I know the carcinoma which we investigated also induced thrombocytopenia. You have shown that, by giving antiplatelet drugs, you cannot prevent this thrombocytopenia and it seems that it is really due more to marrow failure.

Dr. R. Greig (Smith and Kline, Inc., Philadelphia). I want to know the quantitation with regard to the distribution of cells.

Dr. Hilgard. This is one of the pitfalls of these experiments. You remember that the amount of tumor cells injected to produce thrombocytopenia is quite substantial. You get some 10-15 colonies and, assuming that one colony derives from one tumor cell, then where are the other tumor cells? It is more than statistically likely that any morphological observation using that approach observed something which is not of importance.

Dr. M. J. Stuart (SUNY, Upstate Medical Center, Syracuse). I would like to ask you the dose response with regard aspirin.

Dr. Hilgard. We have not repeated the experiments with other doses.

Dr. A. G. G. Turpie (McMaster University, Hamilton). Recently, I had the opportunity to look at some human tissues for intravascular tumor cells. I wonder if you think your presentation agrees with this association between platelets and tumor cells?

Dr. Hilgard. There is very nice experimental work with intravascular fibrin and platelets in a spontaneous model. There were tumor cell emboli in the vessels of the lung, but they were hardly ever associated with platelets and fibrin.

Dr. Mustafa A. Hay (El Safat, Kuwait). I wonder if you have tried treating the tumor cell with platelets before injecting them.

Dr. Hilgard. The answer is no, we have not done that.

Dr. N. Hanna (Frederick Cancer Research Center, Frederick). Tumors are heterogeneous and there are cells that can aggregate platelets and other cells that cannot. There are two stages: one has platelet-tumor interaction which would consume only a few platelets and then platelet-platelet interaction which will consume a lot. You are not measuring platelet-tumor cell interaction alone, your assay measures the whole phenomenon. So even if tumor cells could clump platelets around themselves you might inhibit one-third of the control but that is not going to show the thrombocytopenia.

Dr. Hilgard. I think that is probably correct.

Dr. T. Gasic (University of Pennsylvania, Philadelphia). I would like to ask what kind of cell suspension you inject?

Dr. Hilgard. It is derived mechanically from subcutaneous tumor.

Dr. J. A. E. Halkett (New England Nuclear Corporation, North Billerica). I wonder if anyone has checked the opsonin level of fibronectin to find out whether these levels have an effect on the number of metastases that you find in the lungs following the injection of cells. For example, have you taken particles and decreased the circulating fibronectic levels and then injected the tumor cells to see if you get fewer or greater numbers of metastases?

Dr. Hilgard. What particles are you referring to?

Dr. Halkett. Well, fibronectin will tend to interact with particles or cells that normally home to the spleen or liver but, if this is not so, then they would go to the lungs. I wonder whether anyone has found any connection with opsonin levels.

Dr. Hilgard. I think these experiments have been performed although not with the intention of altering the fibronectin concentration. It is generally known that if, before injecting tumor cells, you give latex particles or any other particles you get a tremendous increase in the number of lung colonies.

Dr. Halkett. I wonder whether some of the variations that have been reported using suspensions of cells could be broken fragments which may decrease the fibronectin concentration and, therefore, send the cells more to the lungs.

ANIMAL MODELS FOR THE STUDY OF PLATELET-TUMOR CELL
INTERACTIONS

M. B. Donati, D. Rotillio, F. Delaini, R. Giavazzi,
A. Mantovani and A. Poggi
Istituto di Richerche Farmacologiche "Mario
Negri", 20157 Milano, Italy

Blood platelets have long been suggested to play a role in cancer cell spread and dissemination (reviewed recently by Donati et al., 1981a and b). Recently, this concept has received further experimental support, although indirect, from animal models and from in vitro studies of platelet-cancer cell interactions.

The main arguments in favor of the involvement of platelets in experimental tumor metastases are the observation that tumor cells can aggregate platelets in vivo, that blood-borne metastases induce thrombocytopenia in the host, that induced thrombocytopenia impairs the development of metastases and that antiplatelet agents may alter the metastatic pattern of tumor spread (Gasic et al., 1968, 1972, 1973, 1978b; Hilgard, 1978). In vitro studies, on the other hand, have drawn attention to the ability of some tumor cells or transformed fibroblast lines to induce blood platelet aggregation (Gasic et al., 1978a; Pearlstein et al., 1979; Marcum et al., 1980).

The aim of this paper is to review briefly some of the experience collected by this laboratory on the involvement of platelets in tumor growth and dissemination. We shall first consider the hemostatic changes that may occur during experimental tumor development, then the indications obtained through pharmacological modulation of the host's platelet function and finally the possibility of using metastatic variants as new tools for more direct evaluation of interactions between platelets and invasive cells.

1. Experimental tumor-associated hemostatic changes

The observation of thrombocytopenia and blood coagulation changes after injection of tumor cells to a healthy host has been considered an argument in favor of the role of platelets and fibrin in cancer cell spread. However, this question needs careful reassessment.

Most in vivo information on cancer cell-platelet interactions has been obtained in artificial models of tumor growth, such as hematogenous dissemination after intravenous injection of large amounts of cancer cells. In these conditions, tumor emboli, more than real metastases, are formed through a process which by-passes the first steps of local tumor invasion, such as development of cell proteolytic and/or migratory properties, detachment from the primary site and penetration through the vascular wall into the bloodstream. This condition of tumor spread appears far removed from the dissemination process occurring in patients bearing solid tumors, considering too that tumor cells are injected abruptly into healthy recipients with none of the complex metabolic and immunological changes generally induced in the host by the presence of a growing tumor (Spreafico and Garattini, 1974; Donati and Poggi,1980).

Data obtained with Walker 256 carcinosarcoma cells indicated that intravenous injection of the cells triggered the acute onset of intravascular coagulation with marked thrombocytopenia (Hilgard and Gordon-Smith, 1974). However, injection under the same conditions of dead cells or of particulate, inert material of a similar size provoked similar effects (Hilgard, 1973). This coagulopathy was presumably due to activation of the coagulation cascade through the contact phase or to entry into the circulation of tissue-derived thromboplastic material.

In our studies, the intravenous injection of Lewis Lung carcinoma (3LL) cells into syngeneic $C_{57}Bl/6$ mice induced a rapid fall in blood platelet count and fibrinogen level within 1-3 minutes and, during the following hour, an increase in serum fibrin(ogen) degradation products (Poggi et al., 1977). This syndrome depended on the size of the inoculum and was rapidly reversible (within less than two hours). No hemostatic changes could be measured in the subsequent hours and days, during lodgement and growth of metastatic cells in the lungs. The intravascular coagulation triggered by cancer cell injection could be prevented by treating the host either with antiplatelet agents or with anticoagulants such as heparin or warfarin, thus suggesting that thrombocytopenia occurred as a consequence of thrombin-induced intravascular platelet aggregation (Poggi et al., 1981).

Table 1 - Occurrence of thrombocytopenia after i.v. injection of 4×10^5 3LL cells into C_{57} Bl/6J mice (mean ± SE of 10 animals per group).

	Blood platelet count ($\times 10^5/\mu l$)	
	Saline	Cancer cells
5 min	9.3+0.5	2.4+0.4
30 min	9.7+0.9	3.9+0.9
120 min	9.0+0.8	8.1+1.0
7 days	10.1+0.4	9.8+0.8
15 days	8.9+0.6	10.4+0.6
21 days	9.5+0.8	9.6+1.2

The role of the activation of clotting at the time of tumor cell injection in the final outcome of disseminating cells and growth of lung colonies is still controversial. According to some reports, drugs which keep platelet aggregation inhibited when the host is challenged with the cancer cell inoculum have a beneficial effect in slowing metastasis growth, just by delaying the very first phases of tumor cell attachment to the vessels of the microcirculation and of the extravasation process. This opinion is not shared by others, who failed to find any change in the pattern and amount of lung colony formation in animals treated with platelet aggregation inhibitors, even throughout the period of lung colony growth (Hilgard and Thornes, 1976; Poggi et al., 1981).

Thus, the existence of a specific link between tumor cell-induced thrombocytopenia and lung colony formation may be questioned, but even less clear is the role of platelet changes in the growth of spontaneously metastasizing tumors.

This has been considered in only a few models.

Thrombocytopenia was observed in all the experimental tumors we studied: murine models, such as 3LL, JW Sarcoma (JWS), mFS6, a benzopyrene-induced fibrosarcoma and rat tumors such as the 256 rat carcinosarcoma, BN-ML myelomonocytic leukemia and 5222 leukemia (Table 2).

Table 2 - Some experimental malignancies the growth of which has been found associated with the development of thrombocytopenia.

Mouse

. Lewis Lung Carcinoma (Poggi et al., 1977)
. JW Sarcoma (Chmielewska et al., 1980)
. mFS6 (Delaini et al., 1981)
. L 1210 leukemia (Hacker et al., 1977)

Rat

. Walker 256 carcinosarcoma (Hilgard et al., 1973)
. BN-ML leukemia (Donati et al., 1977)
. L 5222 leukemia (Donati et al., 1977)

In the 3LL spontaneous metastasis model, kinetic studies with labelled platelets indicated that the progressive thrombocytopenia was associated with normal platelet survival and no evidence of an increased splenic pool. Bone marrow examination indicated a reduced number of megakaryocytes, thus suggesting that impaired platelet production rather than increased platelet consumption was responsible for the observed thrombocytopenia (Poggi et al., 1977). This was confirmed by the observation that neither chronic anticoagulant treatment nor antiaggregation corrected the thrombocytopenia (Mussoni et al., 1978) (Table 3). Thus, both the evolution and the significance of the thrombocytopenia occurring after 3LL cell injection can greatly differ according to the experimental system considered.

Table 3 — Occurrence of thrombocytopenia in different experimental conditions of 3LL growth 21 days after i.m. implantation of 2×10^5 3LL cells into C_{57} Bl/6J mice. Means $\pm$ SE of 10-15 animals per group.

	Blood platelet count ($\times 10^5/\mu l$)
Control	9.8±0.6
Tumor-bearing	3.6±0.3
Tumor-bearing + ditazole	2.9±0.4
Tumor-bearing + warfarin	3.1±0.6
Tumor-bearing + amputation	10.1±0.5

In the 3LL system, thrombocytopenia was accompanied by microangiopathic hemolytic anemia and an increased fibrinogen level with reduced labelled fibrinogen survival and fibrin accumulation within the tumor (Poggi et al., 1976, 1977). Thus, although low-grade intravascular coagulation was suggested in 3LL, the fall in the platelet count occurred through a different mechanism. On the other hand, in the JW Sarcoma, thrombocytopenia was not accompanied by any signs of intravascular coagulation or of fibrin deposition around the tumor (Chmielewska et al., 1980).

In the 3LL model, thrombocytopenia and the other hemostatic changes did not appear to have any clear pathogenetic link with metastasis formation, since they were also observed in animals treated with warfarin, in which metastatic growth was markedly inhibited. Moreover, although they occurred mainly in concomitance with metastasis growth, hemostatic changes in 3LL appeared closely related to the primary tumor, because no changes were detected when lung metastases grew in the absence of the primary, as in the artificial lung colony model or in the spontaneous model after removal of the primary at adequate times after tumor implantation (Table 3).

These data suggest that the presence of the primary tumor uniquely influences the host's hemostatic system. From the experimental models we have studied, it does not appear that tumor-associated necrosis or the immunogenic characteristics of the tumor cells play a major role in this phenomenon. It has been suggested that the primary tumor exerts an inhibitory effect on the growth of metastatic nodules through a still undefined mediator (Yuhas and Pazmino, 1974). It is not known whether a similar mechanism operates in depressing platelet production.

A further argument in support of the concept that hemostatic changes are linked to the presence of the primary tumor, not of metastases is offered by observations with metastatic variants of the mFS6 fibrosarcoma. Sublines from isolated lung nodules of this benzopyrene-induced tumor gave rise, when implanted i.m., to primary tumors of the same weight but to completely different types of metastatic growth, ranging from 0 to 100% metastatic incidence (Giavazzi et al., 1980). With all these sublines the same degree of thrombocytopenia and hyperfibrinogenemia was seen, progressively developing during tumor growth (Delaini et al., 1981). In this model, the occurrence and degree of thrombocytopenia also appeared unrelated to the cells' ability to interact with platelets as studied in vitro. In fact, no difference was found in platelet drop following i.m. implantation of cells capable or not of inducing in vitro mouse platelet aggregation (Table 4).

Table 4 - In vitro induction of mouse platelet aggregation and occurrence of thrombocytopenia at day 26 after implantation of 1×10^5 cells into C57Bl/6J mice in mFS6 metastatic variants (Giavazzi et al. 1980; Delaini et al., 1981).

	Metastasis incidence (%)	Tumor cell-induced platelet aggregation	Platelet count (% of controls)
mFS6	55	−	30
M_4	95	+	35
M_7	100	+	28
M_8	5	−	32
M_9	0	−	35

2. Pharmacological modulation of platelet function in tumor-bearing animals.

Most of the information available on the role of platelets in dissemination has been derived from studies with drugs influencing the host's hemostatic system. The findings to date are highly controversial probably because of differences in the experimental models used, the period of treatment during tumor growth and a number of other interfering factors such as the animal's diet. Above all, the drugs used in these studies, besides their effect on platelet function, all have a number of other pharmacological activities which could by themselves influence cancer cell dissemination.

Several investigations have indicated that inhibition of platelet aggregation prior to the intravenous injection of viable tumor cells reduced the number and incidence of lung nodule formation (Hilgard and Thornes, 1976; Donati and Poggi, 1980). This effect was generally ascribed to longer persistence of cells in the circulation of treated animals, because of better patency of the microcirculatory bed and impairment of tumor cell-platelet-fibrin emboli formation (Hilgard and Thornes, 1976; White and Griffiths, 1976). However, as already mentioned, the approach of mimicking blood-borne metastasis by intravenous injection of tumor cells is highly artificial, since this model reflects only the final steps of dissemination (transport in blood and take by target organs), completely by-passing the initial phases which include detachment from the primary tumor and entry into the bloodstream. Platelets might be involved in cancer cell-endothelium interactions in these initial phases too and it is possible that, like fibrin, they play an opposite role in detachment from the primary and lodgement, the relative importance of these stages depending largely on the experimental conditions.

For these and probably many other reasons, quite different results have been obtained using the same drugs in "spontaneous" or "artificial" metastasis models. Table 5 summarizes our experience when 3LL-bearing mice were treated, throughout the whole period of cancer cell growth, with drugs influencing the host's hemostatic system at various levels. Except for dipyridamole, all the drugs studied had some inhibitory effect on the "artificial" metastasis model, whereas only warfarin also markedly reduced spontaneous metastasis formation (Poggi et al., 1981).

Table 5 – Schematic representation of the effects obtained with various drugs active on the hemostatic system on artificial and spontaneous lung metastasis growth. + stands for statistically significant reduction; – stands for no significant change or increase.

Drug	Lung colonies	Spontaneous metastasis
Warfarin	+	+
Heparin	+	–
Batroxobin	+	–
Aspirin	+	–
Ditazole	+	–
Dipyridamole	+	–
RA 233	+	–

As can be seen in Table 5, different types of platelet aggregation inhibitors were studied, nonsteroidal antiinflammatory agents, such as aspirin and ditazole, and phosphodiesterase inhibitors, such as dipyridamole and RA 233, a pyrimido-pyrimidine derivative. All these compounds were given chronically with a schedule which impaired the host's platelet function as indicated by significant prolongation of the tail transection bleeding time (Table 6). None of them influenced spontaneous metastasis growth.

Table 6 – Effect of various platelet aggregation inhibitors given chronically on the tail transection bleeding time (Mussoni et al.,1977) of 3LL-bearing mice (15 days after tumor cell implantation). Means $\pm$ SE of 4-6 animals per group.

Drug	Daily dose (mg/kg b.w.)	Tail bleeding time (sec.)
Water	–	246$\pm$24
Dipyridamole	40-60	576$\pm$89
RA 233	16-20	786$\pm$69
Aspirin	40-60	576$\pm$46
CMC	–	260$\pm$36
Ditazole in CMC	200	515$\pm$34

CMC = carboxymethylcellulose.

It thus appears that normal platelet function is not a prerequisite for hematogenous metastasis in 3LL-bearing mice. This is in agreement with observations by Hilgard et al. (1976) who found that various inhibitors of platelet aggregation had no effect on spontaneous metastases of 3LL tumors. Using ICRF 159, a drug with antimetastatic effects in the 3LL system, Atherton et al. (1975) have shown that the antimetastatic effect and the inhibition of thrombus formation are not necessarily linked.

It is important to consider that the supposed antimetastatic effect of drugs modulating the host's hemostatic system may be influenced by a number of other factors, because of the multiplicity of pharmacological effects of each drug. Possible modifications of cell metabolism, growth or motility, changes in blood flow, in immune responses or in prostaglandin metabolism should be taken into account. For instance, platelet aggregation inhibitors such as aspirin or ditazole, which are non-steroidal anti-inflammatory drugs, act essentially as inhibitors of prostaglandin synthesis, not only in platelets (where they prevent the generation of cyclic endoperoxides and thromboxanes, potent aggregating agents) but also in other cells, such as vascular cells, macrophages and tumor cells themselves (Flower et al., 1974). Changes in the pattern of prostaglandin (PGs) generation by these cells may have important implications on tumor growth, since some PGs are responsible for immunosuppression, response to inflammatory stimuli, anaphylactic-type reactions and bone-resorbing activity of tumors (Karmali, 1980).

Prostaglandins of the E type were the first to be described in cancer cells (Easty and Easty, 1976); recently, a very potent platelet aggregation inhibitor was found in incubates of tissues from two experimental tumors, the 3LL and the JWS (Poggi et al., 1979). This inhibitory activity was identified as prostacyclin (PGI_2) by various biological, physicochemical and pharmacological criteria. Indeed, it was absent in tissues of animals pretreated with indomethacin (1-10 mg/Kg b.w.) or aspirin (50-200 mg/Kg b.w.); it was also abolished by incubation of the supernatants with an antiserum raised against 6-keto $PGF_{1\alpha}$ which neutralized the antiaggregating activity of synthetic PGI_2. It co-chromatographed with 6-keto $PGF_{1\alpha}$ on thin-layer radiochromatography and was also identified as this arachidonic acid metabolite by mass-fragmentography (Poggi et al., 1982).

In the 3LL model, PGI_2 activity was greater in

metastatic lung than in the primary tumor, but less in isolated metastatic nodules than in normal lung tissue (Table 7). In both primary tissue and metastatic nodules, the ability to generate PGI_2 was somewhat correlated with the relative distribution of cardiac output measured by a radioactive microsphere technique (Raczka et al., 1982, Table 7) and appeared to modulate the response of the tumor vasculature to the vasoactive agent noradrenaline (Quintana et al., 1982).

Table 7 - PGI_2 activity and relative cardiac output distribution (% CO/g tissue) in tissues from 3LL-bearing animals (Poggi et al., 1982). Means $\pm$ SE of 5-8 animals per group.

Tissue	% CO/g	PGI_2 (ng/g wet tissue)
. primary	1.68+0.16	39.9+13.9
. metastatic lung	9.18+1.19	1115+193
. isolated metastases	3.78+0.58	395+183

To rule out the possibility that, in tumor tissue supernatants, PGI_2 could to some extent derive from contaminating cells such as endothelial cells, macrophages or other host cells, we subsequently studied the arachidonic acid metabolism of cultured 3LL cells isolated from either the primary tumor or metastatic nodules. All 3LL cells (devoid of macrophages) were found to generate PGI_2, shown by thin-layer radiochromatography to be 3 times higher in cells from metastases than from the primary (Rotilio et al., manuscript in preparation).

The significance of this observation awaits further clarification. It is in any case worth considering that treatment of animals with prostaglandin inhibitors involves not only inhibition of blood platelet aggregation, but also of a potent vasodilator and platelet aggregation inhibitor produced by cancer cells. We do not know yet whether this dual effect of aspirin-like drugs could account, at least partially, for the controversial results with these agents in experimental tumor models (Hilgard et al., 1976; Lynch et al., 1978; Mussoni et al., 1978; Santoro et al., 1976).

3 - Metastatic variants as new models of platelet-invasive cells interactions.

Evidence for the role of platelets in metastasis formation, based on pharmacological studies, is necessarily indirect and, as we have seen, interpretation of the data is fraught with difficulties. Recent developments in the knowledge of experimental metastasis biology provide tools for more direct evaluation of the role of fibrin in cancer cell invasiveness. It has been suggested that metastases do not result from random survival of cells released from the primary tumor but from the selective growth of specialized subpopulations of highly metastatic cells with specific properties which could affect their arrest pattern (Poste and Fidler, 1980).

This concept has been developed from the study of so-called metastatic variants, i.e. cell sublines which, derived from isolated lung nodules of the same primary tumor, consistently exhibit different metastatic capacity when reinjected in syngeneic animals.

Very recently, sublines were obtained from the already mentioned mFS6, a benzopyrene-induced fibrosarcoma; these were heterogenous as regards several biological properties including the capacity for spontaneous dissemination (Giavazzi et al., 1980). In view of the suggested role of platelets in some steps of dissemination, we used these sublines to study whether the different metastatic potential was associated with differences in cancer cell-platelet interactions.

As mentioned before, platelet aggregating activity was found when cells from the highly metastatic line M_4 were challenged with human or mouse platelets, whereas the poorly metastatic line M_9 showed no such effect (Dejana et al., unpublished results). Moreover, only cells from the poorly metastatic line, when stimulated with arachidonic acid, released in the supernatant a platelet aggregation inhibitor with the biological characteristics of prostacyclin.

We consequently studied arachidonic acid metabolism in the same cells, which were tested for their ability to metabolize either exogenous labelled arachidonic acid or the same substrate previously incorporated into cell membrane phospholipids. The metabolites identified in all cell types by thin-layer radiochromatography were 6-keto $PGF_{1\alpha}$ (stable degradation product of prostacyclin), $PGF_{2\alpha}$, PGE_2, thromboxane B_2 (TxB_2) and PGD_2.

The two sublines were found to differ mainly in the relative amounts of 6-keto $PGF_{1\alpha}$ and of TxB_2. Indeed, M_4, the cell line with higher metastatic incidence, generated a lower amount of 6-keto $PGF_{1\alpha}$ and a higher amount of TxB_2 than the corresponding low metastatic line M_9 (Table 8). A similar discrepancy among the two sublines was observed when 6-keto $PGF_{1\alpha}$ and TxB_2 were measured by specific radioimmunoassays (data not shown).

Table 8 - Percent conversion of ^{14}C-arachidonic acid into 6-keto-$PGF_{1\alpha}$ and TxB_2 by cells from mFS6 variants (Giavazzi et al., 1980). Means $\pm$ SE of 5 values per cell type.

Cell type	Metastatic incidence (%)	6-keto $PGF_{1\alpha}$ (%)	TxB_2 (%)
mFS6	55	3.54+0.64	2.89+0.34
M_4	95	3.90+1.12	5.70+0.64
M_9	0	10.00+0.95	0.96+0.26

It thus appears that the balance between prostacyclin and thromboxane was shifted towards a proaggregatory condition in the highly metastatic line, whereas predominance of PGI_2 appeared to be one of the cell properties associated with low metastatic potential.

In a different model of metastatic variants, derived from the B16 melanoma, it was recently reported that the amount of PGD_2 formed was inversely correlated to their potential to give artificial metastases and that in vitro treatment of the cells with a cyclooxygenase inhibitor prior to their intravenous injection increased the formation of lung colonies (Fitzpatrick and Stringfellow, 1979). PGD_2, however cannot be expected from in vitro studies to play a major role in platelet-cancer cell interactions in mice, since it exhibits virtually no inhibitory activity on mouse platelet aggregation at concentrations more than 50 times higher than the threshold inhibitory concentration of PGI_2 in the same system (Bertelé et al., unpublished observations). Therefore, some other mechanism must be implied in the protective effect of PGD_2 from <u>artificial</u> metastasis formation.

Taken all together, our studies with mFS6 fibrosarcoma sublines seem to indicate that high potential for spontaneous metastasis is associated with a greater ability of the cells to interact with platelets and/or lack of inhibitory activities. This data constitutes direct evidence of the role of blood platelets in <u>spontaneous</u> metastasis models.

Conclusions

There appears to be some experimental evidence that platelets interact with cancer cells in vitro and that platelet changes may occur during tumor cell growth and dissemination. The pathogenetic role of these changes in the formation of metastases appears however still difficult to establish on the basis of pharmacological approaches. It is conceivable that a deeper study of metastatic variants will provide a more direct key to understanding whether cancer cell-platelet interactions do really influence tumor invasiveness.

Acknowledgements

The authors'work mentioned in this survey was partially supported by Grant NIH-PHRB-1R01 CA 12764-01, National Cancer Institute, NIH, Bethesda, Maryland, U.S.A. and by Contracts 80.01621.96 and 80.01466.96 of Italian National Research Council. Judy Baggott and M. Paola Bonifacino helped prepare this manuscript.

References

Atherton A, Busfield D, Hellmann K (1975). The effects of an antimetastatic agent ($\pm$)-1,2-Bis(3,5-dioxopiperazin-1-yl)propane (ICRF 159) on platelet behaviour. Cancer Res 35:953.

Chmielewska J, Poggi A, Mussoni L, Donati MB, Garattini S (1980). Blood coagulation changes in JW Sarcoma, a new metastasizing tumour in mice. Eur J Cancer 16:1399.

Delaini F, Giavazzi R, De Bellis Vitti G, Alessandri G, Mantovani A, Donati M B (1981). Tumor sublines with different metastatic capacity induce similar blood coagulation changes in the host. Br J Cancer 43:100.

Donati M B, Davidson J F, Garattini S (eds) (1981a). "Malignancy and the Hemostatic System".New York: Raven Press.

Donati M B, Mussoni L, Kornblihtt L, Poggi A (1977). Changes in the hemostatic system of rats bearing L5222 or BNML experimental leukemias. Leuk Res 1:177.

Donati M B, Poggi A (1980). Malignancy and haemostasis. Br J Haematol 44:173.

Donati M B, Poggi A, Semeraro N (1981b). Coagulation and malignancy. In Poller L (ed): "Recent Advances in Blood Coagulation". Edinburgh: Churchill Livingstone, p 227.

Easty G C, Easty D M (1976). Prostaglandins and cancer. Cancer Treat Rev 3:217.

Flower R J (1974). Drugs which inhibit prostaglandin biosynthesis. Pharmacol Rev 26:33.

Gasic G J, Boettiger D, Catalfamo J L, Gasic T B, Stewart G J (1978a). Aggregation of platelets and cell membrane vesiculation by rat cells transformed "in vitro" by Rous Sarcoma virus. Cancer Res 38:2950.

Gasic G J, Boettiger D, Catalfamo J L, Gasic T B, Stewart G J (1978b). Platelet interactions in malignancy and cell transformation: functional and biochemical studies. In de Gaetano G, Garattini S (eds): "Platelets: A Multi-disciplinary Approach". New York: Raven Press, p 447.

Gasic G J, Gasic T B, Galanti N, Johnson T, Murphy S (1973). Platelet-tumor-cell interactions in mice. The role of platelets in the spread of malignant disease. Int J Cancer 11:704.

Gasic G J, Gasic T B, Murphy S (1972). Anti-metastatic effect of aspirin. Lancet II:932.

Gasic G J, Gasic T B, Stewart C C (1968). Antimetastatic effects associated with platelet reduction. Proc Natl Acad Sci USA 61:46.

Giavazzi R, Alessandri G, Spreafico F, Garattini S, Mantovani A (1980). Metastasizing capacity of tumor cells from spontaneous metastases of transplanted murine tumors. Br J Cancer 42:462.

Hacker M, Roberts D, Jackson C (1977). Depression of the platelet count after inoculation of mice with L1210 or L5178Y cells. Br J Haematol 35:465.

Hilgard P (1973). The role of blood platelets in experimental metastases. Br J Cancer 28:429.

Hilgard P (1978). Blood platelets and experimental metastases. In de Gaetano G, Garattini S (eds) "Platelets: a Multidisciplinary Approach". New York: Raven Press, p 457.

Hilgard P, Gordon-Smith E C (1974). Microangiopathic haemolytic anaemia and experimental tumour-cell emboli. Br J Haematol 26:651.

Hilgard P, Heller H, Schmidt C G (1976). The influence of platelet aggregation inhibitors on metastasis formation in mice (3LL). Z Krebsforsch 86:243.

Hilgard P, Hohage R, Schmitt W, Köle W (1973). Microangiopathic haemolytic anaemia associated with hypercalcaemia in an experimental rat tumour. Br J Haematol 24:245.

Hilgard P, Thornes R D (1976). Anticoagulants in the treatment of cancer. Eur J Cancer 12:755.

Karmali R A (1980). Review: Prostaglandins and cancer. Prostag Med 5:11.

Lynch N R, Castes M, Astoin M, Salomon J C (1978). Mechanism of inhibition of tumour growth by aspirin and indomethacin. Br J Cancer 38:503.

Marcum J M, McGill M, Bastida E, Ordinas A, Jamieson G A (1980). The interaction of platelets, tumor cells and vascular subendothelium. J Lab Clin Med 96:1046.

Mussoni L, Poggi A, de Gaetano G, Donati M B (1978). Effect of ditazole, an inhibitor of platelet aggregation, on a metastatizing tumour in mice. Br J Cancer 37:126.

Mussoni L, Poggi A, Donati M B, de Gaetano G (1977). Ditazole and platelets. III. Effect of ditazole on tumour-cell induced thrombocytopenia and on bleeding time in mice. Haemostasis 6:260.

Pearlstein E, Cooper L B, Karpatkin S (1979). Extraction and characterization of a platelet aggregating material (PAM) from SV40-transformed mouse 3T3 fibroblasts. J Lab Clin Med 93:332.

Poggi A, Dall'Olio A, Balconi G, Delaini F, de Gaetano G, Donati M B (1979). Generation of prostacyclin (PGI_2) activity by Lewis Lung Carcinoma (3LL) cells. Thromb Haemost 42:339.

Poggi A, Dall'Olio A, Rotilio D, Delaini F, de Gaetano G, Donati M B (1982). Tissues and cells from Lewis Lung Carcinoma generate a platelet aggregation inhibitory activity identified as prostacyclin. Submitted to Cancer Res.

Poggi A, Donati M B, Garattini S (1981). Fibrin and experimental cancer cell dissemination: problems in the evaluation of experimental models. In Donati M B, Davidson J F, Garattini S (eds): "Malignancy and the Hemostatic System". New York: Raven Press, p 89.

Poggi A, Donati M B, Polentarutti N, de Gaetano G, Garattini S (1976). On the thrombocytopenia developing in mice bearing a spontaneously metastasizing tumor. Z Krebsforsch 86:303.

Poggi A, Polentarutti N, Donati M B, de Gaetano G, Garattini S (1977). Blood coagulation changes in mice bearing Lewis Lung Carcinoma, a metastasizing tumor. Cancer Res 37:272.

Poste G, Fidler I J (1980). The pathogenesis of cancer metastasis. Nature 283:139.

Quintana A, Raczka E, Latallo Z S, Donati M B (1982). Cardiac output redistribution induced by noradrenaline in two murine tumor models. Eur J Cancer in press.

Raczka E, Quintana A, Poggi A, Donati M B (1982). Distribution of cardiac output during development of two metastasizing murine tumors. Eur J Cancer in press.

Santoro M G, Philpott G W, Jaffe B M (1976). Inhibition of tumour growth "in vivo" and "in vitro" by prostaglandin E. Nature 263:777.

Spreafico F, Garattini S (1974). Selective antimetastatic treatment - Current status and future prospects. Cancer Treat Rev 1:239.

White H, Griffiths J D (1976). Circulating malignant cells and fibrinolysis during resection of colorectal cancer. Proc R Soc Med 69:467.

Yuhas J M, Pazmino N H (1974). Inhibition of subcutaneously growing line 1 carcinomas due to metastatic spread. Cancer Res 34:2005.

DISCUSSION

Dr. S. J. Scialla (Hematology and Oncology Associates of Northeastern Pennsylvania, Scranton). There is a striking difference in your model compared to the human model. As was mentioned yesterday, thrombocytosis seems to be a key element in patients with solid tumors as compared to the thrombocytopenia. We have routinely done bone marrow on patients with small cell carcinoma in all stages as part of the stage work up. A striking thing is to see enhanced megakaryocyte activity and to see platelets aggregating around tumor cells. I think this is an important difference in your model. The other difference is the population of patients with solid tumors and their environment. The arteriosclerosis that is involved in the vessel is, I think, an important model for us to understand as a unique characteristic in the human.

Dr. M. B. Donati (Instituto di Recherche Farmacologische, Milan). Yes, I agree that thrombocytopenia is a problem in this model. It may be of interest probably to relate this to some other type of human disease where there is a bone marrow depression. Actually, I think it may be a problem as we were discussing before Dr. Hilgard's presentation. Animals living for the whole period of metastasis development can reduce the number of platelets so they have the possibility of developing other mechanisms for metastasis and may not need the platelets as much. On the other hand, I am afraid that many models of experimental metastasis in animals do induce thrombocytopenia. I certainly agree that this is not similar to the human model where there is thrombocytosis but I have not found models where thrombocytopenia does not occur.

Dr. D. Cowan (Case Western Reserve University, Cleveland). I have a question and a comment. The question is, how did you determine the megakaryocyte production. Was it by enumeration?

Dr. Donati. Yes, it was by enumeration.

Dr. Cowan. My second point is that some years ago we

did some studies of megakaryocytes on patients with tumors
and found what appeared to be defects in maturation. We
have recently concluded a study on 250 patients carrying
out determination of platelet content and platelet size
using Dr. Karpatkin's technique and doing monthly or rather
semi-monthly determinations of cell size distribution. In
those patients who were thrombocytopenic we did not find an
increased proportion of megathrombocytes.

Dr. P. Hilgard (Bristol-Meyers International Corporation,
Brussels). In your high and low metastatic cell lines, one
of them aggregates platelets while the other one does not.
Does that correlate with other characteristics of these cell
lines? Do they have higher levels of other enzymes? Is it
just expression of a higher general biological activity or
is it a specific platelet aggregation?

Dr. Donati. There are no major changes in growth kinetics when they grow in vitro and they do not differ markedly
in their immunogenic capacity. As far as the interaction of
the hemostatic system is concerned, these cells do not differ
in their plasminogen activator production. All the sublines
have a higher production than the original line but not different among the cells. Some sublines with lower metastatic
potential have the highest tissue factor activity.

Dr. D. S. Rappaport (University of Minnesota Medical
School, Minneapolis). In your artificial model of metastasis
you had a large number of particles that were injected into
the lungs and I am sure this has an effect on blood flow
which could result in platelet aggregation.

Dr. Donati. I am sure that there are a number of changes.
I think it is also very important that, in the artificial
model, you are injecting cells into a healthy animal. In
the contagious model the animal becomes sick and gradually
undergoes a number of nutritional metabolic changes.

IN VIVO MODELS FOR STUDIES OF HUMAN TUMOR METASTASIS

Nabil Hanna

Cancer Metastasis and Treatment Laboratory
NCI Frederick Cancer Research Center
Frederick, Maryland 21701

An understanding of the host effector mechanisms that control tumor spread is fundamental to the design of new approaches for the prevention and therapy of cancer metastasis. In particular, studies of these mechanisms will help us to find relevant animal models for ascertaining the metastatic potential of tumor cells and for isolating highly metastatic cells present in the heterogeneous cell populations of the primary tumor.

Tumor metastasis is a highly selective process in which specialized subpopulations of cells that pre-exist in the primary neoplasm complete all the steps required for tumor dissemination and the development of organ metastases (Poste and Fidler, 1980). To establish metastases, tumor cells must invade the surrounding tissues and penetrate into the lymphatic and/or vascular systems. Once in the circulation, tumor cells must survive the host defense mechanisms, be arrested in the capillary bed, extravasate into the organ parenchyma, and proliferate to form visible metastatic foci. The outcome of this complex process is determined by both host factors and intrinsic properties of the tumor cells (Fidler et al, 1978). Therefore, it is not surprising that modification of either host factors or tumor cell properties alters the outcome of tumor metastasis. For example, specific immunity against tumor antigens (Fidler et al, 1977), nonspecific stimulation of the immune response by biological response modifiers (Hanna and Fidler, 1980; Jones and Castro, 1977) or anticoagulation therapy (Brown, 1973) may inhibit the development of metastases. In contrast, immunosuppression (Fidler and Kripke, 1980) or pretreatment with cytotoxic drugs (van Putten et al, 1975) enhance metastasis.

There is substantial evidence that cells populating primary malignant neoplasms are heterogeneous with respect to metastatic potential, antigenicity, immunogenicity, production of plasminogen activator and collagenolytic enzymes, radiosensitivity and response to cytotoxic drugs (Fidler and Kripke, 1977; Hart and Fidler, 1980; Poste and Fidler, 1980). Furthermore, tumor cells isolated from primary and metastatic lesions may differ in antigenicity, enzymic production and response to chemotherapeutic drugs (Trope, 1975; Donelli et al, 1977; Baylin et al, 1978; Fogel et al, 1979).

Since only a minor cell population pre-existing in the primary tumor has the potential for invasion and metastasis, the isolation and characterization of such specialized cells is invaluable for the understanding of the pathogenesis of metastasis. One of the in vivo methods used for selection of highly metastatic tumor cell variants involves the isolation of cells from individual metastatic nodules (Fidler, 1973a). This approach has profound implications for the identification of the cellular properties relevant to tumor dissemination and the evaluation of the drug susceptibility of those subpopulations of cells that are endowed with high metastatic potential. Unfortunately, however, similar in vivo methods utilizing animal models for studies of human tumor metastasis have not yet been developed. The athymic T cell-deficient nude mouse has been used successfully for transplantation of xenogeneic normal and malignant human tissues (Nomura et al, 1977; Fogh and Giovanella, 1978; Kindred, 1979). Most studies, however, agree that such xenogeneic tumors, although malignant in the host of origin, rarely metastasize upon transplantation into nude mice (Rygaard and Povlsen, 1969; Povlsen et al, 1973; Fidler et al, 1976).

In this presentation I will discuss the possible role of T cell-independent immune mechanisms mediated by natural killer (NK) cells in host resistance against hematogenous tumor dissemination. I will also propose the use of young nude mice with low levels of NK cell-mediated cytotoxicity as a sensitive in vivo model for determining the metastatic potential of human neoplasms and isolating highly metastatic tumor cell variants that pre-exist within the malignant primary tumors.

RESISTANCE OF THE NUDE MOUSE TO TUMOR METASTASIS

Although the nude mouse lacks functionally mature T lymphocytes, it exhibits consistently high levels of NK cell activity (Herbermann et al, 1975), a normal response to T cell-independent antigens (Manning et al, 1972) and high levels of natural antibodies that react with tumor cells (Martin and Martin, 1974). Moreover, tumoricidal macrophages can be isolated from normal or adjuvant-treated nude mice (Johnson and Balish, 1980). Such T cell independent defense mechanisms may contribute to the natural resistance of the nude mouse against transplanted allogeneic and xenogeneic neoplasms. The in vivo role of natural immune mechanisms in the inhibition of tumor growth and dissemination is demonstrated by the findings that further immunosuppression of nude mice or the use of newborn recipients increases the rate of successful takes, the growth rate and, occasionally, the metastatic spread of transplanted tumors (Lozzio et al, 1980; Ohsugi et al, 1980; Watanabe et al, 1980).

The low incidence of metastases, despite the local growth of transplanted malignant tumors, can be attributed to host defense mechanisms that are more effective in destroying circulating tumor cells than in destroying those residing within the primary tumors. This is not surprising since the progenitors of metastasis circulate as single cells or small clumps and are, therefore, highly accessible and more vulnerable to destruction by immune and nonimmune defense mechanisms. Moreover, in the blood, tumor cells are exposed to destructive forces unique to this environment that might not operate within the solid tumor mass. Indeed, most tumor cells that enter the circulation are destroyed before they extravasate and establish metastases (Fidler, 1973b; Liotta et al, 1974). Such host defense mechanisms are highly active in nude mice, as can be seen by the lower incidence of pulmonary metastases in adult nude mice as compared to that detected in immunocompetent syngeneic recipients (Fidler et al, 1976; Skov et al, 1976).

ROLE OF NK CELLS IN THE CONTROL OF HEMATOGENOUS TUMOR METASTASES

Recent reports from our laboratory (Hanna, 1980; Hanna and Fidler, 1980) indicate that NK cells play a significant role in the destruction of circulating melanoma and fibrosarcoma tumor cells and thus inhibit their hematogenous metastatic spread. The incidence of tumor metastases was studied in hosts that naturally exhibit low (3-week-old mice or adult beige mice) or high (adult nude mice) NK cell activity and in recipients whose NK cell activity was enhanced or suppressed by treatment with interferon inducers or cyclophosphamide (cy), respectively. An inverse correlation between the levels of NK cell activity and the incidence of experimental metastases was observed (Hanna and Fidler, 1980; Hanna, 1980). For example, the in vivo depletion of NK cells in mice treated with cy or β-estradiol resulted in a marked enhancement of pulmonary and extrapulmonary metastases. Similarly, a high incidence of metastases was observed in 3-week-old syngeneic mice and beige mice, both of which exhibit low levels of NK cell activity. On the other hand, activation of NK cells by the administration of bacterial adjuvants (Corynebacterium parvum) or interferon inducers (polyinosinic-polycytidylic acid or statolone) inhibits tumor metastases. Such agents, however, were effective only when injected before the i.v. inoculation of tumor cells. This is in agreement with our recent findings that the adoptive transfer of normal spleen cells into cy-treated recipients inhibited the drug-induced enhancement of metastasis only when injected before tumor cell inoculation (Hanna and Fidler, 1980). The reactive cells in these reconstitution experiments were identified as NK cells (Hanna and Burton, 1981). Thus, in both systems, host destruction of tumor cells coincided with the presence of tumor cells in the circulation. Extravasation into the organ parenchyma occurs by 6-24 hours (Fidler, 1970), and tumor cells then escape destruction by the highly cytotoxic cells present in abundance in the circulation.

THE EXPRESSION OF METASTATIC POTENTIAL OF ALLOGENEIC AND XENOGENEIC TUMORS IN YOUNG NUDE MICE WITH LOW LEVELS OF NK CELL ACTIVITY

The finding that the low NK cell activity in 3-week-old syngeneic mice was associated with a high incidence of tumor metastases suggested a series of experiments aimed at finding a similar correlation between the age dependence of NK cell activity and the incidence of experimental metastases of allogeneic and xenogeneic tumors implanted into nude mice. The natural cell-mediated cytotoxicity of spleen cells obtained from 2- to 3-week-old or 6- to 10-week-old nude mice was evaluated in vitro against (^{3}H)proline-prelabeled murine melanomas and UV-induced fibrosarcomas. Only low levels of natural killing were detected with effector cells obtained from young 2- to 3-week-old mice. In contrast, adult nude mice 6 to 10 weeks old consistently exhibited high NK cell activity (Hanna, 1980; Hanna and Fidler, 1981). However, NK cell-mediated cytotoxicity in young nude mice could be boosted by treatment with bacterial adjuvants and interferon inducers. The effector cells active in the in vitro assay are not macrophages as shown by the fact that removal of nylon-wool-adherent cells did not influence the cytotoxicity of spleen cells obtained from either adult nude mice or 3-week-old mice pretreated with polyinosinic-polycytidylic acid. Moreover, they were sensitive to treatment with anti-NK-1.2 serum and complement known to selectively deplete NK cells (Burton, 1980; Hanna and Burton, 1981).

For in vivo studies of tumor metastasis, cell lines and clones that differ in immunogenicity and/or metastatic potential were injected i.v. or s.c. into 3- and 6-week-old nude mice. All tumors grew s.c. in recipients of both age groups. However, when adult nude mice were challenged i.v. with allogeneic tumor cells, including those of high lung colonization capacity, very few lung colonies were produced. In contrast, equal number of metastatic tumor cells injected into young nude mice produced many pulmonary metastases (Hanna, 1980; Hanna and Fidler, 1981). The in vivo activation of NK cells by injection of interferon inducers 1 to 3 days before tumor inoculation rendered the 3-week-old nude mice resistant to metastasis development. Kinetic analysis of tumor cell arrest and clearance indicated that the higher incidence of lung tumor colonies in young nude mice is not caused by increased efficiency of tumor cell arrest in the vascular bed, but rather to increased survival of the arrested tumor cells during the first 24 hours after tumor cell inoculation (Hanna, 1980). The adequacy of using the young nude mouse model for ascertaining the metastatic potential of tumor cells is supported by the following observations. The metastatic behavior of allogeneic and xenogeneic rat tumors following injection into young nude mice was similar to that observed in syngeneic hosts. All metastatic neoplasms produced pulmonary metastases in young nude mice, whereas nonmetastatic tumors did not. For example, the highly immunogeneic UV-1316 and UV-1591 fibrosarcomas are nonmetastatic tumors that grow progressively in immunosuppressed syngeneic hosts, but do not metastasize spontaneously or experimentally in such recipients. Similarly, neither tumor produced any pulmonary metastases after i.v. injection into 3-week-old nude mice. In contrast, tumor cells that are immunogeneic and also metastatic may fail to metastasize in normal immunocompetent syngeneic recipients but metastasize readily in the immunosuppressed primary host and in young nude mice. Thus, the young nude mouse allows investigators to determine whether the failure of highly immunogenic tumor cells to metastasize in immunocompetent syngeneic hosts is caused by host immune response or the inability of the tumor cell to complete the metastatic process. It is quite striking that the quantitative differences in formation of experimental metastases among tumor cell lines and clones observed in syngeneic hosts were maintained after injection into young nude mice. For example, in 3-week-old nude mice, the B16-F10 melanoma cell variant, selected for its high lung colonization potential in syngeneic C57BL/6 mice, produced significantly more tumor colonies than the B16-F1, or B16-F10^{LR} variants with low lung colonization potential.

As a result of the successful metastasis of allogeneic murine tumors in young nude mice, we examined the applicability of this model to studies of the metastatic potential of zenogeneic rat tumors (Hanna and Fidler, 1980). Injection of chemically induced rat tumor cells MADB-100 and MADB-105, produced more lung colonies in 3-week-old than in 6-week-old nude mice. Moreover, the MADB-105 variant, selected for high lung colonization potential in syngeneic rats, was also more metastatic than its MADB-100 parent cell line. MADB-200 tumor cells formed few lung tumor colonies in young nude mice and syngeneic rats. Thus, the metastatic behavior of xenogeneic rat tumors in young nude mice resembled that observed in syngeneic normal recipients.

Since high levels of NK cell activity are associated with resistance to experimental tumor metastasis, genetic and environmental factors that influence NK cell activity also affect the susceptibility of the nude mouse to metastasis development. Indeed, we have observed that BALB/c nude mice exhibit lower NK cell activity and a higher frequency of metastases than age-matched NIH Swiss nude mice. Moreover, the incidence of tumor metastasis was higher in nude mice raised under pathogen-free conditions in barrier or isolator facilities than in age-matched littermates housed under conventional conditions. Viral infections induce high levels of interferon which activates NK cells and thus renders young nude mice resistant to metastasis development.

Experimental tumor metastasis is not unique to young nude mice. Experimental depletion of NK cells also renders adult 10-week-old nude mice sensitive to metastasis formation. For example, adult nude mice treated with cy or β-estradiol exhibit reduced levels of NK cell-mediated cytotoxicity in vitro and increased incidence of experimental metastasis of allogeneic tumors in vivo. Such results support previous findings that NK cells are indeed responsible, to a great extent, for the resistance of nude mice to tumor dissemination. It should be emphasized, however, that the expression of high NK cell activity as early as 4 to 5 weeks of age renders the 3-week-old mice inadequate for studies of spontaneous metastasis of locally growing tumors. By the time the implanted tumor reaches a critical size, any tumor cells released into the circulation will be destroyed by the highly active NK cells already present in the adult recipient.

THE USE OF YOUNG NUDE MICE FOR ISOLATING HIGHLY METASTATIC TUMOR CELLS

As discussed previously, metastases originate from specialized subpopulations of cells that pre-exist within the heterogeneous primary tumor mass. Therefore, methods were designed to isolate and characterize tumor cell variants capable of invasion and metastasis (Fidler, 1973a; Brunson and Nicolson, 1978). The isolation of such highly metastatic cell variants has important biological and clinical implications for drug sensitivity studies and for evaluations of host-tumor cell interactions relevant to the metastatic process and its control. The validity of using the young nude mouse model for the isolation of highly metastatic tumor cells is supported by the observation that tumor cells isolated from lung tumor colonies of young nude mice injected i.v. with allogeneic melanoma cells are highly metastatic when reinjected into either young nude mice or syngeneic hosts (Pollack and Fidler, submitted for publication). Therefore, we hope that this in vivo model will offer a unique opportunity to distinguish between metastatic and nonmetastatic primary human tumors and to isolate highly metastatic tumor cells from heterogeneous xenogeneic human neoplasms. Also, pulmonary metastases established from such metastatic cell variants in nude mice may prove beneficial for assessing the antimetastatic activity of therapeutic agents.

Recently, we attempted to select metastatic subpopulations of cells from a human amelanotic melanoma cell line, A375. In these studies, 1 X 10^6 cells from the parent A375 cell line were injected s.c. or i.v. into 3-week-old BALB/c nude mice. Eight weeks later, the tumors growing s.c. and the lung metastasis (1 to 10 nodules per lung) were harvested separately, dissociated with collagenase type I, and reestablished in culture. Following several passages in vitro, the lung colonization potential of the parent cell line and cells isolated from tumors growing s.c. or lung metastases was evaluated after i.v. injection into young nude mice. The parent cells passaged in vitro or grown and passaged at a s.c. site produced comparably few lung nodules, indicating that implantation at a s.c. site does not select for cells that are highly metastatic. In contrast, cells isolated from individual pulmonary metastases produced 50 to 100 times more lung colonies than the parent cell line. Collectively, these results demonstrate that the enhanced metastatic potential of cells isolated from lung nodules is a result of selection rather than adaptation to growth in the nude mouse. Several of these highly metastatic cell variants are being used for the evaluation of the efficacy of chemotherapeutic drugs in the treatment of pulmonary metastases in nude mice. Also, the nude mouse model provides ample opportunities to study the significance of platelet-tumor cell interactions in the metastasis of human malignant neoplasms.

Research sponsored by NCI, DHHS, under Contract N01-CO-75380 with Litton Bionetics, Inc.

REFERENCES

Baylin SD, Weisenberg WR, Eggleston JC, Mendelsohn G, Beaven MA, Abeloff MD, Ettinger, DS (1978). Variable content of histaminase, L-dopa decarboxylase and calcitonin in small-cell carcinoma of the lung. Biological and clinical implications. N Eng J Med 299:105.
Brown JM (1973). A study of the mechanism by which anticoagulation with warfarin inhibits blood-borne metastases. Cancer Res 33:1217.
Brunson KW, Nicolson GL (1978). Selection and biologic properties of malignant variants of a murine lymphosarcoma. J Natl Cancer Inst 61:1499.

Burton RC (1980). Alloantisera selectively reactive with NK cells: Characterization and use in defining NK cell classes. In Herberman RB (ed): "Natural Cell-Mediated Immunity Against Tumors." New York: Academic Press, p. 19.

Donelli MG, Colombo T, Broggini M, Garratini S (1977). Differential distribution of antitumor agents in primary and secondary tumors. Cancer Treat Rep 61:1319.

Fidler IJ (1970). Metastasis: Quantitative analysis of distribution and fate of tumor emboli labeled with ^{125}I-5-iodo-2'-deoxyuridine. J Natl Cancer Inst 45:773.

Fidler IJ (1973a). Selection of successive tumor lines for metastasis. Nature 242:148.

Fidler IJ (1973b). The relationship of embolic homogeneity, number, size and viability to the incidence of experimental metastasis. Eur J Cancer 9:223.

Fidler IJ, Caines S, Dolan Z (1976). Survival of hematogenously disseminated allogeneic tumor cells in athymic nude mice. Transplantation 22:208.

Fidler IJ, Kripke ML (1977). Metastasis results from preexisting variant cells within a malignant tumor. Science 197:893.

Fidler IJ, Gersten DM, Riggs C (1977). Relationship of host immune status to tumor cell arrest, distribution and survival in experimental metastasis. Cancer 40:46.

Fidler IJ, Gersten GM, Hart IR (1978). The biology of cancer invasion and metastasis. Adv Cancer Res 28:149.

Fidler IJ, Kripke ML (1980). Tumor cell antigenicity, host immunity and cancer metastasis. Cancer Immunol Immunother 7:201.

Fogel M, Gorelik E, Segal S, Feldman M (1979). Differences in cell surface antigens of tumor metastases and those of the local tumor. J Natl Cancer Inst 62:585.

Fogh J, Giovanella BC (eds) (1978). "The Nude Mouse in Experimental and Clinical Research." New York: Academic Press, Inc.

Hanna N (1980). Expression of metastatic potential of tumor cells in young nude mice is correlated with low levels of natural killer cell-mediated cytotoxicity. Int J Cancer 26:675.

Hanna N, Fidler IJ (1980). The role of natural killer cells in the destruction of circulating tumor emboli. J Natl Cancer Inst 65:801.

Hanna N, Burton RC (1981). Definitive evidence that natural killer (NK) cells inhibit experimental tumor metastasis in vivo. J Immunol, in press.

Hanna N, Fidler IJ (1981). Expression of metastatic potential of allogenic and xenogeneic neoplasms in young nude mice. Cancer Res 41:438.

Hart IR, Fidler IJ (1980). Cancer invasion and metastasis. Quart Rev Biol 55:121.

Herberman RB, Ninn ME, Laurin DH (1975). Natural cytotoxic reactivity of mouse lymphoid cells against syngeneic and allogeneic tumors. I. Distribution of reactivity and specificity. Int J Cancer 16:216.

Johnson WJ, Balish E (1980). Macrophage function in germ-free, athymic (nu/nu), and conventional-flora (nu/+) mice. J Reticuloendothel Soc 28:55.

Jones PDE, Castro JE (1977). Immunological mechanisms in metastatic spread and the antimetastatic effects of C. parvum. Br J Cancer 35:519.

Kindred B (1979). Nude mice in immunobiology. Prog Allergy 26:137.

Liotta LA, Kleinerman J, Saidel GM (1974). Quantitative relationships of intravascular tumor cells, tumor vessels and pulmonary metastases following tumor implantation. Cancer Res 34:997.

Lozzio BB, Machado EA, Lair SV, Lozzio CB (1980). Reproducible metastatic growth of K-562 human myelogenous leukemia cells in nude mice. J Natl Cancer Inst 63:295.

Manning JK, Reed ND, Jutila JW (1972). Antibody response to Escherichia coli lipopolysaccharide and type III pneumococcal polysaccharide by congenitally thymusless (nude) mice. J Immunol 108:1470.

Martin WJ, Martin SE (1974). Naturally occurring cytotoxic anti-tumor antibodies in sera of congenitally athymic (nude) mice. Nature 249:564.

Nomura T, Ohsawa N, Tamaoki N, Fugiwara K (eds) (1977). "Proceedings of the Second International Workshop on Nude Mice." Tokyo: University of Tokyo Press.

Ohsugi Y, Gershwin ME, Owens RB, Nelson-Rees WA (1980). Tumorigenicity of human malignant lymphoblasts: Comparative study with unmanipulated nude mice, antilymphocyte serum-treated nude mice, and x-irradiated nude mice. J Natl Cancer Inst 65:715.

Poste G, Fidler IJ (1980). Pathogenesis of cancer metastasis. Nature 283:139.

Povlsen CO, Fialkow PJ, Klein G, Rygaard J, Wiener F (1973). Growth and antigenic properties of a biosyderived Burkitt's lymphoma in thumus-less (nude) mice. Int J Cancer 11:30.

Rygaard J, Povlsen CO (1969). Heterotransplantation of human malignant tumor to nude mice. Acta Pathol Microbiol Scand 77:758.

Skov CB, Holland JM, Perkins EH (1976). Development of fewer tumor colonies in lungs of athymic nude mice after intravenous injection of tumor cells. J Natl Cancer Inst 56:193.

Trope C, Hakansson L, Dencker H (1975). Heterogeneity of human adenocarcinomas of the colon and the stomach as regards sensitivity to cytostatic drugs. Neoplasma 2:423.

van Putten LM, Kram LK, van Dierendonck HH, Smink T, Fuzy M (1975). Enhancement by drugs of metastatic lung nodule formation after intravenous tumour cell injection. Int J Cancer 15:588.

Watanabe S, Shimosato Y, Juroki M, Sato Y, Nakjima T (1980). Transplantability of human lymphoid cell line, lymphoma, and leukemia in splenectomized and/or irradiated nude mice. Cancer Res 40:2588.

DISCUSSION

Dr. R. Greig (Smith, Kline, and French Inc., Philadelphia). Are you suggesting that the NK cells have a principal effect on cells in the circulation?

Dr. N. Hanna (Frederick Cancer Research Center, Frederick). What do you mean by principal? I would say that these are one of the major cells that kill tumors. The shearing effect, the dynamics, the deformability of the cells or the turbulence of the blood depends upon the system. When you get clearance you can get up to 50% survival as opposed to 1% survival. If you inject 2,000 cells, you get only 100 colonies which means you have a very efficient system which is not alone responsible for the killing of tumor cells. Within the system, it will affect, perhaps, those cells which are attached to the reticuloendothelial cells. NK cells can kill those that have escaped the other destructive mechanisms.

Dr. S.J. Scialla (Hematology and Oncology Associates of Northeastern Pennsylvania, Scranton). I assume that you feel that the NK cells are important in the immune response in humans as a protection against tumors. One interesting clinical study has been in the treatment of breast cancer because cytotoxicity is a key element. It has been shown that if you increase the dosage of chemotherapy, you are apt to get better results. These studies have been done by various institutions and it seems that chemotherapy does have an important element in formal survival and prevention and recurrence of metastases. One would think that this dosage of cytotoxic substances on a continuous basis would be suppressing NK cells in the human.

Dr. Hanna. We were unable to show that the selective activity of NK cells could affect the tumors after it had metastasized into the lung. I believe that one can divide the process into two main steps: 1) in the circulation where platelets, NK cells and macrophages could be very efficient, and 2) where the tumor cells have lodged within the tissue and become inaccessible to these elements.

We have to differentiate between the two mechanisms. If you want to go for chemotherapy, first you have to know what factors or effectors you are depleting and are these important in that stage of the disease.

Dr. E.V. Sugarbaker (Miami Cancer Institute, Miami). I so enticed by what you have already presented, I would like to ask whether you have any information about other tumor cell lines.

Dr. Hanna. With other human cell lines, we did not try very much because we do not know their origin. We are setting up for long-term studies using nude mice and using specimens where we have the history of the patient, the history of the tumor and its recurrence. The more tumors we get the better we can use this model for long-term studies.

CULTURED HUMAN TUMOR CELLS FOR CANCER RESEARCH: ASSESSMENT OF VARIATION AND STABILITY OF CULTURAL CHARACTERISTICS

Jørgen Fogh, Nicholas Dracopoli, James D. Loveless and Helle Fogh
Human Tumor Cell Laboratory, Sloan-Kettering Institute for Cancer Research
Rye, New York 10580

INTRODUCTION

While the use of cells from human tumors after short periods of cultivation (early cultures) may be advantageous for some areas of investigation, such as therapy screening of individual patients' cells, established cell lines, which provide unlimited amounts of cells, are the choice in many aspects of research. Established lines, when employed in specific studies, provide the advantage that many of their characteristics are already known. When different investigations are concerned with the same cell line, valuable comparisons of experimental results are feasible. Recent years have seen a change from the very limited number of cultured human tumor cell lines previously available for research to a great abundance of lines representing numerous types of human tumors. This change has resulted from a switch of emphasis in cancer research from animal systems to an increasing interest in work with human model systems. An important factor has been the recognition of the patience needed to recover long-term cultures from very few cells in the original population. No single major technical breakthrough and no revolutionary development in nutrition can be given the credit for the many lines now established. The interest in research on specific tumor types combined with the knowledge that individual cultures cell lines from human tumors of similar histopathology may vary greatly in their characteristics have increased the need for and stimulated the activity necessary to develop many cell lines from each type of tumor. Thus, we have moved forward from the time when cancer research with cultures human tumor cells was conducted only with HeLa cells.

With the many human tumor cell lines now available, the necessity of thoroughly determining their characteristics, similarities and differences, has become even more obvious. It has also been clearly demonstrated, for example by the generally known problem of HeLa cell contamination, that mixups and mislabelling of cultured cells, with dramatic and tragic consequences for experimental results, are not unusual occurrences.

For more than twenty years our laboratory has worked with cultured human tumor cells and many lines have been established here. Lines have also been obtained from investigators at Sloan-Kettering Institute and other laboratories in the United States and abroad. Our present collection includes 432 cultured human tumor cell lines representing numerous tumor types, as well as several hundred other human cell lines. We have characterized most lines by methods chosen to analyze for malignancy of the cultured cells, consistency with tumor type, special characteristics, evidence of cross-contamination with cells of other species or with human cells, and mycoplasmal, bacterial and fungal contamination. We have provided these lines to other investigators for their experimental studies based upon cultured human tumor cells.

In this paper we will discuss and summarize some of our experience from working with cultured human tumor cell lines, as well as some of our research with these cells. The athymic nude mouse is an important animal for our work. In addition to using tumor growth in nude mice as a criterion for malignancy of cultured human tumor cells, our investigations have included the establishment of the growth and histopathologic characteristics of the spectrum of human tumors that can be directly heterotransplanted to this mouse and maintained as continuous lines. More recent studies have been concerned with metastases in this human/mouse model. Some aspects of these investigations will also be summarized.

METHODS

In this section details of some routine methods, considered to be of general interest, are provided.

(1) Media and Serum Preparation.

Liquid concentrates of routinely used media (i.e. MEM, L-15 and RPMI-1640) are purchased commercially. McCoy's 5A medium is prepared in this laboratory from basic ingredients. Glass distilled water is used in all media preparation. Human serum, used occasionally, is obtained from individual donors; fetal calf serum from commercial suppliers. Fetal calf serum is shipped to us in the frozen state and stored at $-20^{\circ}C$ until filtration. Filtration of all media and serum is done in this laboratory as follows: (a) Media filtered through .22μ Millipore filters; (b) Sera filtered through a series of Millipore filters; 1.2μ, .45μ, and finally .22μ. All filtration is accomplished with pre-sterilized filter units with filters in place. Immediately after filtration all filters are checked. If improper positioning, cracks and irregularities are detected, the medium or serum lot is refiltered. All glassware used is washed with glass distilled water, and sterilized by autoclaving. Media and sera are dispensed separately and not mixed until actual use. Each lot of media or serum and each bottle is numbered individually, sealed in plastic bags and stored at $+4^{\circ}C$ until use.

Sterility and toxicity tests are performed on each batch of medium, serum or other fluid to be used in tissue culture. Every fifth bottle is tested as follows: One ml of the medium or serum is inoculated into a test tube with 7 ml of thioglycollate, which subsequently is incubated at $37^{\circ}C$ for two weeks and repeatedly checked for contamination. Cultures of various human tumor line cells, carried without antibiotics, are carried with the medium or serum to be tested for two weeks. During this period the cultures are repeatedly observed for signs of contamination or toxicity. If contamination or toxicity is observed, the batch is not used for tissue culture. No medium or serum is used with this schedule for tissue culture work prior to two weeks after the filtration for medium and three weeks for serum (Fogh and Fogh, 1969).

(2) Cell Morphology and In Vitro Cytopathology Evaluation

Cells that grow as monolayers are recultured on coverslips in leighton tubes. When cultures are approximately 75% confluent, examination of living cells is made by Nomarksi interference-contrast microscopy (Fogh and Sykes, 1972) at several magnifications and of fixed cells after staining. We routinely use hematoxylin and eosin staining (Fogh and Sykes, 1972); other methods may be useful for specific cytological details or reactions. The Nomarski technique, also applicable to cells growing in suspension, provides a three-dimensional view of the living cells unaffected by preparation and fixation. For the special examination referred to as "in vitro cytopathology," cells grown to 75% confluency on 3 X 1 inch microscope slides in large Leighton culture tubes are rinsed several times with phosphate-buffered saline and fixed with 70% ethanol. In addition, a suspension of cells is prepared by trypsinization of a monolayer culture. After inactivation of trypsin with serum, the cells are sedimented, resuspended in 3.8 ml of phosphate-buffered saline, and fixed with 5.0 ml of 70% alcohol. The fixed preparations are evaluated by a cytologist after Papanicolaou staining to assess the malignancy and nature of the cultured cells (Bean and Hajdu, 1975; Hajdu et al., 1974).

(3) Tests for Tumor Production in Nude Mice

(a) Cell Inoculation of Nude Mice

Cells from actively dividing cultures are scraped from the flask into 2-4 ml of medium and sedimented by centrifugation at 150 g for 10 min. The supernatant is decanted and the cells are resuspended in a volume of 0.2 ml medium/serum and drawn into a 1 ml tuberculin syringe fitted with a 22 or 25 gauge needle. Athymic nu/nu mice, male or female, approximately 6 weeks old, of Rex/Trembler origin, backcrossed to Swiss three times, are inoculated subcutaneously on the dorsal side between the shoulder blades with 10^6 to 20×10^6 cells (Fogh and Hajdu, 1975). Tumors of approximately 1 cm^3 are excised for pathology examination.

(b) Nude Mouse-Grown Tumor Diagnosis

A portion of the tumor specimen is fixed in 10% neutral buffered formalin in preparation for examination by the pathologist. The specimen is processed according to routine tissue methods. After paraffin imbedding, 6 μm-thick sections are cut and stained with hematoxylin and eosin. Additional sections are cut from selected cases and stained with special stains, e.g. PAS, alcian blue, mucicarmine, trichrome and melanin, in order to demonstrate specific features that may have escaped observation at the time of routine microscopic study.

(4) Poliovirus Susceptibility Tests

Monolayer culture tubes of similar cell density are rinsed several times with serum-free culture medium and inoculated as follows: Tubes A; 1.9 ml serum-free medium, 0.1 ml normal rabbit serum, 0.05 ml poliovirus type 1 (Titer=approximately 10^8 $TCID_{50}$/ml). Tubes B; 1.9 ml serum-free medium, 0.1 ml poliovirus antiserum (neutralizing 100 $TCID_{50}$ at a minimum titer of 1:1000), 0.05 ml poliovirus type 1. Tubes C; 1.9 ml serum-free medium, 0.1 ml normal rabbit serum. Cultures are incubated at 37°C and observed at 24 hr intervals. Pronounced cytopathic effect only in Tubes A indicates susceptibility of the cells to poliovirus.

(5) Chromosome Analysis

The steps in the chromosome preparation procedure used are similar for cells derived from various types of tumors. However, reagents, concentrations, and times of exposure may vary among the different cell types. In general, Colcemid is added to the cultures to a final concentration of 0.05 to 0.1 μg/ml, followed by incubation at 37°C for 1 to 18 hrs. Cells are then trypsinized and sedimented at 180 g, the pellet is resuspended in a hypotonic solution consisting of a mixture of potassium chloride and sodium citrate for 15 min at 37°C and fixed at 4°C with alcohol-acetic acid, 3:1, three times. Each fixation period lasts 30 min. Metaphase preparations are prepared by a method based on the suspension technique developed by Moorhead and colleagues (Moorhead et al., 1960). The preparations on the slides are stained with aceto-orcein or Giemsa stain for chromosome counts and observation of abnormalities, or they are banded with trypsin, followed by orcein or Giemsa staining, or with quinacrine dihydrochlorate, for more detailed determinations of individual chromosomes.

(6) Enzyme Analysis

(a) G6PD and Other Polymorphic Enzymes

A monolayer of cells is rinsed 3 times with PBS; cells are removed from the culture flask by trypsinization or by scraping, sedimented by low speed centrifugation at +4°C and resuspended in twice the cell pellet volume of cold distilled water. Cells are lysed by three cycles of rapid freezing (in liquid nitrogen) and thawing, or by sonication. Debris is sedimented by centrifugation at 1300 g for 10 min at 4°C. Supernatant is transferred to a 2 ml cryotube (Nunc) and stored at -80°C until required for analysis. G6PD typing is done in twelve percent starch gels with Tris-EDTA-borate buffer system at pH 8.6 (Ruddle and Nichols, 1971). After applying test material and controls (HeLa, type A; and Detroit 562, type B), electrophoresis is carried out at 5 V/cm for approximately 18 hr. The gels are cooled to 4°C during electrophoresis. The gels are sliced and stained with a standard nitrotetrazolium staining mixture in agar overlay. G6PD type is determined by comparing mobility of bands of test lines with HeLa and Detroit 562 cell controls.

In addition to the determination of the G6PD phenotypes in each of the cell lines to exclude the possibility of HeLa cell contamination, we are also examining the electrophoretic phenotypes of 16 to 20 enzyme systems to exclude the possibility of contamination by other human cells. A variety of electrophoretic techniques, including starch, acrylamide, cellulose acetate, agarose gel electrophoresis and isoelectric focusing, are being used. Our methods for enzyme analysis now includes the characterization for the phenotypes of up to 20 polymorphic enzyme systems: ACP1, ADA, AK1, ESD, FUCA, GLO1, GOT2, G6PD, ME2, PEPA, PEPB, PEPC, PEPD, PGD, PGM1, PGM3, GDH, ACON1, PGP, and αGLU. The electrophoretic methods required for the analysis of the isozyme patterns exhibited by these enzymes are given in Harris and Hopkinson (1976) and Siciliano and Shaw (1976).

(b) Tissue Specific Isozymes

We are also attempting to confirm the tissue of origin of various tumors by examining the patterns of expression of a number of multi-locus enzyme systems which show differential patterns of expression in adult tissue and during development. We are in the process of surveying a relatively large number of candidate enzymes in order to determine the most useful markers for the various tissues of origin. The seventeen multi-locus enzymes, which are determined by at least 38 loci, currently being examined are PFK, CK, HEX, GPD, AMY, CA, ADH, LDH, PK, ALP, ACP, GUK, ALD, ENO, DIA, HK, and ES. The activity of the isozymes will be measured by scanning spectrophotometry of thin layer acrylamide and cellulose acetate gels, and also by standard spectrophotometric techniques for the assays of specific enzyme activity and protein content in whole cell lysate.

(7) HLA Typing

Methods used (in Dr. Marilyn Pollack's laboratory at Sloan-Kettering Institute for Cancer Research) with cultures of human tumor cells prepared in the Human Tumor Cell Laboratory, are those listed in "HLA-A,B,C and DR Alloantigen Expression on Forty-Six Cultured Human Tumor Cell Lines" (Pollack et al., 1981).

(8) Culture Contamination

(a) Tests for Mycoplasma and Other Contaminants

Cultures to be tested for mycoplasma, and/or other contaminants, are carried for at least a week without antibiotics, which may inhibit the growth of the contaminant. Cultivation of mycoplasma in special broths and agars: The following artificial media (agar dispensed in 35 mm plastic petri dishes) are used: BYE agar (Barile et al., 1958), BYE broth (Barile et al., 1958), TC PPLO agar (Carski and Shepard,1961), containing 15% heat-inactivated human serum; in some tests horse or fetal calf serum substituted for the human serum. These three media are used for testing with yeast hemin extract (Baltimore Biological Laboratory, Baltimore, Maryland) added in a concentration of 5% of the total volume. Routinely the inoculated agar plates are incubated at 37°C in an atmosphere of 5% CO_2 in nitrogen.

Method for direct microscopical demonstration of mycoplasma in cultured cells: Cultures of FL cells, seeded at very low density in Leighton tubes with inserted rectangular coverslips, in antibiotic-free LY medium with 20% heat-inactivated human or fetal calf serum, are used as indicator cells. They are inoculated, usually three days later, with the sample to be tested. After additional incubation, the cultures are prepared for microscopic examination as follows: The coverslip is removed from the culture container and placed in a small petri dish containing 3 ml of 0.6% sodium citrate solution. One milliliter of distilled water is added, dropwise, with a 1 ml pipette, to make the concentration of sodium citrate 0.45%. Ten minutes later, 4 ml of Carnoy's fixative (1 part glacial acetic acid, 3 parts absolute ethyl alcohol) is added, dropwise, with a 2 ml pipette, in order to make the fixation gradual. The coverslip is transferred to another petri dish containing 3 ml Carnoy's fixative. Ten minutes are allowed for fixation. The coverslip is taken out and left until absolutely dry. The cells are then stained for 5 min with orcein stain (2% natural orcein and 60% glacial acetic acid), and after three washes in absolute alcohol are mounted in Euparol (Flatters and Garnett, Ltd., Manchester, England). The slides are examined under phase optics at magnifications from X200 to 5100 for recognition of mycoplasma, bacteria and fungi (Fogh and Fogh, 1964, Fogh and Fogh, 1969).

Scanning electron microscopy may be used occasionally in attempts to supplement our information on presence of mycoplasma.

Standard bacteriological and mycological media are used in tests for bacterial and fungal contaminations.

(b) Mycoplasma Elimination

Among various methods which have been found successful in this laboratory are (1) treatment with Kanamycin of contaminated cultures (Fogh and Hacker, 1960); (2) the Gori and Lee method (Gori and Lee, 1964). This latter method includes an initial short treatment with high concentration of Aureomycin, Kanamycin and chloramphenicol under hypotonic conditions, followed by treatment with the same antibiotics at 1/10 the original concentrations for several weeks. Other antibiotics may be considered. There has been evidence, including experience in our laboratory, that growth of tumors from cultured cells in nude mice, followed by *in vitro* cultivation, may eliminate mycoplasmas (Van Diggelen, et al., 1977).

(9) Frozen Storage of Cells

Cultures of cells, actively dividing at the time of harvest, are trypsinized, sedimented by centrifugation at +4°C, resuspended in several milliliters of cooled medium plus serum, and counted. Cell concentration is adjusted to 2.0 to 3.0 X 10^6 per milliliter in medium/serum with 10% glycerol or 5-10% dimethyl sulfoxide as preservative. A sterile syringe with a 22 gauge 1.5 inch sterile needle is used for dispensing 1.0 ml of cell suspension per sterile pre-cooled 1.5 ml glass ampule with aluminum foil cover. Ampules are sealed using the Kahlenberg Globe ampule sealer (Kahlenberg-Globe Equipment Co., Sarasota, Florida), labeled with tape, then held at +4°C for 20 to 30 min to allow the cells to absorb the cryopreservative before being placed in the freezing chamber of the Linde Liquid Nitrogen Biological Freezing System consisting of a CRCF-1 chamber and a CRC-1 console, which is precooled to +4°C. The thermocouple probes are placed in unsealed "dummy" ampules containing the same cell suspension as that being frozen. The flow of nitrogen is regulated to freeze the cells at a precisely controlled rate (1.0-1.5°C/min) until the temperature of the cells is lowered to -100°C. The ampules are then transferred from the freezing chamber to the liquid phase of a liquid nitrogen refrigerator. To assure that the freezing has been successful, the contents of several ampules are tested for viability and any contamination.

We usually freeze 20 to 30 ampules, an amount practical for our work and also the safest amount to handle to secure uniformity and lack of contamination during freezing procedures. We prefer to use a number of portable small liquid nitrogen refrigerators since they can easily be moved in case of emergency. We store only in the liquid phase of the liquid nitrogen to secure the lowest possible storage temperature and the least likelihood of complete depletion of refrigerant which will result in total loss of the contents of the refrigerator. Each freezing is divided and stored in two laboratory locations for obvious safety reasons. As the stock is being reduced, ampules are removed alternately from the two locations. We have avoided any electronic monitoring of the liquid nitrogen level, but fill the refrigerators to the top each day, also over weekends.

CHARACTERIZATION OF CULTURED HUMAN TUMOR CELL LINES

(1) Clinical Data and Confirmation of Tumor Diagnosis

Extensive clinical data on the patient from whose tumor a cell line is derived is usually not of prime importance. However, it should be possible to obtain such information. Thus, the case number must be available for possible future reference, but the patient's name must be protected and not revealed to outside laboratories (it is ethically unacceptable that cell lines are designated according to patient name). The patient's age, sex, race and blood type are considered important and this information must be obtained. Attempts are also made to learn of special therapy prior to surgery, for example chemotherapy, radiation, hormonal or immuno- therapy, since such therapy might affect the cell cultures. A copy of the pathology report relating to the particular surgery from which the tumor specimen for cultivation was obtained is important and a representative slide of the tumor section is obtained whenever possible. For cell lines originated in our tissue culture laboratory the procedure has been that part of the specimen received in this laboratory is processed for examination by the pathologist in order to obtain a "confirming diagnosis". This procedure has not been generally performed in other laboratories. Our procedure is that all material and pathology data obtained from other institutions in connection with a cell line to be added to our collection are reviewed by the expert cancer pathologists in the Department of Pathology at Memorial Hospital. By this approach, in addition to ultimate expertise, consistency in cancer diagnosis is applied to the data recorded in our computerized file. Similar methods, whenever possible, are used for effusions, pleural and ascitic, when these fluids have been the sources of the cultured malignant cells, and in these cases cytologic examinations are used to confirm the presence of tumor cells in the effusions, and their nature.

(2) Confirmation of Malignancy of Cultured Cells

An obvious criterion to fulfill for a culture derived from a human tumor is that it contains malignant cells, and not, as has often been the case, cells originating from other components of the tumor specimen. Although the cell morphology may be indicative it is not infrequent that tumor cells cannot be definitely distinguished from non-tumor cells based only upon the morphology of monolayer cultures. Our approach to confirm malignancy is primarily based upon in vitro cytopathologic examination, test of tumor-producing capacity in athymic nude mice, with subsequent pathology confirmation, and chromosome analysis.

(a) In Vitro Cytopathology

Data from numerous cell lines examined in the past have shown that this method in most cases is highly useful. (Bean and Hajdu, 1975).

(b) Test of Tumor-Producing Capacity

Our program includes tests to determine tumor-producing capacity of cell lines in athymic nude mice (Fogh et al., 1977). Positive result is the ultimate proof of malignancy of a cultured cell line when pathology diagnosis confirms malignancy of the nude mouse-grown tumor. Our work in the past has shown that some cell lines, approximately one-third, may not be tumor-producing by conventional heterotransplantation techniques. However, some of these negative results might be explained by insufficient inoculum. The site of inoculation may also be of importance, and treatment of the mice (for example with ALS, ATS or radiation) may increase their capacity for supporting growth of human cells as tumors.

(c) Chromosome Analysis

Abnormal chromosome numbers or chromosome morphology are good indicators of malignancy of a cultured cell line. This analysis, therefore, is included for the characterization of cultured human tumor cell lines.

(d) Other Methods

Our efforts include examination of living cells by Nomarski interference contrast microscopy (Fogh and Sykes, 1972) and microscopy of fixed cells after staining. Analysis of cellular morphology and culture texture by these methods is used to monitor cellular relationships and changes in cell size and shape, nuclear-cytoplasmic ratio, shape and location of nucleus and number of nucleoli, as helpful indicators to determine malignancy. Other methods are used occasionally, such as determination of plating efficiency and the efficiency of colony growth in soft agar, as well as density dependent inhibition of growth.

(3) Confirmation of Tumor Type of Cultured Cells

Having determined that a cultured human tumor line is malignant, based upon the tests described above, we are still left with the following important problem: does a line presumably derived from a lung carcinoma, for example, actually consist of lung tumor cells? At present, methods to confirm the tissue of origin of cell lines are limited. In certain cases this problem may have started in the clinic, or even in the pathology department. There are tumors located in the organs which may present diagnostic problems in terms of whether these tumors are of primary site or metastasis, and the pathologist not infrequently faces the problem of diagnosing a metastasis as "primary site unknown", although such a tumor can be classified as epithelial or non-epithelial and, in the case of an epithelial tumor as adenocarcinoma, squamous carcinoma, etc. For certain types of tumor metastases the primary tumor is more easily recognized, such as melanomas and alveolar cell carcinoma of the lung. The cultured cell lines, now a giant step removed from their in vivo location and highly modified in terms of cellular morphology, in most cases cannot be identified with tissue of origin on the basis of morphology. Pigmented melanoma lines, again are exceptions. The problem of confirmation of tumor type is serious for two major reasons: clinical and original pathology information may be insufficient for certain cultured human tumor cell lines already established, and extraneous cell contamination of the cultured lines has been proven to be of not infrequent occurrence. Obviously research studies based upon cultured tumor lines with incorrect tumor diagnosis are misleading and improper.

Our approach in attempts to confirm tumor type is based upon (a) in vitro cytopathology diagnosis which in certain cases has proved important in determining, and confirming, the tumor of origin. In many cases, however, this examination can only provide information about the presence or absence of tumor cells, but not about specific type.

(b) <u>Diagnosis of nude mouse-grown human tumors</u> originating from cultured human tumor cell lines. Extensive studies in this laboratory (Hajdu and Fogh, 1978) have shown an amazing similarity between such tumors and the human tumor of origin. Actually, in a certain number of cases there has been increased differentiation in the nude mouse-grown tumors compared to the original human tumor, and there have been cases where this increased differentiation has been helpful in determining the primary site of the human tumor, unknown to the pathologist prior to this test (Sharkey et al., 1978).

(c) <u>Chromosome analysis,</u> with few possible exceptions, is not useful to confirm the type of tumor of a cultured cell line. This conclusion is arrived at from analysis of 130 cultured human tumor cell lines analyzed by chromosomal techniques in this laboratory (Fogh et al., 1980a).

(d) <u>Tissue specific isozymes.</u> Useful data confirming the tissue of origin and tumor type of individual cell lines may be obtained by the analysis of selected multi-locus enzyme systems that either show tissue specific characteristics or whose activity profiles are affected by a cell's transformation to malignancy. We are currently undertaking an extensive study of a number of multi-locus enzyme systems: Pyruvate kinase, Phosphofructokinase, Hexokinase, Creatinine kinase, Alkaline phosphatase, Lactate dehydrogenase, Alcohol dehydrogenase, Aldolase and Enolase with the specific aim of (1) identifying tumor and tissue specific isozyme markers in normal and neoplastic tissues that can be used to confirm the tissue of origin and tumor type of individual cultured cell lines; (2) determining if cultured human tumor cell lines are suitable models for the study of biochemical genetic changes occurring in neoplastic disease by comparing the isozyme patterns and activity profiles of selected enzyme systems with those found in the in vivo tumors of the same types from which the cell lines were initially derived and (3) surveying some of the enzyme systems of the primary metabolic pathways to establish the range of modification that occurs during neoplastic transformation in these systems in cultured cell lines derived from a wide range of tumor types.

In many respects poorly differentiated neoplastic cells, despite their phenotypic variability, are more similar to each other than are the normal in vivo cells from which they are derived (Schapira, 1971). They generally share a number of enzymological features, including the loss of typically adult isozymes with the reversion to fetal phenotypes (Schapira, 1971 and Weinhouse, 1973a), and show a relative increase in the activity of the enzymes of the glycolytic pathway with a corresponding decrease in the activity of the enzymes involved in gluconeogenesis (Weinhouse, 1973a). A recent review (Weinhouse, 1980) has listed 20 enzymes that express fetal isozymes in hepatomas alone. Isozyme changes in neoplastic tissue are not solely the result of random changes in karyotype or loss of genetic material. The reversion to fetal isozyme phenotypes, not under the normal endocrine and dietary influences affecting normal adult tissues, and the increase in anaerobic glycolosis is indicative of the highly specialized and selective changes in the biochemistry and physiology of the transformed cells (Weinhouse, 1973b). Biochemical changes in neoplastic cells appear to be aimed at maximizing the efficiency of the utilization of substrates to facilitate the explosive growth rates of the transformed cells at the expense of the surrounding host tissue.

Much of the present knowledge of biochemical changes in neoplasia is based on the comparison of rat hepatomas with normal adult and fetal rat tissue (Weber, 1974); much less is known about human systems. It has been demonstrated that the loss of adult tissue isozyme profiles and their replacement with isozymes, either absent or only present in very low concentrations in adult tissue, is correlated with a decrease in histological differentiation and an increase in the growth rate of the neoplastic cells (Weber, 1974). Aldolase B, for example, is only found in differentiated liver tissue in man and the rat, and is replaced by Aldolase A in hepatomas. Aldolase B is absent in fetal tissue and only begins to replace the fetal Aldolase A in neonatal liver (Bertoletti and Weiss, 1972). Some isozyme patterns, however, are not affected by the cellular transformation and appear to be stable markers in the tissue of origin and in poorly differentiated neoplastic cells. For example, the LDH-Z isozyme, which in normal tissue is only expressed in trophoblast cells, is also expressed in cultured cell lines derived from choriocarcinomas (Siciliano, et al., 1980).

Many models have been developed describing the changes in patterns of isozyme expression in poorly differentiated cells. The "molecular correlation concept" (Weber, 1974) and other models have been largely developed through studies of non-human tumors. Enzymes of the carbohydrate and purine and pyrimidine metabolism such as hexokinase, pyruvate and uridine phosphorylase are surveyed for their isozyme phenotypes and activity profiles in cultivated tumor cell lines derived from a variety of different tumor types to determine if many human tumors show similar and consistent increases in the rates of anaerobic glycolysis, for example, and to identify specific biochemical strategies of the different tumors.

(e) Monoclonal Antibodies

We are considering the possibility of differentiating between cell lines through the use of monoclonal antibody techniques.

(4) Exclusion of Extraneous Cell Contamination

Interspecies and intraspecies extraneous cell contamination of cultured cells has been apparent for many years (Stulberg, 1973). Since Gartler proposed wide-spread HeLa cell contamination based upon isozyme analysis of other established cell lines, this particular contamination has caused special and deserved attention (Gartler, 1966 and Gartler, 1968). The X-linked G6PD A phenotype occurs in about 30% of American Black males, whereas the A and AB phenotypes occur in 13% and 35% of the females of this population (Boyer et al., 1962 and Porter et al., 1964). The A phenotype is not found in Caucasians. HeLa cells, derived from a Black patient, are G6PD type A. Because this phenotype has been observed in a number of cell lines reported to be derived from Caucasians, it has been suggested that the lines have become contaminated or mixed-up with HeLa cells (Gartler, 1966, Nelson-Rees et al., 1981). Impressive lists of suspected HeLa contaminated lines have been reported. Many of these lines have been considered as in vitro representatives of different human tumors.

Contamination with cells other than HeLa have also occurred. For example, a cluster of 8 cell lines out of approximately 100 cell lines derived from human solid tumors at the Scott & White Clinic were found in our laboratory to have the same enzyme phenotypes when analyzed for 15 polymorphic enzymes. The similarity was confirmed by cytogenetic studies. The chronology of establishing these cell lines, isozyme and cytogenetic studies, indicate that 6 of these lines have cross-contamination (Leibovitz et al., 1979).

Our approach to deal with the problem of extraneous cell contamination is the following:

(a) Prevention

Every precaution to avoid cell contamination, mislabeling or mixups is taken in this laboratory by thorough instruction and supervision of all workers on the severity of the problem, and the necessity of handling cultures of each cell line individually, in such a way that this problem cannot possibly occur. New lines, obtained from cell line originators whenever possible or, if not possible, by investigators with experience and unquestionable reputation in regards to this problem, are cryopreserved in our laboratory shortly after arrival for future reference.

(b) Detection of Extraneous Cell Contamination

Our approach is, in addition to the evaluation of cell lines in terms of growth and morphology, dependent upon the following specific methods: (I) susceptibility to poliovirus, (II) polymorphic enzyme analysis, (III) chromosome analysis and (IV) HLA typing.

Susceptibility to infection with poliovirus is determined to support that the cultured cells are cells of human species, and to exclude that "extraneous cell contamination" with cells of non-primate species has occurred. We have tested numerous human tumor cell lines in the present collection and found them all to be poliovirus Type 1.

Polymorphic enzyme analysis. An important part of this laboratory's program has been to conduct a survey of isozymes in our collection of human tumor cell lines. The rationale for this survey is: (1) to distinguish each cell line from others using polymorphic genetic enzymes and, in as many cases as possible, to show that each line

is genetically unique. (2) To detect possible mixup of cell
lines or contamination of one by another. (3) As a spin-off
of this survey, phenotype frequencies of human tumor cell
lines can be compared with frequencies of normal human populations to determine whether significant deviations occur.
The entire collection of human tumor cell lines is progressively becoming further characterized using up to 20
polymorphic enzymes. The major part of the human tumor
cell line collection has now been studied to various extents.
The total number of typings obtained today is more than
2500. The combinations of phenotypes for the lines are
frequently analyzed to determine similarity, differences
and uniqueness among lines. Those lines not showing unique
combinations of isozyme phenotypes are selected for further
study involving the use of additional polymorphic enzyme
systems. HeLa contamination can be excluded with complete
confidence when a line is G6PD type B. It is expected that
certain cell lines will be type A and still be bona fide
lines when the patient is Black. However, classification
of type A lines as HeLa contaminants when the patient is
recorded as a Caucasian cannot be accepted with 100% assurance. The probability of identity increases as more polymorphic genetic markers are analyzed and are shown to be
identical with HeLa. It is possible that a so-called
Caucasian cell line could in fact be derived from a Black/
Caucasian hybrid with visible Caucasian features but carrying the G6PD type A (Workman et al., 1963). Such a person
could be the result of a single Black/Caucasian mating
many generations ago.

We have recently published detailed data showing that
among 100 characterized cell lines, 59 lines from different
patients and 6 pairs of lines (each pair from the same
patient's tumor) had unique phenotype combinations and were,
therefore, presumed to be authentic, uncontaminated cell
lines. Besides these 71 lines, the remaining 29 lines consisted of several small groups of two to three lines, each
group having a different combination and being among the
more frequent in the normal population. The 29 lines,
therefore, were not suspected to be contaminants. Since
this publication, (Wright et al., 1981), many more of the
presently available human tumor cell lines have been analyzed
for polymorphic enzyme phenotypes. These data are leading
to further definitions of the cell line collection in
terms of uniqueness and probability of the presence of cell
line contamination.

Chromosome Analysis. Whereas chromosome analysis has been used to detect cell line contamination, in particular with regard to HeLa contamination (Nelson-Rees et al., 1981), this method is not generally as useful an approach for this aspect of the program as the analysis of polymorphic enzymes. Our extensive studies have shown that no special chromosomal feature distinguishes most tumor types. Chromosome numbers varied from cell to cell in each cultured cell line. In most samples a modal number, defined as the most frequent occurring count, was determined. However, this number tended to vary with culture passage level; in some cases it increased and in some cases it decreased at increasing passage. Modal chromosome number, therefore, is not a proper characteristic for distinguishing a cultured human tumor cell line. The same conclusion was arrived at when the mean chromosome numbers were determined. We must conclude, therefore, that modal number or mean number or their ranges are of limited value for characterization and identification of most cultured human tumor cell lines, and for making comparison among lines from different tumor categories. An attempt to correlate the modal or mean number with the tumor type clearly showed variation within the tumor type. The modal ranges for chromosome numbers of the many lines we studied varied from diploid to hypertetraploid. Data on more specific karyotypic changes have also been disappointing in terms of serving for comparison, and the background data for statistical tests of significance to evaluate these chromosome changes are still lacking. With few exceptions, therefore, chromosome analysis would not be the method of choice to detect extraneous cell contamination of our cultured human tumor cell lines (Fogh et al., 1980a).

HLA Typing. HLA typing may help in establishing the uniqueness of a cell line in combination with isozyme analysis or sometimes alone. A number of cell lines in our collection have been HLA typed (Pollack et al., 1981). Whereas stability in culture and after conversion to malignancy has been established for many polymorphic enzymes (Wright et al., 1979), such data are still insufficient for HLA antigens. Of 30 lines examined, 15 had previously been shown to have unique enzyme phenotype combination within a total sample of 100 lines. The combinations for the other 15 lines overlapped in many cases. However, HLA typing data defined 14 of these lines as unique within the sample of 30 lines. Thus, our data showed that HLA typing can be an important adjunct to isozyme typing and, conceivably, may serve the purpose of identification in many cases. The frequencies of the combinations observed for the cell lines that would be expected in normal populations varied from 1.5×10^{-2} to 1.9×10^{-6} for the enzyme phenotypes of lines analyzed for 6 or more enzymes, from 7.5×10^{-4} to 4.3×10^{-7} for the HLA typing, and from 5.0×10^{-5} to the extremely low frequency of 8×10^{-13} for the combined data.

(5) <u>Bacterial, Fungal or Mycoplasmal Contamination</u>

It is a well-known fact that bacteria, fungi and mycoplasma may accidentally be introduced into tissue cultures which are highly susceptible. The problems related to mycoplasma only became obvious during the last 25 years. Some apathy towards these problems has prevailed and much work has been carried out with contaminated cell systems, often leading to incorrect interpretations of diagnostic and experimental data.

The practical aspects concern the sources, the prevention, the detection and the elimination of contaminants. The methods of the Human Tumor Cell Laboratory have led to a record which perhaps is unique for a large and diverse tissue culture operation. Although random bacterial or fungal contamination has occurred, this laboratory has not had one accidental tissue culture contamination with mycoplasmas during 18 years of operation, and all cell lines provided to other investigators have been free of mycoplasma contamination. This record has been accomplished while several major research programs carried out within the laboratory have been specifically concerned with effects of mycoplasma on cultured cells.

Our methods for detection of mycoplasma in cultured cells have proven highly efficient and we have detected this type of contamination in a variety of cultured human tumor cell lines received from other laboratories. In many cases our methods for elimination of these contaminants have been successful, so that some of the lines presently in the bank which were previously contaminated with mycoplasmas are now free of such contaminants.

(6) Special Features of Lines.

From characterization of numerous human tumor cell lines it is apparent that lines vary in many of their characteristics; i.e. cell morphology, culture structure, rates of growth, tumor-producing capacity, etc. When parameters such as secretion of distinct products, isozyme phenotype and HLA typing are added to the description of a line, most cultured human tumor cell lines, if not all, can be considered as unique.

(7) Frozen Storage and Records

(a) Freezing and Recovery

This laboratory has successfully frozen approximately 500 cultured cell lines or cell strains and recovered viable cultures from approximately 1000 batches. In addition to the routine maintenance of the cell bank, special experiments have been carried out to determine optimal conditions for the successful cryopreservation, storage, and retrieval to the viable state of many and varied types of cells (Loveless et al., 1979).

(b) Recording of Freezing Data

Years ago this laboratory established a computerized program which provides information on all locations for each cell line or strain stored in liquid nitrogen refrigerators. The computerized file shows data on short supply, on available storage space, and on inventory changes. Each cell line or strain is listed as to passage available, date of freezing, cell concentration per ml, medium and serum requirements, type and concentration of cryopreservative, any contamination present, exact location within the refrigerator, the number of ampules frozen, and the number of ampules left in storage. Changes in this data base are updated constantly by use of a computer terminal in our laboratory connected to our time-shared computer facility. We are able to interact directly with our data base as to detailed information, at any time, on each cell type, the contents of any location within the bank, as well as comprehensive monthly reports reflecting inventory changes.

(c) <u>Recording of Cell Line Characteristics</u>

A computerized catalog of cell line data information is available. This catalog lists cell lines available according to tumor and/or cell type. Information is listed as to cell line designation, clinical data and pathology information and confirmation, cell line originator, from whom the cell line was obtained, date and passage level at which obtained, certain references, and culture conditions, both previous and as cultured in the Human Tumor Cell Laboratory. Characterization data are listed according to individual categories, i.e. in vitro cytopathology, nude mouse tumor-producing capacity, poliovirus susceptibility, chromosome data, contamination previously and/or presently, electron microscopy data and isozyme phenotype data. When updated at regular intervals, this system assures that an up-to-date catalog of all cell line data is readily available. The program contains various listings summarizing individual characteristics of the cell lines.

ASSESSMENT OF VARIATION AND STABILITY OF CULTURAL CHARACTERISTICS OF HUMAN TUMOR CELL LINES

The degree of stability of human tumor cells in culture involves both the similarities between (a) the tumor and the host and the cells in culture and (b) the effects of long-term cultivation. The assessment is difficult, and not all criteria apply equally well. There is a general tendency for the characteristics which are expressions of malignant change to show variation from cell to cell within the population of a line. Cell lines derived from tumors of the same tissue and with similar tumor histopathology often differ in their overall characteristics. Some of the more obvious criteria to consider for the evaluation of variation versus stability include cell morphology, rate of growth, chromosomal changes, capacity to produce tumors and metastases. All demonstrate variation from cell to cell in most cultured human tumor cell populations. In contrast, the characteristics which are not expressions of malignant change such as polymorphic enzyme phenotypes, HLA antigens and poliovirus susceptibility, show relatively consistent characteristics among the cells of a single line.

(1) Cell Morphology

The morphology of the cultured cells may be an important criterion for tentative establishment of malignancy and for distinction between epithelial and non-epithelial cells. This feature, however, does not serve as an absolute criterion for characterization, and even less for identification. In addition to unstable variations within the populations, cell morphology is affected by external factors such as media modifications. The similarity of the in vitro morphological appearance observable among some cell lines derived from different tumors, and the differences which can be observed among cell lines derived from tumors of similar histopathology make distinction by morphology most uncertain. Cells in tumors present a morphology not directly comparable to cells growing in monolayers. Cells in malignant effusions may more readily be compared with cells growing in suspension or with cells released from monolayers by trypsinization. Therefore, the technique referred to as in vitro cytopathology (Bean and Hajdu, 1975) was designed as a means to evaluate cultured cells according to the cytologist's criteria. Changes in morphology often occur between early and established line cultures.

(2) Growth Rate

There is no absolute correlation between the rate of growth of human tumors, or their mitotic frequencies, and the growth in culture of these tumor cells. Differences are often observed between early cultures and established lines and the rate of growth of cells in culture are highly dependent upon nutritional factors. When several cell lines are established from the same tumor the lines may vary in rapidity of growth and different cell lines derived from tumors of similar histopathology have shown great variation in growth rates. These observations indicate variation rather than stability.

(3) Chromosomes

We have referred above to the variation in chromosome metaphases within human tumor cell lines and among lines, also those derived from similar tumors. Chromosomal characteristics may vary with time of cultivation.

(4) Tumor-Producing Capacity

The tumor-producing capacity of many human tumor cell lines has been established in recent studies using athymic nude mice (Fogh et al., 1977.). There have been some surprising observations. Some lines have not been tumor-producing, even at very high inocula and long periods of observation, whereas other lines derived from similar histopathologic types of human tumors have produced tumors in nude mice with confirmed tumor diagnosis. There have been instances where later culture passages have produced tumors in contrast to early cultures derived from the same tumor. The lack of tumor-producing capacity of cells that are apparently malignant according to other criteria, is not fully explained. Similar results have been obtained when human tumors obtained directly from surgery have been implanted into nude mice; certain tumors grew, certain tumors did not. In our experience, a cultured human tumor cell line shown to be tumor-producing at a certain passage level will maintain its tumor-producing capacity also at later passage levels. A remarkable stability has been demonstrated in that the histopathology of tumors produced by cultured human tumor cell lines has demonstrated a picture essentially similar to the human tumor of origin even after extensive periods of cultivation of the cells prior to inoculation into nude mice.

(5) Other Criteria

In a recent study of 58 human tumor cell lines, most of them originating from tissues with low glycogen storage, glycogen was found to be present in all lines (Rousset et al., 1981). However, concentrations varied extensively from one cell line to another. Particularly high values were found in a cell line from breast, two from kidney, two from melanoma, one from uterus, one from urinary bladder, one from glioblastoma and one from ovarian tumor. Glycogen values were higher than those found in cell lines derived from glycogen rich tissue. Thus this study showed considerable variation among cell lines, even those from similar tumors and demonstrates that malignant cells may express highly modified phenotypes in this and many other biochemical characteristics.

It is apparent, therefore, that stability of cultural characteristics of human tumor cells vary according to the criteria applied.

(a) At the cellular level, the conclusion must be that the cell lines are not stable for many characteristics which have changed with malignancy. Characteristics such as the polymorphic enzyme phenotypes, sensitivity to poliovirus and, to a considerable extent HLA antigens, which have not been changed with malignancy, appear to be quite stable. This is true when the cultured cells are compared with normal cells or with the tumor cells in the host and these latter characteristics are stable after short and long-term cultivation as well.

(b) The conclusion is somewhat different when the whole cell population is considered. Although general morphology of cultured human tumor cells is not directly comparable with normal cells or tumor cells in the patient, the cultured cell population often has the capacity to recreate the original tumor, for example in the nude mouse. The capacity to recreate shows an amazing stability even after long-term cultivation. The ability to metastasize appears to be a less stable characteristic (see below). The chromosomal picture in the cultured cell population is obviously different from normal cells in the patient and often from that in the tumor of the patient. Changes are often observed with long-term cultivation. Based upon chromosomal criteria, we must conclude that cultured tumor cell populations are unstable. A similar conclusion is arrived at when rates of growth of cultured human tumor cells are considered. They are usually different from the normal cells in the host; there is no definite relation to the growth of the tumor of the host. Rates of growth may change with long-term cultivation. In contrast, polymorphic enzymes show stability for cultured cell population in all three comparisons.

We may conclude that cultured human tumor cells do not provide an ideal model since many of their characteristics related to malignancy are not stable. However, we do not have a thorough understanding of the morphological and physiological changes occurring during the development and growth of a tumor, and so we should not necessarily expect the in vitro model to be stable in all its characteristics.

HUMAN TUMOR PRODUCTION IN NUDE MICE

Tumor production by a cell line followed by histopathologic confirmation of malignant features of the tumor is the best possible demonstration of malignancy of the cultured cells, and our data have been summarized under CHARACTERIZATION OF CULTURED HUMAN TUMOR CELL LINES. As another model for experimental investigations of human tumor cells we have also studied tumor growth after direct inoculation in nude mice of surgically obtained tumors. Extensive observations of the histopathology showed a strong similarity with the human tumor of origin, although in a number of cases the nude mouse-grown tumors appeared more highly differentiated. This information has been helpful in several clinical cases where the tumor diagnosis was uncertain, for example metastases with primary site of tumor unknown. As with cultured human tumor cell lines, not all surgically obtained tumors grew in nude mice; the rate of success varied with type of tumor. Detailed data have been published on the frequency and rapidity of growth of individual tumor categories (Sharkey et al., 1978; Fogh et al., 1979; Fogh et al., 1980b; Fogh et al., 1980c). Three hundred and eighty-one tumors of 14 tumor categories were attempted including tumors of the breast, lung, head and neck, female genitals, soft tissue sarcomas, gastrointestinal (colon and non-colon), bone tumors, germ cell tumors, kidney tumors, lower urinary tract tumors, malignant melanomas, and a group of tumors with primary site unknown. The highest frequencies of growth were observed for malignant melanoma, adenocarcinoma of the colon, kidney tumors, lung tumors and bone tumors. Takes were higher for recurrent tumors (50.0%) and metastases (38.5%) than for tumors of primary sites (20.5%). The degree of success in establishing continuous tumor lines also varied among the tumor categories. Thus, 100% of the melanomas and kidney tumors growing in primary transplant established as lines, whereas only 17% of such female genital tract tumors and no lymphoreticular tumors continued to grow in passages. The time of growth in primary transplant, based upon a geometric mean diameter of 10 mm tumor size, varied from 4.2 weeks for the female genital tract tumor to 23.9 for the bone tumor category, and for all the tumors averaged 9.8 weeks. For the established tumor lines, passage times varied from 2.3 weeks for lower urinary tract tumors to 7.4 weeks for colon tumors and on average 5.7 weeks for all tumors.

Although many aspects of malignancy are expressed in the nude mouse-grown tumors, invasion and metastases have been only rarely observed in this host. When we studied 106 malignant human tumor lines in a total of 1,045 nude mice, metastasis were only observed in 14 instances (1.3%), involving 11 different tumor lines. Three of the lines showed repeated metastasis. The metastasizing tumors included carcinomas of the breast, kidney, lung and stomach and two metastatic carcinomas with tumor of primary site unknown. Deep penetration of the body wall during growth of the tumor transplant was highly correlated with metastasis. Serial passage in nude mice did not select for a more malignant tumor line and the incidence of metastasis did not differ at various passage levels (Sharkey and Fogh, 1979).

Present efforts in our laboratory have succeeded in establishing sublines of cultured human tumor cells which not only metastasize consistently in the nude mouse but lend themselves to experimental quantitative studies of human tumor cell metastasis and to therapy investigations based upon countable metastatic lung lesions. Most investigations have been concerned with human malignant melanomas; metastases are easily observed and counted macroscopically or with the dissecting microscope.

Four of 12 cultured human melanoma lines produced metastases from subcutaneous or peritoneal tumor sites. Two of these lines (MeWo and SK-MEL-3), metastasizing from both sites, were chosen as potential metastasis models. Tumors from subcutaneously inoculated MeWo cells produced lung metastases in 80% of Swiss nude mice and in 100% in C57 nude mice, 10 to 15 weeks after inoculation. Tumor size and the number of lung metastasis were directly related. By culturing cells from MeWo lung metastasis, we have developed subpopulations with increased metastasizing capacity. Similar experiments have been carried out with cells of the SK-MEL-3 line, which produce more rapidly growing tumors than MeWo cells. Thus, 100% of inoculated mice showed lung metastasis within 5 to 6 weeks. Whereas part of the metastasis developed from MeWo tumors were nonmelanotic, nearly all metastasis resulting from subcutaneous tumors of SK-MEL-3 cells were highly melanotic. Lung colonizing capacity of cultured MeWo and SK-MEL-3 cells after intravenous inoculation into the tail vein of nude mice has also been demonstrated, and lung colonies were easily observed 4 to 6 weeks after inoculation. Sublines cultured from metastases have shown a dramatic increase in the number of colonies produced (Fogh et al., in press; Sordat et al., in press).

Stability of human tumor lines grown in nude mice has been evaluated in a recent publication (Povlsen et al., in press). The term "established tumor line" has been used for a tumor which appears to be capable of continuous passage in the nude mouse and is available in passage or frozen (Fogh et al., 1980c). The following criteria were considered for evaluation of long-term stability: (a) persistent tumor growth during passage, (b) rate of tumor growth, (c) site of pattern of tumor growth, (d) microscopic tumor appearance (e) chromosomal characteristics of tumor, (f) biochemical tumor characteristics, (g) host reactions to tumors. In many respects the human tumor/nude mouse model appears stable and, perhaps, more than other human tumor/animal models. However, the final evaluation of stability of this model must await large-scale studies of a series of chosen standard tumors which should be examined by standardized methods and with microbiologically uniform, genetically well-defined animals.

SUMMARY

Coinciding with a greater general emphasis on experimental investigations with human tumor models, lines of cultured human tumor cells increasingly have been established and used in many disciplines of cancer research. The necessity to determine certain basic characteristics of the many lines, as they relate to the host and tissue of origin and the expression of malignancy, has become obvious as has the need to exclude extraneous cell contamination or mixups with other cultured cell lines. Our studies have characterized approximately 400 cultured human tumor cell lines, derived from many different human tumor types, as to their cell morphology, growth, nutritional requirements, tumor-producing capacity in nude mice, chromosome abnormalities, polymorphic enzyme phenotypes and HLA types. The characteristics which are expressions of malignant change, typically show variation from cell to cell within the population of each cultured tumor cell line, and among cell lines derived from tumors of the same tissue and with similar histopathology. Thus, individual types of cancer should be studied on the basis of more than one cell line. The variability, as expected, includes the malignancy-related chromosomal changes. Although chromosome analysis may be useful to confirm species of cultured cell lines, modal chromosome number or mean number or their ranges vary and are of limited value for identification of a human tumor cell line and for comparison of lines from different tumor categories. The background data for statistical tests of significance to evaluate specific karyotypic changes are also lacking. In contrast, markers relating to the host, such as polymorphic enzyme phenotypes, have shown a remarkable stability in the cultured cells, even after long-term cultivation, and have been most useful for cell line identification. With enzyme phenotype analysis (up to 20 enzymes in our laboratory) combined with HLA typing, the probability of contamination can be shown to be extremely low (8×10^{-13} for one tumor cell line). Present efforts include attempts to characterize the cell lines based upon tissue specific and malignancy altered isozymes. Mycoplasma contamination of cultured cell lines has been avoided during many years with simple methods which are adaptable in any laboratory.

Even after numerous culture passages, cultured human tumor cells maintain their capacity to produce tumors in nude mice with re-creation of structure and pattern of the original human tumor. Increased differentiation of nude mouse-grown tumors is not uncommon. Metastases may occur, although variably and rarely. We have established sublines of cultured human melanoma cells which consistently metastasize to and colonize in the lungs of nude mice, with countable lesions. This system is valuable for special therapy studies and for quantitative and experimental investigations of metastases of human cancer.

References

Barile MF, Yaguchi R, Eveland WC (1958). A simplified medium for the cultivation of pleuropneumonia-like organisms and the L-forms of bacteria. Am J Clin Path 30:171.
Bean MA, Hajdu SI (1975). Cytological characterization of human tumor cells from monolayer cultures. In Fogh J (ed): "Human Tumor Cells In Vitro," New York: Academic Press, p. 333.

Bertolotti R, Weiss MC (1972). Expression of differentiated functions in hepatoma cell hybrids. II. Aldolase. J Cell Physiol 79:211.

Boyer SG, Porter IH, Weilbacher RG (1962). Electrophoretic heterogeneity of glucose-6-phosphate dehydrogenase and its relationship to enzyme deficiency in man. Proc Natl Acad Sci USA 48:1868.

Carski TR, Shepard CC (1961). Pleuropneumonia-like (Mycoplasma) infections of tissue culture. J Bacteriol 81:626.

Fogh J, Fogh H (1964). A method for direct demonstration of pleuropneumonia-like organisms in culture cells. Proc Soc Exptl Biol Med 117:899.

Fogh J, Fogh H (1969). Procedures for control of mycoplasma contamination of tissue cultures. Ann NY Acad Sci 172:15.

Fogh J, Hacker C (1960). Elimination of pleuropneumonia-like organisms from cell cultures. Exptl Cell Res 21:242.

Fogh J, Hajdu SI (1975). The nude mouse as a diagnostic tool in human tumor cell research. J Cell Biol 67:117a.

Fogh J, Sykes JA (1972). A comparison of methods for morphological studies of cultured cells. In Vitro 7:206.

Fogh J, Fogh H, Daniels WP (1980a). A critical review of chromosome numbers in cultured human tumor cell lines. 18th Somatic Cell Genetics Conf. Tucson, Arizona.

Fogh J, Fogh JM, Orfeo T (1977). One hundred and twenty-seven cultured human tumor cell lines producing tumors in nude mice. J Natl Cancer Inst 59:221.

Fogh J, Orfeo T, Tiso J, Sharkey FE (1979). Establishment of human colon carcinoma lines in nude mice. Exptl Cell Biol 47:136.

Fogh J, Orfeo T, Tiso J, Sharkey FE, Fogh JM, Daniels WP (1980b). Twenty-three new human tumor lines established in nude mice. Exptl Cell Biol 48:229.

Fogh J, Tiso J, Orfeo T, Fogh JM, Daniels WP, Sharkey FE (in press). Analysis of human tumor growth in nude mice. in "Proceedings of the Third International Workshop on Nude Mice." New York: Gustav Fischer, Inc.

Fogh J, Tiso J, Orfeo T, Sharkey FE, Daniels WP, Fogh JM (1980c). Thirty-four lines of six human tumor categories established in nude mice. J Natl Cancer Inst 64:745.

Cartler SM (1966). Genetic markers as tracers in cell culture. Natl Cancer Inst Monogr 26:167.

Gartler SM (1968). Apparent HeLa cell contamination of human heteroploid cell lines. Nature 217:750.

Gori GB, Lee DY (1964). A method of eradication of mycoplasma infections in cell cultures. Proc Soc Exptl Biol Med 117:918.

Hajdu SI, Bean MA, Fogh J, Hajdu EO, Ricci A (1974). A Papanicolaou smear of cultured human tumor cells. Acta Cytol 18:327.

Hajdu, SI, Fogh J (1978). The nude mouse as a diagnostic tool in human tumor cell research. In Fogh J, Giovanella BC (eds): "The Nude Mouse in Experimental and Clinical Research," New York: Academic Press, p 235.

Harris H, Hopkinson DA (1976). "Handbook of Enzyme Electrophoresis in Human Genetics". Amsterdam and Oxford: North-Holland.

Leibovitz A, Wright WC, Pathak S, Siciliano MJ, Daniels WP, Fogh H, Fogh J (1979). Detection and analysis of G-6-PD phenotype B cell lines contamination. J Natl Cancer Inst 63:635.

Loveless JD, Orfeo T, Hernandez R, Fogh J (1979). Liquid nitrogen freezing and storage of cultured human cells. In Vitro 15:200.

Moorhead PS, Nowell PC, Mellman WJ, Battigs DM, Hungerford DA (1960). Chromosome preparations of leukocytes cultured from human peripheral blood. Exptl Cell Res 20:613.

Nelson-Rees WA, Daniels DW, Flandermeyer RR (1981). Cross contamination of cells in culture. Science 212:452.

Pollack MW, Heagney SD, Livingston PO, Fogh J (1981). HLA-A,B,C and DR alloantigen expression on forty-six cultured human tumor cell lines. J Natl Cancer Inst 66:1003.

Porter IH, Boyer SH, Watson-Williams EJ, Adam A, Szeinberg A, Siniscalco M (1964). Variation of glucose-6-phosphate dehydrogenase in different populations. Lancet 1:895.

Povlsen C, Rygaard J, Fogh J (in press). Long-term growth of human tumors in nude mice. Evaluation of stability. In Fogh J, Giovanella BC (eds): "The Nude Mouse in Experimental and Clinical Research, Vol. II," New York: Academic Press, Inc.

Rousset M, Zweibaum A, Fogh J (1981). Presence of glycogen and growth related variations in fifty-eight cultured human tumor cell lines of various tissue origins. Cancer Res 41:1165.

Ruddle FH, Nichols EA (1971). Starch gel electrophoretic phenotypes of mouse X human somatic cell hybrids and mouse isozyme polymorphisms. In Vitro 7:120.

Schapira F (1971). Foetal enzyme patterns in cancer tissues. Rev Europ Etudes Clin Et Biol XVI:205.

Sharkey FE, Fogh JM, Hajdu SI, Fitzgerald PJ, Fogh J (1978). Experience in surgical pathology with human tumor growth in the nude mouse. In Fogh J, Giovanella BC (eds): "The Nude Mouse in Experimental and Clinical Research," New York: Academic Press, p. 187.

Sharkey FE, Fogh J (1979). Metastasis of human tumor in athymic nude mice. Int J Cancer 24:733.

Siciliano MJ, Shaw CR (1976). Separation and visualization of enzymes on gels. In Smith I (ed): "Chromatographic and Electrophoretic Techniques, Ed. 4, Vol. 2," London: W. Heinemann Medical Books Ltd, p. 185.

Siciliano MJ, Bordelon-Riser ME, Freedman RS, Kohler PO (1980). A human trophoblastic isozyme (lactate dehydrogenase-Z) associated with choriocarcinoma. Cancer Res 40:283.

Sordat BCM, Ueyama Y, Fogh J (in press). Metastasis of tumor xenografts in the nude mouse. In Fogh J, Giovanella BC (eds): "The Nude Mouse in Experimental and Clinical Research, Vol. II," New York: Academic Press, Inc.

Stulberg CS (1973). Extrinsic cell contamination of tissue cultures. In Fogh J, Giovanella BC (eds): "The Nude Mouse in Experimental and Clinical Research," New York: Academic Press, p. 1.

Van Diggelen OP, Shin S, Phillips DM (1977). Reduction in cellular tumorigenicity after mycoplasma infection and elimination of mycoplasma from infected cultures by passage in nude mice. Cancer Res 37:2680.

Weber G (1974). Molecular correlation concept. In Busch H (ed): "Molecular Biology of Cancer," New York: Academic Press, p. 488.

Weinhouse S (1973a). Metabolism and isozyme alterations in experimental hepatomas. Fed Proc 32:2162.

Weinhouse S (1973b). Isozyme patterns of hepatomas and tumour progression. Neoplasma 20:559.

Weinhouse S (1980). New dimensions in the biology of cancer. Cancer 45:2975.

Workman PL, Blumberg BS, Cooper AJ (1963). Selection, gene migration and polymorphic stability in a US white and Negro population. Am J Hum Genet 15:429.

Wright WC, Daniels WP, Fogh J (1979). Stability of polymorphic enzyme phenotypes in human tumor cell lines. Proc Soc Exptl Biol Med 162:503.

Wright WC, Daniels WP, Fogh J (1981). Distinction of seventy-one cultured human tumor cell lines by polymorphic enzyme analysis. J Natl Cancer Inst 66:239.

DISCUSSION

Dr. G. J. Gasic (University of Pennsylvania, Philadelphia). How do you get rid of fibroblasts?

Dr. J. Fogh (Sloan-Kettering Institute for Cancer Research, Rye). I don't think that there is any one method that is uniformly successful. We have used mechanical methods to separate tumor cells from fibroblasts by lifting out the tumor cells and transferring them. We have used high concentrations of serum in a number of cases; if you want a magic concentration then it has been 26% in our laboratory.

If I had to identify one thing that is more important than anything else, that would be patience. It may take months and months to remove the fibroblasts; I think that we have had some lines which took 18 months or so.

Dr. D. S. Rappaport (University of Minnesota Medical School, Minneapolis). There are properties of tumors that may be important in man or animals but may be different in the nude mouse.

Dr. Fogh. I do not know anything excitingly different in that area.

Dr. P. Hilgard (Bristol-Myers International Corporation., Brussels). In terms of the main stability of cell lines, how long did you keep them without their behavior changing?

Dr. Fogh. That is a very difficult subject. We are talking about a heterogeneous changing population. We know that anything that concerns cancer is heterogeneous variation. It is as though you are dealing with a box of ants. You may have a box with the same number of ants as in the beginning but they are all in different positions now - is this stability? Some characteristics are stable: tumors in nude mice, histopathology, etc. What do we really know about the stability of the tumor in the patient: how much work has been done on this, how many people have checked the same patients? So, the whole thing becomes an enormous problem. We do not know enough about it.

　　　Dr. J. L. Ambrus (Roswell Park Memorial Institute, Buffalo). I suppose it all depends upon what you are looking for. We looked for interferon sensitivity in many cell lines and we found that some of them are exquisitely sensitive and some are very resistant to interferon. In fact, these characteristics remain stable over many, many years and over many generations. At least from that one point of view, there is a remarkable all-round stability, but I am sure we can look for other characteristics and you will find a lack of stability.

Acknowledgment

Supported in part by PHS Contract N01-CB43854, PH grants CA-08748, and CA25548 from the National Cancer Institute.

POSITIVE AND NEGATIVE AGGREGATION RESPONSES TO CULTURED
HUMAN TUMOR CELL LINES AMONG DIFFERENT NORMAL INDIVIDUALS

E. Bastida, A. Ordinas & G. A. Jamieson
American Red Cross Blood Services
Blood Research Laboratories
9312 Old Georgetown Road
Bethesda, Maryland 20814

ABSTRACT

Platelets from approximately 50% (7/16) of normal individuals have been shown to have greater sensitivity to aggregation induced by critical threshold concentrations of three human tumor cell lines. These results may have implications for the genetics and epidemiology of human neoplastic disease.

Introduction

The primary interaction of platelets and tumor cells has usually been studied by aggregometry (Gasic et al, 1976; Holme et al, 1978; Pearlstein et al 1979, Hara et al, 1980) and a rough correlation has been established between the ability of tumor cells to aggregate platelets and their ability to induce lung metastases in mice (Gasic et al, 1976). The basis for this aggregation by tumor cells has not been determined but has generally been ascribed as being due to the release of ADP·from platelets (Gasic et al, 1976; Holme et al, 1978) to the thromboplastic activity of the tumor cell (Gastpar, 1970) or to a lipoglycoprotein complex of the tumor cell surface (Hara et al, 1980). This complex may be shed as microvesicles (Gasic et al, 1976) and may be identical to the platelet aggregating material (PAM) which has been isolated from mouse tumor cells (Pearlstein et al, 1979).

We have now found differences between different donors in the ability of their platelets to aggregate in response to various human tumor cell lines. These different individual susceptibilities appear to be largely independent of the cell line involved and may have implications in the evaluation of mechanisms of human neoplastic disease.

Results

For the present experiments, platelet-rich plasma was prepared from blood collected in sodium heparin (5 u/ml) from healthy volunteers who had not taken medications for the previous seven days: platelet counts (∿300,000/ul) and recoveries (∿80%) were in the normal range. An aliquot of washed tumor cells from each cell line, to final concentrations as noted in the Table, was then added to an aliquot of platelet-rich plasma from each donor in the aggregometer (Chronolog Corp., Broomall, Pa.) and the extent of aggregation determined: aggregation induced by 10uM ADP was taken as 100% response. We are indebted to Dr. Jørgen Fogh, Sloan-Kettering Institute, New York, for providing us with the U87MG (glioblastoma), HT29 (adenocarcinoma) and SKNMC (neuroblastoma) lines, and to Dr. Adi Gazdar, Veterans Administration Hospital, Washington, D.C. for the Hut 28 (mesothelioma) and A549 (epithelial lung carcinoma) lines. Following culture under standard conditions, cells were harvested in the absence of proteases by decanting the culture medium, washing the monolayers twice with Hank's balanced salt solution (HBSS) and then treating them for 5 min with HBSS free of Ca^{2+} and Mg^{2+} and containing 5 mM EGTA. The cell suspension was centrifuged at 800 xg for 10 min, the supernatant solution was removed for subsequent testing and the cell pellets were washed twice with a solution of HBSS free of Ca^{2+} and Mg^{2+} but containing bovine serum albumin and 0.25 mg/ml apyrase, and were finally resuspended in the same solution but without apyrase. Cells were counted in a hemocytometer and viability determined by exclusion of trypan blue. The range of viable cells was between 90-97%. None of the cell lines showed contamination with fibroblasts and the effects observed were not affected by treatment of the cultures with collagenase.

The results are summarized in the Table. Individual donors were examined 2-4 times over a period of 4 months. Results were constant with regard to aggregation response in comparison with 10uM ADP as control and lag times did not vary more than $\pm$ 10 sec in repeat experiments. At a concentration of 10^5 cell/ml the U87MG line caused the aggregation of PRP from each of the sixteen donors examined. With HT 29 cells at 10^6 cells/ml seven of the donors showed full platelet aggregation response while 9 donors failed to show aggregation. At 5×10^6 cells/ml, the Hut 28 line showed zero aggregation with the same nine donors who had shown zero aggregagation response with HT 20 at 10^6/ml. For SKNMC, the same nine donors did not show an aggregation response as in the case of the HT 29 line at 10^6 cells/ml. None of the donors responded to the A549 at tumor cell concentrations as high as 10^7/ml.

The maximum aggregation response was not the same with each cell line: the U87MG and HT 29 lines gave maximum values equal to that observed with 10 uM ADP while the responses of the SKNMC and Hut 28 lines were equal to 60% and 50%, respectively, of the ADP value. However, these values were reproducible and probably reflect differences in the sizes of the platelet-tumor cell aggregates in these two cases.

Discussion

These results show, not surprisingly, that certain tumor cell lines do not cause platelet aggregation (A549) while others (U87MG) cause aggregation at all tumor cell concentrations examined. However, for several lines, there are critical tumor cell concentrations within which donors can be grouped into responders and non-responders. More significantly, this division appears to be largely independent of the particular tumor cell line examined. The tumor cell line used in this study show different aggregation profiles and, as reported elsewhere in this conference, different effects are observed following treatment with apyrase, hirudin and phospholipases suggesting that several different mechanisms may be involved. However, the responses of individual donors to the various cell types found in the present work are remarkably constant. Of the three cell types where differential responses were observed (HT 29, Hut 28 and SKNMC), nine of the donors (3,4,5,7,9,14,15,16) showed no response with any of the lines. Further studies are necessary to determine whether a graduated response can be shown at low concentrations of U87MG cells although aggregation by this line appears to proceed by a unique mechanism.

These observations suggest a greater sensitivity to platelet aggregation by human tumor cells, in general, on the part of certain individual donors. This may explain the fact that certain types of tumors can be metastatic in one individual but not in another (Sugarbaker and Ketchman, 1977) and the results may have more general significance in their application to the epidemiology and genetics of human neoplastic disease.

ACKNOWLEDGEMENTS

This work was supported, in part, by USPHS Grants HL 14697, HL 20971 and Biomedical Research Support Grant RR 05737. Publication No. 546 from the American Red Cross.

REFERENCES

Baumgartner HR (1973). The Role of Blood Flow in Platelet Adhesion, Fibrin Deposition, and Formation of Mural Thrombin. Microvasc Res 5, 167.

Gasic GJ, Kock PAG, Hsu B, Gasic TB, Niewiarowski S (1976). Thrombogenic activity of mouse and human tumors: Effects of platelets, coagulation, and firbinolysis and possible significance for metastases. Z Krebsforsch. 86, 263.

Gastpar H (1970). Stickiness of Platelets and Tumor Cells Influenced by Drugs. Thromb Diath Haemorrh Supple 42: 291.

Hara Y, Steiner S, Baldini MG (1980). Platelets as a Source of Growth-promoting Factor(s) for Tumor Cells. Cancer Res 40, 1212.

Holme R, Oftebro R, Hovig T. (1978). In vitro interactions between cultured cells and human blood platelets. Thrombos Haemostas (Stuttg) 40, 89.

Marcum JM, McGill M, Bastida E, Ordinas AS, Jamieson GA (1980). The interaction of platelets, tumor cells and vascular subendothelium. J Lab Clin Med 96, 1046.

Pearlstein E, Cooper LB, Karpatkin SJ (1979). Extraction and characterization of a platelet aggregating material from SV-40 transformed mouse 3T3 fibroblasts. J Lab Clin Med, 93, 332.

Sugarbaker EV, Ketcham AS (1977). Mechanisms and prevention of cancer dissemination: an overview. Sem Oncol 4, 19.

Table

Aggregation Responses of Individual Donors to Six Human Tumor Cell Lines

Cell line	U87MG 10^5/ml		HT 29 10^6/ml		Hut 28 5×10^6/ml		SKNMC 5×10^6/ml		A549 10^7/ml	
Cell conc.	%*	T**	%	T	%	T	%	T	%	T
Donors										
1	100	2.7	100	2.2	50	4.0	60	3.0	0	–
2	100	2.2	100	3.0	50	4.5	60	3.2	0	–
3	100	3.1	0	–	0	–	0	–	0	–
4	100	3.5	0	–	0	–	0	–	0	–
5	100	2.6	0	–	0	–	0	–	0	–
6	100	2.0	100	4.1	50	3.5	60	2.7	0	–
7	40	4.2	0	–	0	–	0	–	0	–
8	100	2.2	100	2.5	50	3.2	60	3.0	0	–
9	100	3.0	0	–	0	–	0	–	0	–
10	100	3.2	0	–	0	–	0	–	0	–
11	100	2.8	100	3.0	50	3.0	60	2.5	0	–
12	100	2.3	100	3.0	50	3.8	60	3.0	0	–
13	100	2.9	100	3.5	50	4.2	60	3.5	0	–
14	100	3.5	0	–	0	–	0	–	0	–
15	100	2.6	0	–	0	–	0	–	0	–
16	100	3.5	0	–	0	–	0	–	0	–

* % Aggregation responses compared with 10uM ADP
** Lag time (min) from addition of tumor cells to onset of aggregation

DISCUSSION

Dr. S. J. Scialla (Hematology and Oncology Associates of Northeastern Pennsylvania, Scranton). Did you use pork or beef heparin in your PRP? Clinically it appears that pork heparin can cause thrombocytopenia which may be calcium dependent and complement mediated and this may not be true for beef heparin. Hyperaggregability in smokers might also affect the response to tumor cells.

Dr. E. Bastida (Universidad de Barcelona, Spain). We have not found any differences using different heparins. We have performed these studies several times in the same donors and the results were entirely reproducible. We did not take smoking habits into account. The donors were part of our staff and were people not taking any kind of drugs. My impression is that few of these people are smokers.

Dr. G. J. Gasic (University of Pennsylvania, Philadelphia). What do you think causes this effect?

Dr. Bastida. We do not know but it is due to a plasma factor as determined by crossover studies with platelets and plasma from good responders and non-responders.

Dr. S. Karpatkin (New York University Medical Center, New York). Did you use washed platelets?

Dr. Bastida. Washed platelets and gel-filtered platelets were not able to support the aggregation even with added fibrinogen.

Dr. J. Estrada (M. D. Anderson Hospital and Tumor Institute, Houston). How do the results compare between a patient's platelets and his own tumor and with other tumor cells?

Dr. Bastida. It would be very interesting to test that but we do not have the information at the present time.

Dr. J. C. Hoak (University of Iowa, Iowa City). How does the response of platelet aggregation with tumor cells correlate with the response to other aggregating agents?

Dr. Bastida. We would not find a correlation. Tumor cells gave half the extent of aggregation found with the usual aggregating agents with 10 μm ADP.

Dr. M. B. Donati (Instituto di Recherche Farmacologische, Milan). Does aspirin treatment of either the cell or the platelets affect the reaction?

Dr. Bastida. In the tumor cells that have a double wave of aggregation only the second wave is affected by aspirin. This will be better explained in the presentation by Dr. Jamieson.

A MITOGENIC FACTOR FOR TRANSFORMED CELLS FROM HUMAN
PLATELETS

Allan Lipton, Nancy Kepner, Cheryl Rogers,
Elizabeth Witkoski and Kim Leitzel
The Milton S. Hershey Medical Center of The
Pennsylvania State University
Dept. of Medicine, Hershey, Pennsylvania 17033

The concept that platelets contain growth factors developed from the observation of Balk that chicken plasma was less effective than chicken serum for promoting the growth of chicken fibroblasts (Balk, 1971). He postulated either that the serum mitogenic factors are released from precursors in plasma or from thrombocytes when blood is clotted during the preparation of serum. Kohler and Lipton (1974) and Ross et al, (1974) extended this observation and actually demonstrated that platelets contain growth factor. Indeed, platelet extracts can promote the multiplication of fibroblasts (Kohler and Lipton, 1974), smooth muscle (Ross et al, 1974) and human glial cells (Heldin et al, 1977).

Purification of a platelet-derived growth factor ($PDGF_1$) was facilitated by the observation that this macromolecule withstood heating for 10 minutes at $100°C$ and was a basic protein with an isoelectric point about 9.8 (Antoniades et al, 1979). This polypeptide growth factor that stimulates the proliferation of connective tissue cells has been purified by several groups (Antoniades et al, 1979; Heldin et al, 1979; Vogel et al, 1978; Heldin et al, 1981). It can be obtained from either fresh or outdated human platelets. A large scale purification procedure not involving the use of sodium dodecyl sulphate (SDS) has recently been published. Purification steps include CM-Sephadex, Blue Sepharose, Bio-Gel P-150 and Sephadex G-200 column chromatography. $PDGF_1$ is a 30,000-33,000 dalton protein as determined by both SDS gel electrophoresis and sedimentation-equilibrium analysis (Antoniades et al, 1979;

Heldin et al, 1979; Vogel et al, 1978; Heldin et al, 1981). In the presence of reducing agents the biological activity is lost and the factor molecule is converted into two distinct components of lower molecular weight (17,000 and 14,000 respectively) (Heldin et al, 1981). This would suggest that $PDGF_1$ consists of two different polypeptide chains linked by disulphide bridges. Amino acid analysis reveals that this platelet-derived growth factor contains all the common amino acids, except tryptophan, but no hexosamine (Heldin et al, 1981).

Subcellular localization of $PDGF_1$ has been accomplished. Two types of granules have been identified in platelets. The majority, known as a α granules, are of moderate electron density, and contain platelet factor 4 (PF4), beta thromboglobulin (B-TG), and platelet fibrinogen. A second type of granule, of higher electron density (dense bodies) stores serotonin, calcium, ADP and ATP. $PDGF_1$ is located in the platelet α granules. It was found that the dose-response pattern of release and subcellular localization of $PDGF_1$ was similar to that of both PF4 and BTG and different from that of ATP_1, ADP, and serotonin (Witte et al, 1978; Kaplan et al, 1979). Recent studies in patients with storage pool disease support this conclusion. The platelets of patients with this disorder have diminished amounts of ATP, ADP, serotonin and calcium and show decreased numbers of dense bodies. In the majority of patients, the number of α-granules and the content of PF4, BTG and $PDGF_1$ have been found to be normal (Weiss et al, 1977). Thus, present evidence suggests that $PDGF_1$ is stored in platelet α-granules and released along with other α-granule constituents following stimulation with thrombin, arachidonic acid and collagen.

Megakaryocytes are platelet precursors in the bone marrow that are known to be active in protein synthesis. Growth factor activity as determined by the stimulation of ^3H-thymidine incorporation into the DNA of quiescent 3T3 cells in culture has recently been found in guinea pig bone marrow. Quantitative dilution studies demonstrated that, of the cells present in the guinea pig bone marrow, only the megakaryocyte possessed significant amounts of activity similar to $PDGF_1$ (Chernoff et al, 1980). The amount of activity present in one megakaryocyte was equivalent to that present in 1,000-5,000 platelets, approximately the number of platelets shed from a single megakaryocyte (Chernoff et al, 1980). Thus, $PDGF_1$ is highly likely to have its origin in the bone marrow megakaryocyte.

$PDGF_1$ plays a critical role in the cell cycle of non-transformed fibroblasts and other mesenchymal cells. Cells exposed to $PDGF_1$ are rendered "competent" - i.e. they are potentially able to leave Go and enter the cell cycle (Ross et al, 1974; Pledger et al, 1978). Fibroblast growth factor (FGF) derived from bovine brain or pituitary gland (Gospodarowicz, 1974; Gospodarowicz and Moran, 1974) can also render cells "competent". A factor recently found in the spinal cord can also recruit quiescent cells into the cell cycle (Jennings et al, 1979).

Progression of cells through G_1 and S requires the continual presence of platelet-poor plasma (Vogel et al, 1978; Pledger et al, 1978). Cells exposed to just $PDGF_1$ will not synthesize new DNA. The concentration of plasma determines, in part, the rate of entry of cells into the S phase. Factors in plasma that can mediate the progression of cells through G_1 and S phase include insulin and somatomedin C (Clemmons and Van Wyk, 1981).

The first evidence that there might be several growth factors present in platelet extracts came from attempts to characterize the platelet mitogen (Kepner et al, 1978; Kepner and Lipton, 1981). As mentioned, growth promoting activity in crude platelet extracts for 3T3 cells ($PDGF_1$) was stable after heating to 100°C for 10 minutes. In contrast, growth promoting activity for SV40 virus transformed 3T3 (SV3T3) cells in crude platelet extract ($PDGF_2$) was destroyed by heating at 100° C for 5 minutes (Figures 1A and 1B).

These factors can be partially separated by gel filtration on a Sephadex G-100 gel column at pH 7.4 (Figure 2). The component that can selectively promote the growth of SV3T3 cells ($PDGF_2$) has a molecular weight of 72,000 daltons. Its effect appears to be quite specific in that the SV3T3-active fractions do not promote the growth of 3T3 cells (Figure 2). The platelet factor that promotes the growth of SV3T3 cells ($PDGF_2$) is stable on exposure to 4 M quanidine hydrochloride, but is destroyed by treatment with sodium metaperiodate (0.05 M at 4° C for 48 hours) and is partially inactivated by trypsin treatment. The

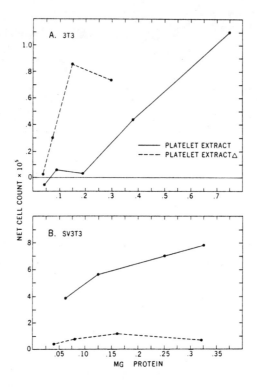

Figure 1. Heat sensitivity of mitogenic activity for 3T3 and SV3T3 cells from human platelets. Thirty to 80 units of outdated human platelet concentrates were pooled and spun at 163 x g for 10 min. at room temperature to remove RBC. They were respun at 5875 x 6 for 10 min. The platelet button was washed and centrifuged 5 times at 5875 x g. Each wash was with 400 to 800 ml 0.85% NaCl solution. The final platelet button was resuspended in 40 to 80 ml 0.85% NaCl solution. Washed platelets were incubated at 37° for 30 min. and then centrifuged at 16,319 x g. The supernatant (crude platelet extract) was divided into 2 equal portions, and one portion was heated at 100° for 5 min. and then both portions were centrifuged at 500 x g. Both samples [heated (---) and nonheated (___)] were sterilized using Millipore filters (0.22 µM) and tested in the standard assay for growth using the indicated amount of protein in a 0.5 ml volume. In this assay, 3T3 and SV3T3 cells (10^5) were plated in 60-mm Falcon plastic dishes in 5 ml of Dulbecco's and Vogt's modification of Eagle's medium containing 0.4 and 0.15% FCS, respectively, and cultures were incubated in 12% CO_2 at 37°C. This concentration of serum did not induce a significant increase in cell number per dish over a period of 4 days. The sample to be tested was added after 2-4 hr. of incubation. Numbers of cells were determined in a Coulter Counter 4 days after the start of the experiment. Net cell count was obtained by subtracting the number of cells in control plates (no additions) after 4 days of incubation. All cell counts were performed in duplicate; all experiments were repeated at least twice.

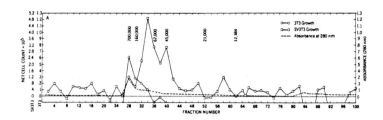

Figure 2. Sephadex G-100 gel filtration of extract of human platelets. Platelet extracts were prepared as described in the legend to Figure 1. Six ml of extract were placed on a 3- x 80-cm Sephadex G-100 column containing 0.01 M Tris-HCL, pH 7.4 at 4°. Five ml fractions were collected and sterilized using 0.22 μM Millipore filters. The growth assays for 3T3 and SV3T3 cells were performed as described in Legend to Figure 1 using 0.5 ml of each column fraction. Markers used were thyroglobulin (7.0 x 10^5), human γ-globulin (1.6 x 10^5), bovine albumin (6.7 x 10^4), ovalbumin (4.5 x 10^4), soybean trypsin inhibitor (2.1 x 10^4), and cytochrome c (1.2 x 10^4 daltons).

active material can be adsorbed onto a Concanavalin A-Sepharose column and specifically eluted with methyl-α-D-glucopyranoside. Ninety-five percent of the original activity was recovered from this Con A Sepharose column. The specific activity is increased about 100-fold at this stage when compared to fetal calf serum (Figure 3). The mitogenic activity for SV3T3 cells from human platelets is apparently due to a glycoprotein.

$PDGF_2$ can also be adsorbed on DEAE-Sephacel at pH 8.0 and then eluted with a sodium chloride gradient (0.01 M - 0.16 M). Twenty-five percent of the original activity is recovered from this column. Active fractions are approximately 100 times more active than fetal calf serum in promoting SV3T3 growth. When the active fractions are pooled and subjected to column preparative isoelectrofocusing, the SV3T3 growth promoting activity ($PDGF_2$) has an isoelectric point between pH 7.8 - 8.3 (Figure 4). In this regard, $PDGF_2$ differs from $PDGF_1$ which focuses between 9.6 and 10.2, with a peak of activity at pH 9.8. It should also be mentioned that

partially purified fractions from Figures 2-4 that contain
$PDGF_2$ activity do not induce competence in quiescent 3T3
cells.

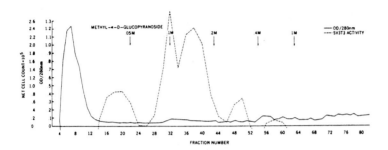

Figure 3. Con A-Sepharose Chromatography of $PDGF_2$ Activity. Crude platelet extract was prepared as described in the legend to Figure 1. This material was then concentrated four-fold by lyophilization and six ml. was placed on a 1 x 5 cm Con A-Sepharose column, .01 M Tris- .01 M NaCl, pH 8.0 at 4°C. After 100 ml were collected, a step wise gradient using .05, .1, .2, .4, and 1 M methyl-α-D-glucopyranoside as indicated was employed. 2.5 ml fractions were collected and sterilized. 0.3 ml each column fraction was added to SV3T3 cells in a standard growth assay (Legend to Figure 1).

The preceding work was performed using cells transformed by the SV40 viral genome. Another means by which transformation can be induced is by incubation of cell cultures with chemical carcinogens. These cells, too, no longer grow in orderly fashion, have a decreased serum requirement, and result in tumors when injected into animals. Treatment of an epithelial rat liver cell line, K-16, with N-acetoxy-2-acetylamino-fluorene produced a transformed cell line, W-8. This line is stimulated by fractions which contain $PDGF_2$ activity. Similarly, cell line NQ-T-1, derived from Balb-c 3T3 cells by treatment with 4-Nitroquinoline 1-oxide, was stimulated to multiply by fractions containing both $PDGF_1$ and $PDGF_2$ activity while the parental Balb-3T3 line is stimulated only by fractions containing $PDGF_1$ activity.

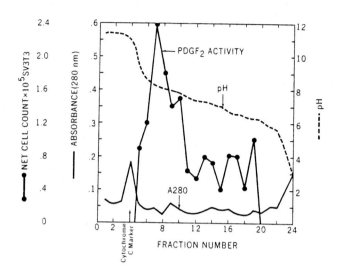

Figure 4. Preparative isoelectrofocusing column of PDGF$_2$ activity from DEAE-Sephacel. Four ml of PDGF$_2$ activity from a DEAE-Sephacel column were dialyzed in 1% glycine and focused on a LKB electrofocusing column (110 ml) according to the manufacturer's instructions. One % ampholine, pH range 7 to 10, was used; and cytochrome c with a pI of 10.6 was used as a marker. The samples to be tested were collected in 5-ml fractions, dialyzed with 0.85% NaCl to remove sucrose, and sterilized using 0.22-µM Millipore filters. The SV3T3 cell growth assay to determine PDGF$_2$ activity was performed as described in Legend to Figure 1 using 0.5 ml of each fraction.

Human platelet lysate has also been shown to stimulate the multiplication of rat epithelial mammary tumor cells (Eastment and Sirbasku, 1978) and many cancer-derived cell lines (Eastment and Sirbasku, 1980; Hara et al, 1980). Similar to our results described above these authors describe the platelet growth factor for these transformed cells as a heat labile, non-dialyzable glycoprotein (Eastment and Sirbasku, 1980; Hara et al, 1980).

We next performed experiments to learn if the megakaryocyte was a source of PDGF$_2$ activity. Adult Fisher rats were used as the source of bone marrow. Marrow was obtained by flushing the marrow cavity of the femur. Marrow extracts were made by freezing and thawing the marrow suspension three times in order to lyse the cells. Such marrow extracts contain a factor(s) that promotes the growth of SV3T3 cells. The amounts of activity obtained from whole marrow extracts appears to be similar to that obtained from platelets and whole serum on a milligram of protein basis.

The next question was whether such activity present in marrow extracts could be atrributed to the megakaryocyte fraction. Enriched megakaryocyte preparations can be obtained using velocity sedimentation in an isokinetic gradient of Ficoll and tissue culture medium (Pretlow and Stinson, 1976). Fractions containing 1-2 per cent of the total cell number as megakaryocytes can be obtained (Figure 5). Individual fractions from this gradient were pooled into 5 sections. The megakaryocyte-rich protein portion (Figure 5, Pool II) could easily be separated from the bulk of the marrow cells (Pool IV). When pooled groups were assayed there was little mitogenic activity for SV3T3 cells ($PDGF_2$ activity) in the "megakaryocyte-rich" Pool II (Table 1). Thus, we have not been able to show that megakaryocytes contain or are an enriched source of $PDGF_2$ activity.

TABLE 1

$PDGF_2$ ACTIVITY IN BONE MARROW FRACTIONS

	SV3T3 (cells x 10^5)
Control	0.78
Pool I	1.10
Pool II	1.12
Pool III	1.92
Pool IV	2.87
Pool V	3.53

Legend. Velocity Sedimentation of Rat Marrow on Isokinetic Ficoll gradients was carried out as described in legend to Figure 5. Various fractions of gradients from 12 rat femurs were pooled. 0.02 mgm protein from each pool was added to Standard Growth Assay for SV3T3 cells as described in the Legend to Figure 1.

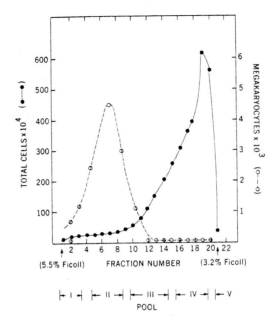

Figure 5. Preparation of "Enriched-Megakaryocyte" pool from Rat Marrow Suspension by Velocity Sedimentation on an Isokinetic Ficoll gradient.[23] Marrow was obtained from Fisher Adult Rat Femur. This was layered on a 3.2% - 5.5% Ficoll gradient. The gradient was then centrifuged at 300 RPM (16 x g) for 10 minutes. 1.5 ml fractions were collected.

It is thus likely that $PDGF_2$ is made elsewhere in the body - perhaps in the liver (Lipton et al, 1975; Witkoski et al, 1979) - and transported by the circulating platelet. $PDGF_2$ activity is found in both the α and dense granules of the platelet when these granules are isolated by sedimentation through a sucrose gradient (Salganicoff and Lipton, unpublished observations).

All the experiments described to this point were performed with cells grown in tissue culture. We next asked the question, can platelet extracts accelerate tumor growth in the whole animal? Oncogenic transformation of hamster embryo fibroblast (HEF) cells has been accomplished using ultraviolet (UV)-irradiated herpes simplex virus type 2 (*HSV-2) (Duff and Rapp, 1971; Duff and Rapp, 1976). Such viral transformed cells regularly produce tumors when injected into random-bred weanling Syrian-hamsters (3-4 week old). We chose this model system because HSV-2 is a putative human tumor virus and the regularity with which tumors can be produced.

1×10^2 HSV-2 transformed HEF cells were injected into the cheek pouch of fifty random-bred weanling female Syrian Hamsters (3-4 week old) on Day 1 only. The viability of injected cell suspensions was checked by trypan blue exclusion and was greater than 90 percent. This dose of cells was chosen for injection because in control experiments 85% of the animals receiving 10^4 cells developed tumors by week 4 after injection. In addition to tumor cells the initial injection contained either (a) 0.2 ml Krebs-Henseleit buffer (10 animals), (b) 0.2 ml Dulbecco's medium (10 animals) and (c) 0.2 ml crude platelet extract (15 animals). Repeat injections of 0.2 ml of a) to c) were made into the appropriate animal's cheek pouch on Days 2, 4, and 6 without cells.

Platelet extract for these experiments was prepared by pooling thirty to eighty units of outdated human platelet concentrates. This pool was spun at 163 g for 10 minutes to remove RBC's. They were respun at 5875 g for 10 minutes. The platelet button was washed and centrifuged 5 times at 5875 g. Each wash was with 400-800 ml isotonic saline. The final platelet button was resuspended in 40-80 ml isotonic saline. Washed platelet button were incubated at 37°C for 30 minutes, then centrifuged at 16,319 g. The supernatant (crude platelet extract) was concentrated four-fold by lyophilization and sterilized using a 0.22 μ Millipore filter.

Figures 6A and 6B summarize the results in hamsters injected with 1×10^2 HSV-2 transformed HEF cells in the cheek pouch. Each experiment was performed three times. Animals that received concomitant injections of either Krebs-Henseleit buffer or Dulbecco's medium did not develop tumors over the 15 week period of observation. Animals that were not injected with transformed cells and received only $PDGF_2$ also did not develop tumors. Thirty-three percent of hamsters injected with platelet extract developed palpable tumors by week 7 (Figure 6A). At the termination of the experiment at week 14, sixty percent of hamsters injected with platelet extract developed readily visible tumors at the injection site (Figure 6B). Tumors were

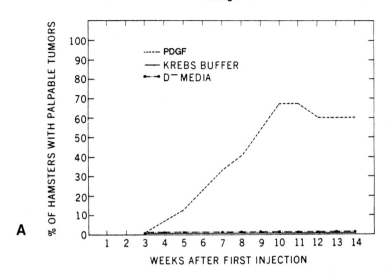

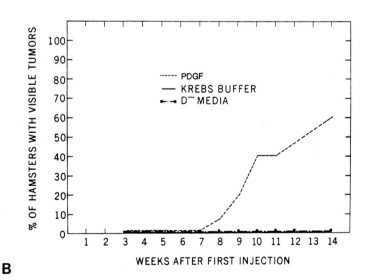

Figure 6. Effects of $PDGF_2$ on Hamster Tumors Formed by HSV-2 Transformed HEF Cells.
A. Percentage of hamsters with palpable tumors after injection with the transformed cells and $PDGF_2$, Krebs-Henseleit Buffer, or Dulbecco's Medium.
B. Percentage of hamsters with visible tumors after injection with the transformed cells and $PDGF_2$, Krebs-Henseleit Buffer and Dulbecco's Medium.

anaplastic spindle cell tumors with some epitheloid differentiation. These results were highly significant when compared to animals injected with Krebs buffer or Dulbecco's medium, p <.001. At the time they were killed, all hamsters were free of disseminated tumors.

The mechanism of enhancement of tumor growth is speculative at present. We do feel that it is due to a direct stimulative effect resulting in enhanced multiplication of tumor cells and not due to immunosuppression of the host. The reason for this is that the hamster cheek pouch is an immunologically privileged sanctuary.

In conclusion, human platelets contain discrete macromolecules that promote the growth of 3T3 ($PDGF_1$) and SV3T3 ($PDGF_2$). Both appear to be proteins. The factor that promotes the growth of SV3T3 and a variety of other transformed cell lines is heat labile, non-dialyzable and has an isoelectric point of 7.8 - 8.3. Platelet extracts that contain $PDGF_2$ activity can promote the in vitro as well as in vivo growth of transformed cells. These observations would appear to offer several new strategies for cancer therapy such as (a) selective inhibition of the multiplication of cancer cells by removal or inhibition of the growth stimulatory glycoprotein $PDGF_2$ - e.g. specific antibody to inhibit action or removal on appropriate ion exchange or adsorption resin, and (b) use of the stimulatory features of $PDGF_2$ to promote tumor cells to enter a dividing state in which all the cells would be susceptible to cycle-active chemotherapy. At present many human tumors are refractory to chemotherapy because cells are in a dormant or non-dividing state.

REFERENCES
  Antoniades HN, Scher CD, Stiles CD (1979). Purification of human platelet-derived growth factor. Proc Natl Acad Sci (USA) 76:1809.
  Balk SD (1971). Calcium as a regulator of the proliferation of normal, but not of transformed, chicken fibroblasts in a plasma-containing medium. Proc Natl Acad Sci (USA) 68:271.
  Chernoff A, Goodman DS, Levine RF (1980). Origin of platelet-derived growth factor in megakaryocytes. J Clin Invest 65:926.
  Clemmons DR, Van Wyk JJ (1981). Sometomedin-C and platelet derived growth factor stimulate human fibroblast replication. J Cell Physiol 106:361.

Duff R, Rapp F (1971). Oncogenic transformation of hamster cells after exposure to herpes simplex virus type 2. Nature 233:48.

Duff R, Rapp F (1976). Properties of hamster embryo fibroblasts transformed in vitro after exposure to ultraviolet-irradiated herpes simplex virus type 2. J Virology 8:469.

Eastment CT, Sirbasku DA (1978). Platelet derived growth factor(s) for a hormone-responsive rat mammary tumor cell line. J Cell Physiol 97:17.

Eastment CT, Sirbasku DA (1980). Human platelet lysate contains growth factor activity for established cell lines derived from various tissues of several species. In Vitro 16(8):694.

Gospodarowicz D (1974). Localization of a fibroblast growth factor and its effect alone and with hydrocortisone on 3T3 cell growth. Nature 249:123.

Gospodarowicz D, Moran J (1974). Stimulation of division of sparse and confluent 3T3 cell populations by fibroblast growth factor, dexamethasone and insulin. Proc Natl Acad Sci (USA) 71:4584.

Hara Y, Steiner M, Baldoni MG (1980). Platelets as a source of growth-promoting factor(s) for tumor cells. Cancer Res 40:1212.

Heldin CH, Wasteson A, Westermark B (1977). Partial purification and characterization of platelet factors stimulating the multiplication of normal human glial cells. Exp Cell Res 109:429.

Heldin CH, Westermark B, Wasteson A (1979). Platelet-derived growth factor purification and partial characterization. Proc Natl Acad Sci (USA) 76:3722.

Heldin CH, Westermark B, Wasteson A (1981). Platelet derived growth factor. Isolation by a large-scale procedure and analysis of subunit composition. Biochem J 193:907.

Jennings T, Jones R, Lipton A (1979). A growth factor from spinal cord. J Cell Physiol 100:273.

Kaplan KL, Broekman MJ, Chernoff A, Lesznik GR, Drillings M (1979). Platelet α-granule proteins: Studies on release and subcellular localization. Blood 53:604.

Kepner N, Lipton A (1981). A mitogenic factor for transformed fibroblasts from human platelets. Cancer Res 41:430.

Kepner N, Creasy G, Lipton A (1978). Platelets as a source of cell-proliferating activity. In G. de Gaetano, S Garattini (eds.). Platelets: A Multidisciplinary Approach, New York Raven Press, p 205.

Kohler N, Lipton A (1974). Platelets as a source of fibroblast growth promoting activity. Exp Cell Res 87:297.

Lipton A, Roehm CJ, Robertson JW, Dietz JM, Jefferson LS (1975). Release of a growth factor for SV40 virus transformed cells by rat liver and hemicorpus perfusions. Exp Cell Res 93:230.

Pledger W, Stiles C, Antoniades H, Scher C (1978). An ordered sequence of events is required before Balb/c 3T3 cells become committed to DNA synthesis. Proc Natl Acad Sci (USA) 75:2839.

Pretlow TG, Stinson AJ (1976). Separation of megakaryocytes from rat bone marrow cells using velocity sedimentation in an isokinetic gradient of ficoll and tissue culture medium. J Cell Phy 88:317.

Ross R, Glomset B, Kariya B, Harker L (1974). A platelet-dependent serum factor that stimulates the proliferation of arterial smooth muscle cells in vitro. Proc Natl Acad Sci (USA) 71:1207.

Vogel A, Raines E, Kariya B, Rivest M-J, Ross R (1978). Coordinate control of 3T3 cell proliferation by platelet-derived growth factor and plasma components. Proc Natl Acad Sci (USA) 75:2810.

Weiss HJ, Lages BA, Wittle LD, Kaplan KL, Goodman DS, Nossel HL, Baumgartner HR (1977). Storage pool disease:Evidence of clinical and biochemical heterogeneity. Thromb Haemost 38:3.

Witkoski E, Schuler MF, Feldhoff RC, Jacob SI, Jefferson LS, Lipton A (1979). Liver as a source of transformed cell growth factor. Exp Cell Res 124:261.

Witte LD, Kaplan KL, Nossel HL, Lages BA, Weiss HJ, Goodman DS (1978). Studies of the release from human platelets of the growth factor for cultured human arterial smooth muscle cells. Circ Res 42:402.

DISCUSSION

Dr. Mustafa A. Hay (El Safat, Kuwait). Have you tried the mitogen on T cells?

Dr. A. Lipton (The Milton S. Hershey Medical Center, Hershey). No, we have not tried those experiments. I have no idea at this point of the relationship of this particular material to other described T cells mitogenic factors.

Dr. L. R. Zacharski (VA Medical Center, White River Junction). Would you comment on the role and efficacy of common platelet inhibitory drugs in affecting growth factors?

Dr. Lipton. I don't know of anybody who has done those experiments. They would be relatively difficult since you are dealing with all the biological vagaries in terms of cell differentiation and platelet differences. These are very important experiments which may be best saved until one has a specific assay for each of the factors.

Dr. J. L. Ambrus (Roswell Park Memorial Institute, Buffalo). We are talking about the possibility of agonizing these platelet factors and yet, I think, it is something we are doing all the time when we apply chemotherapy to our patients. We push chemotherapy for six weeks, produce thrombocytopenia and thereby inhibit the delivery of platelet growth factors.

Dr. Lipton. You may be right but I am not sure; it needs to be looked at very carefully in clinical settings.

Dr. R. Greig (Smith, Kline and French, Philadelphia). Did you compare the vascularization in the tumors in animals?

Dr. Lipton. We have noticed no difference in vascularization. The extract does not cause tumors unless you coinject it with cancer cells and we saw no inflammation in animals that were treated with that particular factor. We feel that these growth factors are present normally in certain tissues in the body such as the platelet, the liver, etc. I think that we are probably looking at a spectrum where certain cancer cells produce, perhaps, their own factor at times while other cancer cells may be totally dependent on extraneous factors.

Dr. Zacharski. I have one comment: Handin's group looked for platelet factor 4 by immunofluorescence. There were no intact platelets and yet the platelet had been there and deposited its contents so that they detected it immunologically. I wonder if this kind of thing could not be happening in the present case? There is no compelling necessity for having platelets present in order to contribute to growth.

Dr. Lipton. Your comment is well taken. We are certainly way out on a limb when we begin to speculate about mechanism. We would like to have you believe that the platelet contains at least two growth factors and that they may be more specific for transformed cells, that is all.

EFFECT OF PLATELET GROWTH FACTOR(S) ON GROWTH OF HUMAN
TUMOR COLONIES

Dale H. Cowan and Joyce Graham

St. Luke's Hospital and Case Western Reserve
University School of Medicine
Cleveland, Ohio 44104

A fundamental property of neoplastic cells is the capacity for uncontrolled proliferation. The possibility that blood platelets may influence the growth of tumor cells derives from a series of observations indicating that platelets release one or more growth factors after appropriate stimulation. Balk (1971) observed that virally transformed chick embryo fibroblasts proliferated rapidly in low calcium, platelet-poor plasma medium, whereas normal chick fibroblasts failed to proliferate. Replacement of serum for plasma resulted in proliferation of the normal fibroblasts. Balk and co-workers (1973) suggested that serum contains growth-promoting activity that is lacking in plasma.

Kohler and Lipton (1974) observed that platelets were a source of the growth-promoting activity for 3T3 mouse fibroblasts that was present in serum but not plasma. At the same time, Ross and co-workers (1974) reported that platelets were a major source of growth-promoting activity for monkey aortic smooth muscle cells. Such activity was lacking in serum prepared from platelet-poor plasma.

Westermark and Wasteson (1976) were the first to report that a platelet-derived growth factor (PDGF) stimulated the proliferation of human cells, specifically, glial cells of astrocytic origin. Castor and co-workers (1977) subsequently reported that a similar factor stimulates the growth of human synovial cells in vitro.

The platelet growth factor reportedly responsible for stimulating normal target cells was extensively purified

and characterized by Ross and co-workers (1978) and by
Antoniades and co-workers (1979). It is a heat-stable,
cationic polypeptide having a molecular weight in the range
of 13,000 - 16,000 daltons (Table 1). Its activity is lost
after digestion with trypsin or reduction with 2-mercapto-
ethanol.

TABLE 1

CHARACTERISTICS OF PLATELET-DERIVED
GROWTH FACTOR(S)

PROPERTIES	PDGF - 1	PDGF - 2	PDGF-3 (?)
TARGET CELL	NORMAL	TRANSFORMED	MALIGNANT
MOL WT	13,000-16,000	~72,000	30,000-50,000
IEP	9.8	7.8 - 8.3	?
ACTIVITY LOST AFTER:			
HEATING			
56°	No	No	No
70°	No	Yes	Yes
100°	No	Yes	Yes
TRYPSIN DIGESTION	Yes	PARTIAL DECREASE	PARTIAL DECREASE
PERIODATE OXIDATION	N.D.	Yes	Yes

In his reports, Balk (1971, 1973) had noted that chick
embryo fibroblasts transformed by Rous sarcoma virus did not
appear to need PDGF for long-term growth in culture since it
proliferated equally well in medium supplemented with plasma
or serum. Kohler and Lipton (1974) observed that SV40-trans-
formed Swiss 3T3 cells also grew equally well in medium supplemented with plasma or serum. Scher and co-workers (1978)
reported that a large number of clonal lines of both mouse
3T3 fibroblasts and human fibroblasts transformed by SV40 or
retroviruses had a similarly reduced requirement for PDGF.

Kepner and Lipton (1981) recently reported that, in
contrast to the PDGF for normal cells, the PDGF for trans-

formed cells is destroyed by heating at 100°C for 5 minutes, has an isoelectric point of 7.8 - 8.3, and has a molecular weight of 72,000 daltons (Table 1). It is destroyed by treatment with sodium periodate, partially destroyed by trypsin digestion, and not affected by exposure to 4 M guanidine hydrochloride. It has chemical properties of a glycoprotein and appears to be distinguishable from the factor that stimulates normal cells. They termed PDGF with activity for normal cells $PDGF_1$ and that with activity for transformed cells $PDGF_2$.

Eastment and Sirbasku (1978) showed that extracts prepared from pooled outdated human platelets stimulated the growth of MTW9/PL rat mammary tumor cells in culture. In a later publication, they reported that platelet growth factor(s) supported the growth of 24 of 29 malignant cell lines obtained from human, monkey, mouse, rat, chicken, Chinese hamster, and Syrian hamster sources (Eastment and Sirbasku, 1980). Hara and co-workers (1980) also observed that human platelet lysate promoted the proliferation of cells of well-established cell lines from rodents.

The factor responsible for supporting the growth of the malignant cell lines in culture is nondialyzable, partially destroyed by trypsin digestion, and completely destroyed by heating at 100°C for two minutes and by exposure to periodate (Eastment and Sirbasku, 1980, Hara and co-workers, 1980). It has a molecular weight of 30,000 - 50,000 daltons (Eastment and Sirbasku, 1978)(Table 1).

Based on these observations and those reported above, several authors have suggested that there are at least three and possibly more growth factors derived from platelets (Scher and co-workers, 1979, Eastment and Sirbasku, 1980). Antoniades and co-workers have suggested that the different sizes reported for the growth factors studied by different investigators may reflect variations in the techniques used in their preparation and purification. However, Eastment and Sirbasku (1980) showed that the factors having different molecular weights displayed distinctive target cell activity: the epithelial tumor cell growth activity was separable from the 3T3 cell activity in crude heated extracts.

The effect of PDGF on quiescent, density inhibited BALB/c3T3 cells is to induce these cells to become competent to respond to a second factor or factors (Pledger and co-workers,

1977). These second factors are termed progression factors and induce the cells to progress through G_0/G_1, enter S phase, and divide (Pledger and co-workers, 1978). A similar coordinated effect stimulates cell proliferation of density-inhibited Swiss/3T3 cells. At least one of the potent mediators of progression that acts with platelet growth factor to stimulate cell proliferation of 3T3 cells is somatomedin C (Stiles and co-workers, 1979). Although hormones promote the growth of malignant cells in culture (Eastment and Sirbasku, 1978), there are as yet no reported studies of a coordinate effect of PDGF and hormones on the proliferation of malignant cells in vitro.

To date, there are no published reports of the effects of platelet-derived growth factor(s) on the growth in culture of human tumor cells prepared from fresh surgical specimens. This may have been due in part to difficulties and limitations affecting techniques for culturing human tumor cells in vitro. The development of an in vitro assay for human tumor stem cells by Hamburger and co-workers (1977, 1978, 1979) working in Salmon's laboratory in Arizona (Salmon, 1980) created the opportunity for studying the effect of PDGF on human tumor cells. Their assay was based on earlier work done at the Ontario Cancer Institute (see, for example, Park and co-workers, 1971) and the concept that there is a subpopulation of stem cells in every renewal tissue in an adult. The hypothesis underlying the assay is that only a small proportion of cells in primary tumors are potential stem cells. (Buick, 1980).

Evidence that the colonies that grow in the semisolid two-layer agar culture system described by Hamburger and Salmon (1977) are derived from tumor stem cells was provided by Salmon (1980) and Von Hoff and co-workers (1980). They used histologic features, cytogenetic markers, and the detection of tumor-associated products to show that the colonies growing in the semisolid matrix were of neoplastic origin. Presently, absent a specific interest in identifying a particular marker in the tumor cell colonies, cytologic analysis is the major method used to certify the neoplastic origin of the colonies growing in the cultures (Von Hoff, personal communication).

In the present study, advantage was taken of the fact that one can expose tumor cells to various factors continuously after plating them in culture. The studies described here were done with a crude platelet lysate present continuously in the underlayer of the two-layer culture system.

MATERIALS AND METHODS

Patient Samples

Tissue specimens were obtained at surgery from patients undergoing routine diagnostic or therapeutic procedures. Standard histopathological examinations were conducted in the Pathology Department to determine the histologic identity of the tissues.

Preparation of Cell Suspensions

Pieces of solid tumor were placed in sterile, iced Hank's balanced salt solution (HBSS) with 10% fetal calf serum (Microbiological Associates, Rockville, MD) and taken to the laboratory where they were minced to approximately 1mm cubes with scissors or scalpel and pushed through a fine mesh wire screen (Cell Separator, Bellco, Vineland, NJ). The resulting suspensions were passed through needles of decreasing size to 25-gauge and washed by centrifugation as described by Hamburger and co-workers (1978). Smears of the resulting cell suspensions were prepared on glass slides, stained by the Papanicoulau method (Hamburger and co-workers, 1978) and examined to determine the presence of cell clumps. Only single cell suspensions were used in this study; suspensions containing cell clumps were discarded. The viability of the cells in suspension was determined in a hemacytometer using trypan blue. Calculations of the number of cells plated were always based on the number of viable cells in suspension.

Assay for Colony Formation

Cells were cultured using the method described by Hamburger and Salmon (1977). The cells were suspended in 0.3% Bacto agar (Difco) in CMRL 1066 supplemented with 15% horse serum, 100u/ml penicillin, 2mg/ml streptomycin, 4mM glutamine, 4mM $CaCl_2$, 0.3mM ascorbic acid, and 2u/ml insulin. Immediately before the cells were plated, 0.1mg/ml asparagine, 0.375mg/ml DEAE dextran, and 5×10^{-5} M 2-mercaptoethanol were added. The final concentration of viable cells ranged from 1-5 x 10^5 cells/ml. One ml of the cell suspension was layered onto one ml feeder layer in 35mm plastic petri dishes.

The feeder layer consisted of McCoy's Medium 5A with or without platelet-derived growth factor at a final concentration of 200µg/ml, plus 10% heat-inactivated fetal calf serum, 5% horse serum, and a variety of nutrients as described by Pike and Robinson (1970). McCoy's medium without PDGF was used for control cultures. Immediately before use 0.2% tryptic soy broth (Grand Island Biological Co.), 0.1mg/ml asparagine, 0.375 mg/ml DEAE-dextran, and 0.5% agar (final concentrations) were added to the enriched medium. The underlayers were poured into the petri dishes promptly after addition of the agar.

After preparing the final 2-layer culture system, the plates were examined under an inverted microscope to assure the presence of a good single-cell suspension. The plates were then incubated at 37°C in a 7.5% CO_2 humidified atmosphere.

Scoring of Cultures

At the end of the culturing period, Hank's balanced salt solution was layered on top of the cultures for 10 minutes at room temperature. After it was removed, 10% formaldehyde in HBSS, pH 7.2, was placed on the cultures for 10 minutes, followed by deionized water for 10 minutes. The top layer containing the cell colonies was floated off into a cup of deionized water, then onto a glass slide, covered with strips of methylcellulose and air dried at room temperature. The cell layer was stained by the Papanicoulau method.

Colonies were counted using a Zeiss mecroscope. Colonies consisted of coherent groups of 30 or more cells (Figure 1). They were readily distinguishable from cell aggregates or clumps. Colony formation in vitro was defined as the presence of 5 or more colonies per plate. To determine the variance in colony formation, cells from 3 different tumors were plated in 10 replicate plates without PDGF. After 3 or 6 days of culturing, the colonies were counted. The variance in colony formation in the replicate plates was less than 10%. The plating efficiency was defined as the number of colonies formed divided by the number of viable cells plated. Stimulation or inhibition of colony formation in the presence of PDGF was defined as a 20% or greater change in the number of colonies present compared to control cultures.

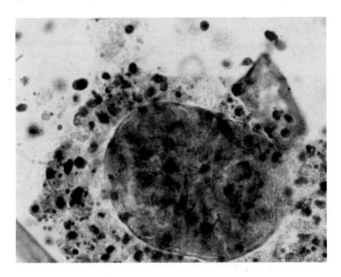

Figure 1. Photomicrograph of a 3-day colony of human colon carcinoma cells (Papanicoulau stain) grown in soft agar (x 640).

Preparation of Human Platelet Lysate

Platelet lysate was prepared either from freshly drawn blood of healthy human donors anticoagulated with 0.55 M sodium citrate or from outdated platelet concentrates obtained from the Northern Ohio Red Cross Blood Program, Cleveland, Ohio. Plastic equipment was used in all stages of the preparation. Red cells were removed by centrifugation. The platelets in the supernatant were sedimented by centrifugation at 1250 x g for 30 minutes at 4°C, and washed twice in phosphate buffered saline (PBS), pH 7.2, containing 0.1% EDTA. The platelets were resuspended in PBS and lysed by 5 freeze and thaw cycles using liquid nitrogen. The lysate was centrifuged at 100,000 x g for 1 hour at 4°C, heat inactivated at 56°C for 20 minutes, dialyzed overnight against PBS with 0.1% EDTA at 4°C, and stored frozen (-20°C) until use. The protein concentration of the lysate was measured by the method of Lowry and co-workers (1951).

Modification of Platelet Lysate

The platelet lysate was subjected to several modifying procedures to examine its growth-promoting activity. These were described by Hara and co-workers (1980) and consisted of (1) dialysis against 3 changes of 40 or more volumes of 0.9% NaCl for 24 hours at 4°C, (2) incubation at 56°C, 70°C, or 90°C for 30 minutes, (3) incubation with trypsin (EC 3.4.21.4, specific activity, 8,000 units/mg protein; Sigma Chemical Co., St. Louis, MO), 0.5mg/mg platelet protein for 3 hours at pH 8.5 and 37°C, stopped by soybean trypsin inhibitor in a weight ratio of 1:1.5 (enzyme:inhibitor), and (4) incubation with 0.05M sodium metaperiodate (Sigma) at pH 5.5 in the dark for 3 days at 4°C, followed by dialysis to remove periodates.

Release of Growth-Promoting Activity from Platelets

Citrate anticoagulated platelets were washed twice with Ca^{++}-free Tyrode's solution, resuspended in Tyrode's solution containing 1.8mM Ca^{++} and treated with 200μg/ml acid soluble collagen (Ethicon Corp. Somerville, NJ) or 2u/ml thrombin (bovine topical thrombin, Sigma Chemical Co., St. Louis, MO) with constant stirring at 1100 rpm at 37°C for 10 minutes, as described by Hara and co-workers (1980). Cellular material was sedimented by centrifugation, and the supernatant was tested for growth-promoting activity. The supernatant obtained after centrifuging the thrombin treated platelets was heat-inactivated at 56°C for 30 minutes, and frozen at -20°C as with the freeze-thaw lysate. The protein concentration was determined by the Lowry method before freezing.

Biological Assay for Platelet-Derived Growth Factor(s)

The growth-promoting activity of the modified and unmodified platelet lysates and of the supernatants prepared after stimulation with collagen and thrombin was assayed using Balb/c3T3 mouse fibroblasts and SV40 transformed 3T3 fibroblasts (SV3T3) as described by Vogel and co-workers (1978). The cells were suspended in the McCoy's 5A enriched medium used for the underlayer of the 2-layer agar system. They were plated in 35mm wells in plastic dishes and grown to confluence in 7.5% CO_2 at 37°C. After reaching confluence, the medium was removed and replaced by McCoy's 5A modified to contain only 0.4% fetal calf serum.

The cells sat in a resting stage for 48 hours after which the preparations of platelet growth factor(s) were added at final concentrations of 100 and 200µg/ml in HBSS. Control cultures were treated with HBSS without platelet factors. The cells were then grown in 7.5% CO_2 at 37°C for 20 hours, 50µCi tritiated thymidine (^{3}HTdR, 50-60 mCi/M, New England Nuclear NET-027 Z, Boston, MA) was added, and the cells were incubated for 4 additional hours. Growth was stopped by adding 5% trichloroacetic acid (TCA). Sodium hydroxide, 0.5M, was added to dissolve the cells off the plates. The solution was placed in Scintiverse (Fisher Scientific Co.) and radioactivity was counted in a liquid scintillation counter (Beckman).

RESULTS

Assay and Characterization of PDGF

The activity of growth factor(s) prepared by different methods is shown in **Table 2**. Freeze-thaw lysis and collagen stimulation yielded preparations having similar stimulatory activity for 3T3 cells. The activity obtained after thrombin stimulation was approximately one-half that obtained by the other methods.

PDGF obtained with collagen and thrombin stimulation promoted the uptake of ^{3}HTdR by SV40 transformed 3T3 fibroblasts to an extent similar to that observed with non-transformed cells. By contrast, the stimulation of transformed fibroblasts by PDGF obtained from freeze-thaw lysis was approximately twice that of non-transformed cells.

The effect of modifying PDGF by different treatments is shown in Table 3. Similar results were observed using PDGF obtained by freeze-thaw lysis and collagen stimulation. The activity of PDGF was unaffected by dialysis against 0.9% sodium chloriade but was completely destroyed by heating to 70°C or 90°C and by periodate oxidation. Trypsin digestion resulted in one-third less activity using 3T3 cells and two-thirds less activity using SV40-transformed 3T3 cells.

TABLE 2

ASSAY OF PLATELET-DERIVED GROWTH FACTOR(S) OBTAINED BY DIFFERENT METHODS*

	INCREASE IN UPTAKE OF ^{3}HTdR RELATIVE TO CONTROL	
	%	
PREPARATION	3T3	SV3T3
FREEZE-THAW LYSATE	80	167
COLLAGEN RELEASED	96	77
THROMBIN RELEASED	41	45

*Results obtained using 100µg platelet protein.

TABLE 3

EFFECT OF MODIFYING PROCEDURES ON ACTIVITY OF PLATELET-DERIVED GROWTH FACTOR(S)

TREATMENT	ACTIVITY (% OF CONTROL)	
	3T3	SV3T3
DIALYSIS	110	114
HEATING		
56°C	100	100
70°C	0	0
90°C	0	0
TRYPSIN DIGESTION	63	33
PERIODATE OXIDATION	0	0

Growth of Tumor Colony Forming Units (TCFU)

Cells from 207 malignant neoplasms were cultured in the 2-layer agar system. Five or more colonies grew from 147 (71%) of the tumors plated. The source of the tumors and the number of each type which grew colonies are listed in Table 4. The viability of the tumor cells varied from 10% to 90%. Large differences in viability were observed with tumors of similar cell types obtained from different patients. The viability of cells from primary and metastatic sites was similar and did not correlate with the presence or absence of colony formation in vitro.

The plating efficiency of cells from which colonies grew varied from 0.001-0.35%. There was no relationship between the plating efficiency, the type of neoplasm, or the source of the cells (i.e. primary or metastatic site).

TABLE 4

COLONY FORMATION OF MALIGNANT TUMOR
CELLS IN A TWO-LAYER SOFT AGAR CULTURE SYSTEM

	PLATED	GROWN*
BREAST	42	32
LUNG	28	22
COLON	24	16
RENAL	20	15
OVARIAN	19	15
SARCOMA	11	7
UTERUS	9	3
HEAD & NECK	7	7
STOMACH	6	3
MELANOMA	5	3
OTHER SITES†	36	24
TOTAL	207	147

*Five or more colonies per plate.

†Includes tumors from prostate, testis, bladder, thyroid, brain, and esophagus, lymphomas, and from unknown primaries.

Effect of PDGF Obtained by Freeze-Thaw Lysis on Human Tumor Colony Formation

The effect of platelet lysate (200µg protein per ml in the underlayer) on human tumor colony formation is shown in Table 5.

TABLE 5

EFFECT OF PLATELET LYSATE
ON TUMOR COLONY GROWTH*

	INCREASE	DECREASE	NONE
BREAST	19	6	7
LUNG	13	8	1
COLON	9	5	1
RENAL	7	7	1
OVARIAN	9	4	2
HEAD & NECK	5	2	0
SARCOMA	4	3	0
UTERUS	3	0	0
OTHER SITES	15	10	0
TOTAL	**84**	**45**	**12**

*The numbers denote the number of cultures of each cell type showing the designated change.

As compared to cultures grown without lysate in the underlayer, the number of tumor colony forming units (TCFU) increased 20% or more in the presence of platelet lysate in 84 (60%) of 141 tumors. The extent of the increases varied from 20 to 175%. There was no distinctive pattern of response for tumors of specific cell types. Platelet lysate did not induce colony formation from tumors which failed to grow without lysate.

The number of TCFU decreased 20% or more below control values in 45 (32%) of the tumors. The extent of the decreases was 20 to 100% and did not correlate with specific tumor types. Twelve (8%) of the tumors which grew in culture showed no response to platelet lysate.

There was no correlation between the viability of the tumor cells before plating and the type or extent of response to the platelet lysate.

In order to determine the time required for maximum response to PDGF, replicate cultures were plated when sufficient numbers of viable cells were present in the tumor cell suspensions. Culturing was stopped and colonies were counted three, six, and ten days after plating. The maximum effect (increase or decrease) was observed after three, six, and ten days in 75 (58%), 51 (40%), and 3 (2%) of 129 tumors, respectively. Cultures having maximum positive responses to platelet lysate in three or six days often showed fewer colonies after ten days compared to control plates. In addition, the colonies in these ten day cultures generally contained more necrotic-appearing cells.

Effect of PDGF Obtained by Collagen and Thrombin Stimulation on TCFU

To determine whether PDGF obtained by collagen stimulation induced the same response as that seen using platelet lysate, 20 tumors were plated in duplicate using PDGF prepared by freeze-thaw lysis and by collagen stimulation. The concentration of platelet-derived protein in the underlayers was 200µg/ml. The tumor cells for these experiments were obtained from lymphomas and from carcinomas of the breast, ovary, prostate, urinary bladder, colon, lung, and kidney. In each case, the types of responses observed using the two sources of PDGF were identical. The extent of the responses observed with PDGF in platelet lysate exceeded that of PDGF from collagen stimulation by 20-80% except in the case of three tumors where the opposite occurred.

Similar studies were done using four tumors to compare the effect of PDGF obtained by thrombin stimulation with that of PDGF obtained by freeze-thaw lysis. The concentration of platelet-derived protein in the underlayer of each test culture was 200µg/ml. The cells tested were prepared from cancers of the breast, ovary, and head and neck. The effect of PDGF obtained from thrombin stimulation was identical in type and extent to that observed using PDGF obtained by platelet lysis.

Effect of Modifying PDGF

Twenty-six tumors were cultured in parallel using untreated platelet lysate and platelet lysate that was dialyzed overnight against 0.9% NaCl at $4^{\circ}C$. The response of TCFU to the dialyzed lysate was similar in type and extent to that observed with untreated lysate.

Sixteen tumors were cultured in parallel with untreated platelet lysate and with lysate that had been heated to $70^{\circ}C$ or $90^{\circ}C$ for 5 minutes. In each case, the effect of the platelet lysate was not observed after heating. Similarly, the effect of the lysate on TCFU was abolished by periodate oxidation. Studies using four different tumors showed that treatment of the lysate with trypsin resulted in a 60% or greater decrease in the stimulatory effect of the untreated lysate.

DISCUSSION

The results have demonstrated that PDGF stimulates the formation of human tumor colonies in a two-layer soft agar culture system. Responsive tumors represented over 20 different cell types and included cells of epithelial and mesothelial origin. Positive responses were evidenced by larger numbers of colonies and, in some instances, by greater colony size.

These results were obtained in a system in which a solution of PDGF was added to a highly enriched feeder layer that included both horse and fetal calf serum. It is possible that platelet factors present in these sera may have obscured the effect of the added human platelet-derived proteins. This possibility is presently being explored using culture systems in which human plasma-derived serum is substituted for the horse and fetal calf serum in the feeder layer.

The time to peak effect in response to the PDGF was generally much shorter than the time usually required for maximal colony formation in this system (Hamburger and coworkers, 1978, Von Hoff and co-workers, 1980). This is compatible with the observations by Pledger and co-workers (1977, 1978) that PDGF is a competence factor which renders cells responsive to progression factors. These lead to the rapid mobilization of cells from G_O and their entry into the replicative cycle. The effect of PDGF on tumor cells may be, therefore, to increase the number of colony forming units and/or accelerate their rate of replication.

The relative stimulatory effects on human tumor cells of PDGF prepared by freeze-thaw lysis and by collagen stimulation were similar to those observed using SV40-transformed 3T3 cells. In both settings, PDGF prepared by platelet lysis was approximately twice as active as PDGF obtained after collagen stimulation. This may reflect the availability after lysis of a second class of platelet-derived factors having mitogenic activity (Zetter and Antoniades, 1979). Presumably, these factors are not released from storage granules by collagen stimulation.

PDGF prepared after thrombin stimulation was equivalent to that obtained after lysis in stimulating human tumor cells but only one-fourth as effective in stimulating SV40-transformed mouse fibroblasts. It is possible that the procedures used to inactivate the thrombin (heating at $57^O C$ for 30 minutes) were inadequate or failed to inactivate another as yet undetermined factor. If so, the findings may support the suggestion of Zetter and Antoniades (1979) that thrombin potentiates the mitogenic response to growth factors and/or has direct mitogenic activity.

The apparent inhibitory effect of PDGF on tumor colony formation may be indicative of a factor having true cell growth inhibiting activity. Alternatively, it may reflect greater sensitivity of cells from specific tumors to PDGF and a dose-response relationship in which relatively high concentrations of PDGF inhibit rather than stimulate colony growth. Preliminary studies in our laboratory of the response of tumor cells to varying doses of PDGF suggest that relatively high concentrations of PDGF may inhibit colony formation.

The growth-promoting activity for human tumor cells was non-dialyzable, heat resistant only to $56^O C$, partially trypsin-sensitive, and totally destroyed by periodate oxidation. These characteristics are similar to those reported by Kepner and Lipton (1981) in a study of the effect of PDGF on SV40-transformed 3T3 cells and to those reported by Eastment and Sirbasku (1978) and Hara and co-workers (1980) using malignant cell lines from a variety of animals. They are different from the properties of PDGF reported by Ross and co-workers

(1978) and Antoniades and co-workers (1979) that stimulated non-transformed 3T3 fibroblasts. It is of interest that the non-transformed and transformed fibroblasts responded qualitatively similarly to the PDGF modified by the various procedures in the present study. The findings provide additional evidence to support the proposition that multiple growth factors are produced by platelets which have distinctive target cell activity. The activity of the factor(s) responsible for stimulating human tumor cells appears to reside in a glycopeptide or a glycoprotein. The isoelectric point(s) and molecular weight(s) of the factor(s) that stimulate human tumor colony formation are presently being investigated.

The human tumor cells studied here displayed considerable variation in response to PDGF. However, despite the extreme heterogeneity of biologic characteristics of human tumor cells, the present study suggests that PDGF can stimulate the proliferation of human tumor colony forming units. To the extent that these observations reflect in vivo phenomena, they suggest that platelets may promote the growth of cells in primary and metastatic tumors. It is possible that more refined documentation of this effect may lead to the development of an assay which selects patients having tumors whose growth would be retarded by antiplatelet agents.

REFERENCES

Antoniades HN, Scher CD, Stiles CD (1979). Purification of human platelet-derived growth factor. Proc Natl Acad Sci USA 76:1809.

Balk SD (1971). Calcium as a regulator of normal but not of transformed, chicken fibroblasts in a plasma-containing medium. Proc Natl Acad Sci USA 68:271.

Balk SD, Whitfield JF, Youdale T, Braun AC (1973). Roles of calcium, serum, plasma, and folic acid in the control of proliferation of normal and Rous sarcoma virus-infected chicken fibroblasts. Proc Nat Acad Sci USA 70:675.

Buick RN (1980). In vitro clonogenicity of primary human tumor cells: Quantitation and relationship to tumor stem cells. In Salmon SE (ed): "Cloning of Human Tumor Stem Cells" New York: Alan R. Liss, p 15.

Castor CW, Ritchie JC, Scott ME, Whitney SL (1977). Connective tissue activation. XI. Stimulation of glycosaminoglycan and DNA formation by a platelet factor.

Arth Rheum 20:859.

Eastment CT, Sirbasku DA (1978). Platelet derived growth factor(s) for a hormone-responsive rate mammary tumor cell line. J Cell Physiol 97:17.

Eastment CT, Sirbasku DA (1980). Human platelet lysate contains growth factor activities for established cell lines derived from various tissues of several species. In Vitro 16:694.

Hamburger AW, Salmon SE (1977). Primary bioassay of human myeloma stem cells. J Clin Invest 60:846.

Hamburger AW, Salmon SE, Kim MB, Trent JM, Soehnlen BJ, Alberts DS, Schmidt HJ (1978). Direct cloning of human ovarian carcinoma cells in agar. Cancer Res 38:3538

Hamburger AW, Kim MB, Salmon SE (1979). The nature of cells generating human myeloma colonies in vitro. J Cell Physiol 98:371.

Hara Y, Steiner M, Baldini MG (1980). Platelets as a source of growth-promoting factor(s) for tumor cells. Cancer Res 40:1212.

Kepner N, Lipton A (1981). A mitogenic factor for transformed fibroblasts from human platelets. Cancer Res 41:430.

Kohler N, Lipton A (1974). Platelets as a source of fibroblast growth-promoting activity. Exp Cell Res 87:297.

Lowry OH, Rosebrough NJ, Farr AL, Randall RJ (1951). Protein measurement with the Folin phenol reagent. J Biol Chem 193:265.

Park CH, Bergsagel DE, McCulloch EA (1971). Mouse myeloma tumor cells: a primary cell culture assay. J Nat Cancer Inst 46:411.

Pike B, Robinson W (1970). Human bone marrow colony growth in vitro. J Cell Physiol 76:77.

Pledger WJ, Stiles CD, Antoniades HN, Scher CD (1977). Induction of DNA synthesis in BALB/c3T3 cells by serum components: Reevaluation of the commitment process. Proc Natl Acad Sci USA 74:4481.

Pledger WJ, Stiles CD, Antoniades HN, Scher CD (1978). An ordered sequence of events is required before BALB/c3T3 cells become committed to DNA synthesis. Proc Natl Acad Sci USA 75:2839.

Ross R, Glomset J, Kariya B, Harker L (1974). A platelet-dependent serum factor that stimulates the proliferation of arterial smooth muscle cells in vitro. Proc Natl Acad Sci USA 71:1207.

Ross R, Glomset J, Kariya B, Raines E (1978). Role of platelet factors in the growth of cells in culture. J Natl Cancer Inst 48:102.

Salmon SE (1980). Background and overview. In Salmon SE (ed) "Cloning of Human Tumor Stem Cells." New York: Alan R. Liss, p 3.

Scher CD, Pledger WJ, Martin P, Antoniades HN, Stiles CD (1978). Transforming viruses directly reduce the cellular growth requirement for a platelet derived growth factor. J Cell Physiol 97:371.

Scher CD, Shepard RC, Antoniades HN, Stiles CD (1979). Platelet-derived growth factor and the regulation of the mammalian fibroblast cell cycle. Biochim Biophys Acta 560:217.

Stiles CD, Capone GT, Scher CD, Antoniades HN, VanWyk J, Pledger WJ (1979). Dual control of cell growth by somatomedins and platelet-derived growth factor. Proc Natl Acad Sci USA 76:1279.

Vogel A, Raines E, Kariya B, Rivet M-J, Ross R (1978). Coordinate control of 3T3 cell proliferation by platelet-derived growth factor and plasma components. Proc Natl Acad Sci USA 75:2810.

Von Hoff DD, Harris GJ, Johnson G, Glaubiger D (1980). Initial experience with the human tumor stem cell assay system: Potential and problems. In Salmon SE (ed): "Cloning of Human Tumor Stem Cells." New York: Alan R. Liss, p 113.

Westermark B, Wasteson A (1976). A platelet factor stimulating human normal glial cells. Exp Cell Res 98:170.

Zetter BR, Antoniades HN (1979). Stimulation of human vascular endothelial cell growth by a platelet-derived growth factor and thrombin. J Supramol Structure 11:361.

ACKNOWLEDGEMENTS

This work was supported in part by a grant from the Saint Luke's Hospital Association and by the Dr. I.J. Goodman and Ruth Goodman Blum Memorial Fund.

DISCUSSION

Dr. D. S. Rappaport (University of Minnesota Medical School, Minneapolis). I am curious as to whether there is something like tumor-derived PDGF that might affect the primary growth.

Dr. D. H. Cowan (Case Western Reserve University, Cleveland). One of the problems is that PDGF presumably has human tumor cells as the target, and presumably, one would have to use the human tumor cells as the model in order to test it.

(Unidentified). Did you resuspend one of your colonies that had already been formed to see if they exhibited this heterogeneity?

Dr. Cowan. There are techniques which have been described for taking the colonies out of cell lines or subculturing them. There are concerns as to whether the responsiveness of the cells to the PDGF will be retained after several passages or whether the biological characteristics might be altered.

Dr. N. Hanna (Frederick Cancer Research Center, Frederick). Did you use mechanical disruption of the tumor?

Dr. Cowan. These were all done by mechanical disruption. The very extensive series of studies reported at the American Association for Cancer Research last month demonstrated the validity of using mechanical disruption although it has been shown that the viability is somewhat impaired.

Dr. M. Steiner (The Memorial Hospital, Pawtucket). Were there any changes in morphology?

Dr. Cowan. No, we were unable to detect any difference in morphology comparing these to stimulated cells, the inhibited cells agains the controlled cells.

Dr. J. C. Hoak (The University of Iowa, Iowa City). I was going to ask whether your growth factor modified the respond to chemotherapeutic agents.

    Dr. Cowan. At this point we do not know. It's a matter of getting cooperation from surgeons and from pathologists and getting material. It has taken us a couple of years to educate pathologists not to dump the tumors in the formalin.

HUMAN PLATELET CHEMOTAXIS CAN BE INDUCED BY LOW MOLECULAR
SUBSTANCE(S) DERIVED FROM THE INTERACTION OF PLASMA AND
COLLAGEN

Rosalin Wu Lowenhaupt

Department of Physiology
University of Cincinnati, College of Medicine
Cincinnati, Ohio   45267

ABSTRACT

Previously, we reported that collagen induces chemotaxis of human platelets as revealed by a capillary-tube method for qualitative assessment of platelet migration (Lowenhaupt, 1978). Chemotaxis was presumably stimulated by a substance(s) generated from collagen incubated in plasma. We have now developed a new quantitative method for evaluating chemotaxis by using ^{111}indium-oxine labeled platelets. Blood from healthy donors was drawn into plastic syringes containing acid-citrate-dextrose. Platelet-rich and platelet-free plasma were prepared as described previously. Platelets were labeled with ^{111}In-oxine according to Goodwin et al (1978) Chemotaxis was studied in a specifically constructed 7-compartment chamber partitioned with nitro-cellulose membranes (3 μm, 1 μm, and 0.45 μm). Chemotaxis could be induced with collagen from either bovine or rat-tail tendon. Molecular sieving of normal plasma on Sephadex G-50 showed a peak (A 220nm) of low molecular weight material which was shifted to the right in preparations preincubated with either bovine tendon or rat-tail tendon collagen. Aliquots from the control plasma peak did not produce chemotaxis. In contrast, aliquots from the collagen-plasma peak markedly stimulated chemotaxis. These findings suggest that plasma releases a low molecular moiety(s) from collagen which promotes chemotaxis. (Supported in part by NIH Grant HL 21988.)

INTRODUCTION

The interaction between platelets and collagen or connective tissue has been studied extensively (see review, Jaffe, 1976). Most studies have been concerned with the adhesion, aggregation, and release reaction, which are responses elicited in vitro by the rapid mixing of platelets and collagen preparations. However, during studies on the mechanisms of acute rejection of transplanted kidneys in dogs, we noticed an immediate accumulation of platelets in the capillaries of the transplanted kidneys (Lowenhaupt and Nathan, 1968). In the electron micrographs of these studies, we also noted that some platelets with extended pseudopods had penetrated through the endothelial lining and the thin basement membrane of the capillaries into the subendothelial tissue in close contact with the collagen fibrils (unpublished results). This led us to speculate on the consequence of a more remote interaction between platelets and collagen. We therefore undertook a series of studies of the motility and chemotaxis of platelets which had been placed at a distance (5-6 mm) from a small amount of collagen.

Human blood platelets exhibit chemotaxis when exposed to collagen bathed in autologous platelet-free plasma in vitro (Lowenhaupt et al, 1973; Lowenhaupt, 1978). Chemotaxis can be induced only in actively migrating platelets. Metabolic inhibitors which impair the plasma membrane transport system, microfilament function or membrane enzyme activity, affect the migratory ability of platelets (Lowenhaupt et al, 1977). Chemotaxis of human platelets requires the presence of normal plasma and does not take place in artificial media alone. Chemical modification of collagen does not seem to alter its ability to promote chemotaxis (Lowenhaupt and Setzler, 1979). Mild digestion of collagen with enzymes, such as trypsin or thrombin, but not bacterial collagenase, increases slightly its effectiveness as an inducer. A significant decrease in the degree of chemotaxis was observed if digested collagen was dialyzed thoroughly with phosphate buffered saline. This suggests that some small molecular substance(s) may be responsible for the induction of chemotaxis. We now report that a small molecule(s) released from collagen which has been incubated in plasma for 24 hr, is equally as capable of inducing chemotaxis as is collagen, itself.

We have specifically designed and constructed a 7-compartment chamber to assay the chemotaxis of ^{111}indium-oxine (^{111}In-oxine) labeled platelets (Lowenhaupt et al, 1981). In this system the migration of platelets to platelet-free plasma (PFP) and a testing substance can be compared. Results are expressed as a chemotaxis ratio. In control studies, we have shown that the chemotaxis ratio is 1.0 when PFP was placed in the compartments at both ends of the 7-compartment chamber (results included in Figure 3).

^{111}Indium-oxine is a lipid soluble complex. It has been shown to be useful for the labeling of leukocytes and platelets (Thakur et al, 1976; Thakur et al, 1977; Heaton et al, 1979). It is superior to ^{51}chromium as a radio-label because of its γ-emission characteristics and high cell labeling efficiency. Its half life of 67 hr is ideally suited to kinetic and distribution studies. In general, ^{111}In has been found to be nontoxic in terms of its radiobiological consequences or the effects of the oxine and ethanol on platelet structure and function.

Blood from healthy donors was drawn into plastic syringes containing acid-citrate-dextrose (ACD, 1.5ml to 10ml of blood). (The study was approved by the Human Research Committee of the University of Cincinnati Medical Center.) Platelet-rich plasma (PRP) and platelet-free plasma (PFP) were prepared as described previously (Lowenhaupt et al, 1977). Briefly, PRP was obtained by slow centrifugation of 180g for 15 min in sterile plastic tubes. The top layer of PRP was transferred into another plastic tube and centrifuged for the second time at the same speed to remove contaminating red and white cells; PRP was transferred to another tube. For each 10ml PRP 1ml ACD was added. The addition of ACD reduces spontaneous platelet aggregation during the subsequent labeling process. PRP was centrifuged at 1000g to remove excess plasma. Platelets were resuspended in 2ml physiological saline for labeling with 500-700 μCi of ^{111}In-oxine (Medi-Physics, Inc., CA) for 30 min. 10ml PPP was added to the radioactive platelet suspension to remove the excess ^{111}In-oxine. The radioactive PRP was centrifuged for 15 min at 1000g and PPP was removed. Another 2ml cold PPP was layered on top of the radioactive platelet pellet and then removed. The labeled platelets were suspended finally to an appropriate concentration with PFP to about 300,000 platelets/μl and used immediately in the chemotaxis assays.

Platelet-free plasma (PFP) was obtained by centrifuging the remaining blood at 1000g for 30 min. The top layer of plasma was then filtered through 0.2 µm Millex filter unit (Millipore) to obtain PFP. All procedures from the drawing blood to the preparation of the final radioactive platelets were carried out with sterile technique at room temperature (21-22°C).

Acid-soluble bovine collagen (ASBC) was prepared from type I, insoluble bovine Achilles tendon collagen (Sigma C 9879), according to the method reported by Day and Holmsen (1972). Essentially, ASBC is the acid-soluble fraction of bovine tendon collagen extracted in acetic acid at 0-4°C. The final concentration of the preparation is 1mg/ml.

Rat-tail tendon collagen was extracted with acetic acid (ASRC). 1.0gm (wet weight) of tail-tendon was dissected from tails of 250gm female Sprague-Dawley rats and cut into 2-3mm lengths and extracted in 150 ml of 17.4mM acetic acid by gentle shaking at 4°C for 48 hr. Insoluble tissue was sedimented and discarded. The concentration of this preparation was determined to be 5mg/ml by lyophilizing and weighing aliquots of the sample.

A sketch of the 7-compartment chamber (Lowenhaupt et al, 1981) is shown in Figure 1. When assembled, the compartments A through G are separated by different pore-size membranes and clamped tightly together to form the 7-compartment chamber. From previous studies, we have found that it is best to use HA 0.45 µm membranes (Millipore) between compartments A & B (memb. I) and F & G (memb. VI); 1 µm Nuclepore membranes between B & C (memb. II) and E & F (memb. V); and 3 µm Nuclepore membranes between C & D and D & E (memb. III & IV respectively). High vacuum grease (Dow Corning) is applied between compartments to prevent leakage. PFP is placed in all compartments to prevent leakage. PFP is placed in all compartments except compartment D, which is filled with ^{111}In-PRP at time zero. Incubation times are recorded from this time point. Each compartment holds about 100 µl of liquid. In control studies, compartment G is filled with PFP. For other studies, compartment G is filled with equal volume of PFP and the testing substances. A 10 min interval was allowed for the interaction of plasma and collagen or the testing substance before the introduction of ^{111}In-PRP into compartment D. The chamber is placed on a level surface for 3 hr at room temperature (lead shielded). After incubation, the contents of compartments D, C, B, A, and E, F, G (in order) are carefully withdrawn and dispensed into individual counting vials. The chamber is then disassembled. Each membrane is carefully removed and layed flat at the bottom of the vial. All 13 samples of each chamber are counted with a gamma counter (Nuclear Chicago). 8-12 such chambers can be processed during each experimental day.

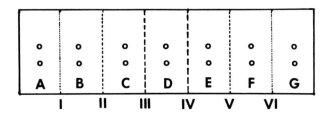

Figure 1. A diagram of 7-compartment chamber. Compartments are indicated with A through G. Membranes are represented with I through VI. Detailed description of membranes and contents in compartments are listed in text.

Before proceeding to the study of chemotaxis, both PRP and ^{111}In-PRP were enumerated with a Coulter Counter, examined under phase-contrast light microscopy (Zeiss) and tested for their ability to aggregate when stimulated by adenosine diphosphate plus $ASBC$ or ASRC by using a BIO/DATA Platelet Aggregation Profiler^{TM}. The ultrastructures of platelets after labeling were only checked occasionally and showed only slightly increased pseudopods with normal distribution of granules. The platelets were discoid before and after labeling. The experiment was discontinued if small aggregates were noted in PRP or ^{111}In-PRP. The labeling procedure did not impair platelet function. The labeling efficiency at our hands is about 65-70%. ASRC preparations do not stimulate platelets to aggregate before or after labeling even with an excess amount of ASRC or if allowed to react for 20-30 min. However, ASBC always stimulates platelet aggregation.

A 2 x 120 cm Sephadex G-50 column (Bio-Rad Econocolumn) was prepared in PBS and standardized with blue dextran, aldolase, ovalbumin, chymotrypsinogen A and ribonuclease A before plasma or the supernatant from the reacted mixture of plasma and collagen was layered on top of the column. The ASRC was first dialyzed against glycyl glycine buffered saline (GGBS, 0.14M NaCl, 0.01M glycyl glycine, pH 7.2) at 4°C and then mixed with equal volume of PFP and reacted at room temperature for 24 hr with constant shaking mixing. The mixture was centrifuged for 30 min at 1000g to sediment the collagen. One ml of plasma or the supernatant of the plasma-collagen reaction was layered on top of the column which had been equilibrated with phosphate-buffered saline (PBS, 0.14M NaCl, 0.01M PO$_4$, pH 7.4).

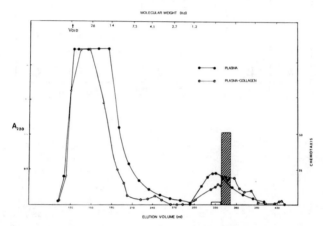

Figure 2. Sephadex G-50 chromatograph of plasma and plasma-collagen. The optical density (●—● and ○—○) of fractions was measured at 220nm. Chemotaxis (mean value of ratio) is indicated by vertical bars (open bar-plasma, hatched bar-plasma-collagen supernatant). Molecular weight shown in upper margin of figure is calculated from the markers (detail in text).

Elution was continued by the addition of PBS. Fractions were collected at a flow rate of 40 drops/min. When the optical density of the fractions was monitored at A 280nm, a single large peak appeared from 100ml to 210ml of elution volume (result not shown) with both plasma or plasma-collagen supernatant. However, when the fractions were monitored at A 220nm, two peaks appeared (figure 2). The large peak was associated with the same fractions as in A 280nm. additional, smaller peak appeared between 300ml and 400ml of elution volume. The low molecular weight peak from plasma-collagen supernatant was slightly shifted to the right of that with plasma alone. The mean values from the results in the chemotaxis studies (Fig. 4) are included in figure 2 and plotted as vertical bars at the regions of those fractions tested ($p < 0.05$).

Our previous studies (Lowenhaupt, 1978; Lowenhaupt et al, 1981) showed that collagen from either bovine or rat-tail tendon could induce human platelet chemotaxis. However, we did not test ^{111}In-oxine labeled platelets which were labeled in platelet-rich plasma. Therefore, their ability to exhibit chemotaxis was evaluated with the present preparation. Figure 3 shows the results of the study. The data are expressed as a chemotaxis ratio of the radioactive counts (CPM) of both end compartments of the chamber and plotted in semilogarithmic scale. When PFP was placed in both end compartments, chemotaxis was not observed since little or no CPM was noted in compartments A, B, and F, G. Random migration of the platelets occurs in both directions since the chemotaxis ratio was one. This control study indicated the usefulness of the present system. Each point in the figure represents the value obtained from one set of chambers. With the addition of ASBC or ASRC to the compartment G, a significant induction of directional migration was observed. ASRC was a consistently strong inducer of platelet chemotaxis. The mean chemotaxis ratio for ASRC induced chemotaxis was 94.5, whereas that of ASBC was only 32.8.

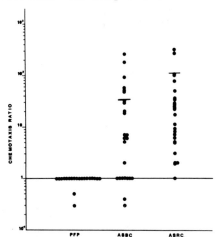

Figure 3. ^{111}In-PRP chemotaxis induced in the presence of collagen. PFP serves as control. Each dot indicates the result of one set of chambers. Refer to the text for details. Bar indicates the mean values.

Figure 4 shows the results of platelet chemotaxis induced by 3 of the low molecular weight fractions (representing peak activity) from plasma or the plasma-collagen supernatant. It is again plotted in the semilogarithmic scale. The calculation of the chemotaxis ratio for each point is the same as described for figure 3. The addition of the low molecular weight fractions from plasma into compartment G appeared to have induced a slight but not **significant** directional migration as demonstrated by the mean chemotaxis ratio of 2. However, when similar fractions from the plasma-collagen reaction were added to G, a significantly greater chemotaxis ratio of 51 was observed ($p < 0.05$). These findings strongly support our suggestion that plasma normally contains no biologically active chemotaxis factors. However when collagen is exposed to plasma or blood as may occur when there is injury to a blood vessel, a reaction results in the formation or activation of a chemotactic factor(s).

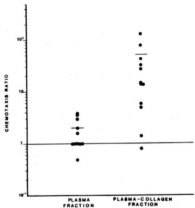

Figure 4. ^{111}In-PRP chemotaxis induced by low molecular weight fractions of plasma or the plasma-collagen reaction. Each point indicates one set of chambers. Studies performed in duplicate with labeled platelets. Bars represent the mean values.

Although platelet adhesion and aggregation have long been considered the primary steps in hemostasis, our investigations over the past several years suggested that plasma can first react with exposed collagen to produce a chemotactic signal for platelets. In support of this hypothesis we have demonstrated platelets move in a specific direction in response to plasma-treated collagen in vitro by 3 independent assay systems (Lowenhaupt et al, 1973; Lowenhaupt, 1978; Lowenhaupt et al, 1981). Our present study demonstrates that: (1) indium labeling can be used to quantitatively follow platelet migration in a 7-compartment chamber assay, and (2) plasma treated collagen releases a low molecular chemotactic signal that may be resolved on Sephadex G-50. It is not known if the signal is a peptide or whether sugars or sugar-peptides are involved in the signal.

We have also observed platelets migrating through the endothelial lining whose morphology of extended pseudopods and cellular granules would suggest that they have not been involved in release reaction and aggregation. However, since the distance across the capillary in an injured area is so small, and there are other physiological influences, it is not possible to suggest that the migratory platelets in vivo are the result of chemotaxis. We hope to examine the in vivo validation of our theory using platelets from Factor XI-deficient patients, since their platelet chemotaxis is repressed in vitro (Lowenhaupt, 1978).

The importance of platelet chemotaxis would be clarified by understanding the relationship of platelet chemotaxis to other blood cells which are active in wound healing, defense and other disease states. In this regard, it might seem profitable to examine whether defined chemoattractants of leukocytes would stimulate the platelet system.

Special thanks to Stephen J. Keller, Ph.D. and John R. Lymangrover, Ph.D. for their valuable advice in preparing this manuscript.

REFERENCES

Day H, Holmsen H (1972). Laboratory tests of platelet function. Ann Clin Lab Sci 2:63.
Goodwin DA, Bushberg JT, Doherty PW, Lipton MJ, Conley FK, Diamanti CI, Meares CF (1978). Indium-111-labeled autologous platelets for location of vascular thrombi in humans. J Nucl Med 19:626.

Heaton WA, Davis HH, Welch MJ, Mathias CJ, Joist JH, Sherman LA, Siegel BA (1979). Indium-111: A new radionuclide label for studying human platelet kinetics. Br J Haematol 42:613.

Jaffe RM (1976). Interaction of platelets with connective tissue. In Gordon JL (ed): "Platelets in Biology and Pathology." New York: North-Holland Publishing Company, p 261.

Lowenhaupt RW (1978). Human platelet chemotaxis: requirements for plasma factor(s) and the role of collagen. Am J Physiol 253:H23.

Lowenhaupt R, Nathan P (1968). Platelet accumulation observed by electron microscopy in the early phase of renal allotransplant rejection. Nature 220:822.

Lowenhaupt RW, Setzler LM (1979). The effects of collagen modifications on platelet chemotaxis. Fed Proc 38:799.

Lowenhaupt RW, Miller MA, Glueck HI (1973). Platelet migration and chemotaxis demonstrated in vitro. Thromb Res 3:477.

Lowenhaupt RW, Glueck HI, Miller MA, Kline DL (1977). Factors which influence blood platelet migration. J Lab Clin Med 90:37.

Lowenhaupt RW, Sperling MI, Silberstein EB (1981). A quantitative method to measure human platelet chemotaxis using ^{111}Indium-oxine labeled gel-filtered platelets. Fed Proc 40:774.

Thakur ML, Colman RE, Welch MJ (1977). Indium-111-labeled leukocytes for the localization of abscesses: preparation, analysis, tissue distribution, and comparison with gallium-67 citrate in dogs. J Lab Clin Med 89:217.

Thakur ML, Welch MJ, Joist JH, Colman RE (1976). Indium-111-labeled platelets: Studies on preparation and evaluation of in vivo functions. Thromb Res 9:345.

DISCUSSION

Dr. K. V. Honn (Wayne State University, Detroit). Have you tried type IV collagen?

Dr. R. W. Lowenhaupt (University of Cincinnati Medical Center, Cincinnati). In my previous experiments, I tried type III but I have not tried type IV as yet.

Dr. E. Pearlstein (New York University Medical Center, New York). Have you considered the possibility that fibronectin may be involved?

Dr. Lowenhaupt. Fibronectin on the platelet could be the receptor for the signal but that is just speculative. Chemotaxis is not very good in smokers or immediately after exercise.

Dr. J. C. Hoak (The University of Iowa, Iowa City). Are there any smokers who would like to rebut?

Dr. R. P. Warrell, Jr. (Memorial Sloan-Kettering Cancer Center, New York). Your first assay was really a very neat little trick on the old MIF assay. I wonder if you looked to see if aspirin inhibits the migration of platelets out of your tubes?

Dr. Lowenhaupt. Aspirin is very tricky. In vivo aspirinated platelets give variable results but in vitro inhibit the migration on a dose-dependent basis.

Dr. J. Halkett (New England Nuclear, Corporation, North Billerica). Do the platelets seem to have a chemotactic memory? That is, if you remove the collagen and put all of the platelets back into the little cubicle and add collagen again, would they move out more rapidly?

Dr. Lowenhaupt. I would not think so, but I have not done that experiment.

Dr. W. Eisert (University of Hannover, West Germany).

Do you observe that all the platelets that show the chemotaxis are actually moving or just a small subpopulation?

    Dr. Lowenhaupt. Just a small subpopulation is chemotactic but all the platelets move out from the tube in a random migration.

ROLE OF THE VASCULAR ENDOTHELIUM

John C. Hoak, William M. Parks, Glenna L. Fry,
Abigail A. Brotherton, and Robert L. Czervionke
Division of Hematology-Oncology, Department of
Medicine, University of Iowa, Iowa City, Iowa
52242

The intact vascular endothelial surface is considered non-thrombogenic and blood platelets usually fail to adhere to it. Likewise it serves as a barrier to the spread of tumor cells and the development of metastatic lesions. Both tumor cells and platelets exhibit a preferential adherence to the subendothelial extracellular matrix. Whether this phenomenon is purely a function of the non-thrombogenic properties of the endothelium is unknown.

A number of possible factors and mechanisms have been proposed to explain the non-thrombogenic properties of the vascular endothelium. Both the intact platelet and endothelial cell at physiologic pH have negatively charged membranes and are mutually repelled by each other. It is also believed that the surface of the endothelium is influenced by the presence of heparans and glycosaminoglycans which might facilitate the inactivation of thrombin and factor Xa by antithrombin III. An enzyme with ADPase activity has been found to be associated with the endothelial surface and would have an obvious role in preventing platelet aggregates from forming near intact endothelium (Heyns et al., 1974). The endothelium is known to produce an activator of plasminogen and carries the potential for activation of the fibrinolytic system to promote lysis of fibrin in thrombi.

An interesting new role for the endothelium has been suggested by the recent work of Lollar and Owen (1980). These workers have demonstrated that the binding of thrombin to the endothelium is an important primary mechanism for rapid removal of thrombin from the circulation and facilitates its subsequent association with antithrombin III to form a thrombin - AT III complex.

A most important non-thrombogenic property of the endothelium relates to its ability to produce and release prostacyclin. Moncada, et al. (1976) have shown that arachidonate metabolic mechanisms exist in endothelial cells to convert arachidonic acid and cyclic endoperoxides to prostacyclin, a most potent inhibitor of platelet aggregation and adhesion.

In relationship to the importance of prostacyclin in thrombosis, several key questions warrant attention.
1. Is prostacyclin responsible for the non-thrombogenic properties of the vascular endothelium?
2. What other anti-thrombotic mechanisms operate at the level of the blood vessel wall?

Results

To delineate the role of prostacyclin, we have performed studies using a platelet adherence system with cultured vascular cell monolayers (Czervionke et al., 1978). Prostacyclin was measured in these studies using a radioimmunoassay for 6-keto-PGF$_{1\alpha}$, the stable end product of prostacyclin (Czervionke, et al., 1979).

Table 1
Effect of Bovine Thrombin on Adherence of Platelets
(Percentages)

	Control (i.m.)	Bovine Thrombin 0.5 U
Endothelium		
venous	1.5 ± 0.2	4.0 ± 0.5
hemangioendothelioma	2.6 ± 0.7	66.9 ± 1.9
Arterial smooth muscle	2.4 ± 1.0	79.4 ± 0.3
Arterial fibroblasts	1.3 ± 0.4	79.4 ± 0.9
Empty dish	2.1 ± 0.1	77.7 ± 2.7

(Monolayers were rocked with 1 ml incubation medium (i.m.) with or without 0.5 U bovine thrombin, and 0.5 ml ^{51}Cr-labelled platelets for 30 min at 37°C. Adherence was determined by the method of Czervionke, et al. (1978). Values given are means ± s.e. for at least six dishes). These data are taken from Fry, et al., 1980.

To test the effect of prostacyclin in the platelet adherence system we used four different approaches to eliminate it from the endothelium.

Aspirin (ASA) is known to acetylate the cyclo-oxygenase of the platelet and to inhibit thromboxane A2 formation. Therefore, we chose to treat the endothelial monolayer with aspirin so that in a similar way the endothelial cyclooxygenase would be inhibited and prostacyclin production would cease. We tested this possibility by using our assay for 6-keto-PGF$_{1\alpha}$ and the thrombin-induced platelet adherence system.

The results are shown in Figure 1. In the absence of thrombin, there was no increased platelet adherence despite inhibition of prostacyclin formation. In the presence of thrombin and in the absence of aspirin, 6-keto-PGF$_{1\alpha}$ increased and there was little platelet adherence. When the endothelium was treated with 0.01 mM aspirin, thrombin still caused significant release of 6-keto-PGF$_{1\alpha}$, and no increase in platelet adherence occurred. However, treatment of the endothelium with 1 mM aspirin prevented the formation of 6-keto-PGF$_{1\alpha}$ even in the presence of thrombin, and platelet adherence increased to 44 per cent.

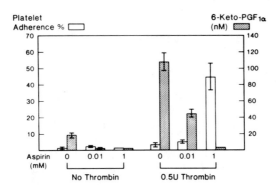

Figure 1. Platelet adherence (open columns) to untreated and aspirin (ASA)-treated endothelium compared with 6-keto-PGF$_{1\alpha}$ release (shaded columns). ASA or buffer control was incubated with the monolayer for 30 min at 37°C, with

rocking. The preincubation solution was removed and the dish was washed twice. Thrombin or buffer control was added, followed immediately by ^{51}Cr-labelled platelets (for adherence) or unlabelled platelets (for PGI_2 determinations). The monolayer was rocked for 30 min at 37ºC. The percentage adherence was calculated by dividing counts per minute of cells attached to the monolayer, multiplied by 100, by total counts per minute added to the dish. 6-Keto-$PGF_{1\alpha}$ released into the supernatant was determined by radioimmunoassay.

In additional studies we demonstrated that the effect of aspirin on the endothelium was temporary. When the aspirin-treated endothelium was removed from contact with the aspirin and restored to culture conditions 2 hours later, thrombin caused significant 6-keto-$PGF_{1\alpha}$ release, and platelet adherence returned to normal. When cycloheximide, an inhibitor of protein synthesis, was added to the endothelial culture during the recovery period, little 6-keto-$PGF_{1\alpha}$ was released, and platelet adherence remained abnormal.

Prostacyclin was also removed from the platelet adherence system by using an incubation system containing a rabbit antibody against 6-keto-$PGF_{1\alpha}$, which cross-reacts with prostacyclin. In the presence of the antibody, thrombin-induced platelet adherence to endothelium increased from 7.7 to 39.1 per cent.

Studies with Hemangioendothelioma cells

We have reported earlier on the use of a murine model of the Kasabach-Merritt syndrome in which 129/J strain mice developed thrombocytopenia, microangiopathic hemolytic anemia and a localized consumption coagulopathy in transplanted hemangioendotheliomas (Hoak et al., 1971; Warner et al., 1971). It has been possible to grow these hemangioendothelioma cells in culture and to study their surface properties and their ability to produce prostacyclin (Fry et al, 1980). The hemangioendothelioma monolayers did not produce prostacyclin in response to thrombin, and in comparison with venous endothelium, produced little prostacyclin upon addition of arachidonic acid or the endoperoxide, PGH_2.

In order to study the effect of exogenous prostacyclin on platelet adherence to different types of vascular cells, the monolayers were pre-treated with 1 mM aspirin to block endogenous production of prostacyclin by thrombin. In order to determine whether some cell types were more sensitive to prostacyclin, concentrations from 25 to 150 mM were used. Fibroblasts were chosen as a representative cell type from the subendothelium, since values for smooth muscle and fibroblasts were not significantly different. Mouse fibroblasts were used as a control for the hemangioendothelioma.

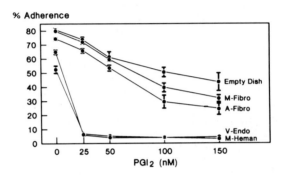

Figure 2. Inhibition by exogenous PGI_2 of 0.5 U thrombin-induced platelet adherence to 1 mM aspirin-treated cell layers. All monolayers were incubated 30 min with 1 mM aspirin in incubation medium. After rinsing twice, platelet adherence was determined as described in Figure 1. PGI_2 was added just before thrombin and platelets to achieve the concentrations shown. Abbreviations are A-Fibro, arterial fibroblasts; V-Endo, venous endothelium; M-Heman, mouse hemangioendothelioma; M-Fibro, mouse L929 fibroblasts. From Fry et al. (1980), published with permission of Grune and Stratton.

As can be seen in Figure 2, thrombin-induced platelet adherence to the venous endothelium and hemangioendothelioma cells decreased dramatically with as little as 25 mM prostacyclin. In contrast, platelet adherence to fibroblasts in the presence of thrombin was extremely resistant to the effect of prostacyclin. Prostacyclin appears to play a key role in the prevention of platelet aggregate formation and their adherence to vascular wall cells when a thrombogenic stimulus such as thrombin is present. It appears to be an important component of the non-thrombogenic effect exhibited by the endothelium. Removal of prostacyclin from the endothelium did not increase baseline platelet adherence, but did increase thrombin-induced platelet adherence from 4 to 60 per cent. Addition of exogenous prostacyclin, at low concentrations, reversed the enhanced thrombin-induced platelet adherence under these conditions.

The failure of high concentrations of prostacyclin to completely block thrombin-induced platelet adherence to smooth muscle cells and fibroblasts suggests that these cells either lack a component normally found in endothelium or produce a substance that promotes adherence. Possible differences include collagen production (type and amount) by smooth muscle and fibroblasts or in the case of endothelium, interactions at the surface that enhance the effect of prostacyclin. Monolayers derived from normal endothelium or from neoplastic endothelium (hemangioendothelioma) exhibit some property in addition to prostacyclin that prevents thrombin-induced platelet adherence. Therefore, despite its ability to decrease platelet adherence to all of the cell types tested, it is unlikely that prostacyclin is the sole factor regulating platelet adherence.

Control Mechanisms of Prostacyclin Formation and Release

Our studies to date suggest a significant role for prostacyclin in the maintenance of a normally functioning endothelium. We have been concerned with control mechanisms for prostacyclin production and release.

Preincubation of cultured endothelial cells with 1 mM TMB-8, an antagonist of intracellular calcium ions (Malogodi and Chiou, 1974), or with 4 mM 3-isobutyl-1-methylxanthine (IBMX), an inhibitor of cyclic nucleotide phosphodiesterase activity, blocked prostacyclin release induced by thrombin or the calcium ionophore A23187, decreased arachidonic acid-induced release by about 50%, but had no effect on PGH_2-induced release (Brotherton and Hoak, in press). Radioimmunoassay of cyclic AMP in the endothelium showed that the basal level (2.16 ± 0.26 pmol of cyclic AMP/4.5×10^5 cells) was increased by an average of 2.6-fold with 4 mM IBMX. As seen in Table 2, prostacyclin (0.4 µM) had no significant effect on cyclic AMP levels in the absence of IBMX, but caused a 2 fold increase with 4 mM-IBMX.

Table 2
Effect of PGI$_2$ and 3-Isobutyl-1-Methylxanthine (IBMX) on the
Concentration of Cyclic AMP in Endothelial Cell Monolayers

Treatment	No. of Exps.	Conc. of cyclic AMP (pmol/4.5 x 10^5 cells)	
		Control	IBMX (4 nM)
Control	7	2.16 $\pm$ 0.26	5.55 $\pm$ 0.57
PGI$_2$ (400 nM)	7	2.86 $\pm$ 0.46	11.81 $\pm$ 1.93

Published with permission of the Royal Society of London from Hoak JC, et al., Philosophical Transactions of the Royal Society, in press.

These findings suggest that an increase in the intracellular concentration of cyclic AMP antagonizes the effects of agents that require calcium ions for the induction of prostacyclin release. In addition, high cyclic AMP phosphodiesterase activity in the endothelium may protect against a negative feedback mechanism involving activation of adenylate cyclase by release prostacyclin. A diagram of hypothetical reactions involved in prostacyclin control mechanisms is shown in Figure 3.

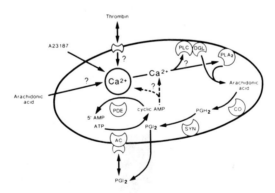

Figure 3. Schematic representation of some of the mechanisms that may be involved in the regulation of PGI$_2$ production by the vascular endothelium. PLA$_2$, phospholipase A$_2$; PLC, phospholipase C; DGL, diglyceride lipase; CO, fatty acid cyclooxygenase; SYN, PGI$_2$ synthetase; AC, adenylate

cyclase; PCE, cyclic AMP phosphodiesterase. Stimulatory pathways are indicated by solid lines and inhibitory pathways are indicated by solid lines and inhibitory pathways by broken lines. From Brotherton and Hoak (in press). Published with permission of the National Academy of Sciences.

Adherence of Lymphoma Cells to Endothelium

In earlier studies, Maca et al. (1978) studied the adherence of cultured lymphoma cells (Raji) to cultured human endothelial monolayers. Significant adherence of the Raji cells (44.6 per cent) was observed. This process appeared not to be dependent upon intact microtubular or microfilament function. Likewise, removing surface sialic acid did not alter this process. In contrast, incubating the endothelial cells for 24 or 48 hours with dexamethasone decreased adhesiveness of the Raji cells to the endothelial cells by approximately 40 per cent.

We have performed additional studies with Raji cells to determine whether prostacyclin plays a significant role in the adherence of these lymphoma cells to endothelium. In the first set of experiments, the endothelial cells were incubated with 1 mM aspirin to inhibit prostacyclin production prior to their incubation with ^{51}Cr-labeled Raji cells. The results, shown in Figure 4, indicate that lack of prostacyclin did not increase the adherence of the Raji cells.

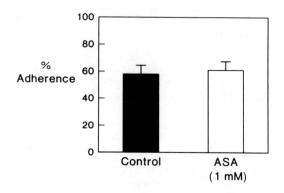

Figure 4.

In additional studies, exogenous prostacyclin was added to the incubation system containing ^{51}Cr-labeled Raji cells and the cultured endothelial monolayers. Again, as shown in Figure 5, prostacyclin did not decrease the adherence of the Raji cells to the endothelium.

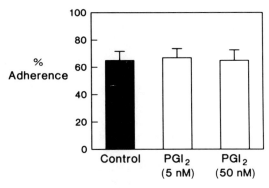

Figure 5.

We confirmed the findings of Maca et al. (1978) in observing that treatment of the endothelium with 1 µM dexamethasone for 48 hours decreased Raji cell adherence from 52 ± 9.4 per cent to 21 ± 4.5 per cent.

Summary

A possible explanation for the non-thrombogenic effect of the endothelium is the presence of prostacyclin, (PGI_2), the potent inhibitor of platelet aggregation and adherence, which is produced and released by the endothelium in response to various stimuli. Removal of PGI_2 from the endothelium did not increase baseline platelet adherence, but did increase thrombin-induced platelet adherence from 4 to 60%. Additions of exogenous PGI_2 at low concentrations reversed the enhanced thrombin-induced platelet adherence under these conditions. Although it is unlikely that prostacyclin is the sole factor regulating platelet adherence to the endothelium, it appears to play a major role in the interaction of platelets with components of the blood vessel wall. Conditions which predispose to adherence of platelets to the vessel wall may involve entrapment of tumor cells and lead to metastasis formation. Whether prostacyclin and other factors involved in the non-thrombogenic character of the vascular endothelium provide a significant defense against attachment of tumor cells is not known, but the potential for such a primary or secondary role clearly exists. In our studies, prostacyclin did not appear to influence the adherence of Raji lymphoma cells to the endothelium. Additional studies are indicated to correlate adherence of tumor cells to the vascular wall with their potential for formation of metastatic lesions.

References

Brotherton AFA, Hoak JC (in press). Proc Natl Acad Sci USA.
Czervionke RL, Hoak JC, Fry GL (1978). Effect of aspirin on thrombin-induced adherence of platelets to cultured cells from the blood vessel wall. J Clin Invest 62:847.
Czervionke RL, Smith JB, Hoak JC, Fry GL, Haycraft DL (1979). Use of a radioimmunoassay to study thrombin-induced release of PGI_2 from cultured endothelium. Thromb Res 14:781.
Fry GL, Czervionke RL, Hoak JC, Smith JB, Haycraft DL (1980). Platelet adherence to cultured vascular cells: Influence of prostacyclin. Blood 55:271.
Heyns ADuP, Potgieter GM, Retief FP (1974). The inhibition of platelet aggregation by an aorta intima extract. Thromb Diath Haemorrh 32:417.
Hoak JC, Warner ED, Cheng HF, Fry GL, Hankenson RR (1971). Hemangioma with thrombocytopenia and microangiopathic anemia (Kasabach-Merritt Syndrome): An animal model. J Lab Clin Med 77:941.
Lollar P, Owen WG (1980). Clearance of thrombin from circulation in rabbits by high affinity binding sites on endothelium. J Clin Invest 66:1222.
Maca RD, Fry GL, Hakes AD (1978). Effects of glucocorticoids on the interaction of lymphoblastoid cells with human endothelial cells in vitro. Cancer Res 28:2444.
Malogodi MH, Chiou CY (1974). Pharmacological evaluation of a new Ca^{2+} antagonist, 8-(N,N-diethylamine)-octyl-3,4,5-trimethoxybenzoate hydrochloride (TMB-8): Studies in smooth muscles. Eur J Pharmacol 27:25.
Moncada S, Gryglewski R, Bunting S, Vane JR (1976). An enzyme isolated from arteries transforms prostaglandin endoperoxides to an unstable substance that inhibits platelet aggregation. Nature 263:663.

Warner ED, Hoak JC, Fry GL (1971). Hemangioma, thrombocytopenia and anemia. The Kasabach-Merritt syndrome in an animal model. Arch Pathol 91:528.

DISCUSSION

Dr. H. J. Day (Temple University Medical Center, Philadelphia). I was intrigued by your feedback mechanism. Is there any evidence that different endothelial cells have abnormal or different levels of susceptibility?

Dr. J. C. Hoak (University of Iowa, Iowa City). In different types of endothelium, we find different levels of phosphodiesterase activity. So, I think the answer is going to be "yes".

Dr. S. Karpatkin (New York University Medical Center, New York). Can you tell us any more about the effects of phospholipase?

Dr. Hoak. It does not affect prostacyclin synthetase because the endoperoxide is not affected. We just do not have the answers yet.

Dr. Day. I think you showed that PGI_2 is partially affected in the presence of collagen. Have you used any antithrombin in your system?

Dr. Hoak. No, not in the way you would like.

Dr. A. Scriabine (Miles Institute for Preclinical Pharmacology, New Haven). What is known about thrombin interaction? Does it activate complement?

Dr. Hoak. It certainly activates intracellular stores of calcium and this appears to be part of the mechanism. The results are not as clean as we would like, so I am not going to say too much about it, but thrombin, as you know, releases calcium into platelets. Some people have proposed that thrombin can cause endothelial cells to contract.

Dr. M. Steiner (The Memorial Hospital, Pawtucket). Have you tried platelet agents other than thrombin?

Dr. Hoak. ADP-induced aggregates will not adhere to

normal endothelium. Collagen will do so occasionally, but nothing has been as potent as thrombin.

DISCUSSION

Dr. H. J. Day (Temple University Medical Center, Philadelphia). I was intrigued by your feedback mechanism. Is there any evidence that different endothelial cells have abnormal or different levels of susceptibility?

Dr. J. C. Hoak (University of Iowa, Iowa City). In different types of endothelium, we find different levels of phosphodiesterase activity. So, I think the answer is going to be "yes".

Dr. S. Karpatkin (New York University Medical Center, New York). Can you tell us any more about the effects of phospholipase?

Dr. Hoak. It does not affect prostacyclin synthetase because the endoperoxide is not affected. We just do not have the answers yet.

Dr. Day. I think you showed that PGI_2 is partially affected in the presence of collagen. Have you used any antithrombin in your system?

Dr. Hoak. No, not in the way you would like.

Dr. A. Scriabine (Miles Institute for Preclinical Pharmacology, New Haven). What is known about thrombin interaction? Does it activate complement?

Dr. Hoak. It certainly activates intracellular stores of calcium and this appears to be part of the mechanism. The results are not as clean as we would like, so I am not going to say too much about it, but thrombin, as you know, releases calcium into platelets. Some people have proposed that thrombin can cause endothelial cells to contract.

Dr. M. Steiner (The Memorial Hospital, Pawtucket, RI). Have you tried platelet agents other than thrombin?

Dr. Hoak. ADP-induced aggregates will not adhere to normal endothelium. Collagen will do so occasionally, but nothing has been as potent as thrombin.

CONTROL OF TUMOR GROWTH AND METASTASIS WITH PROSTACYCLIN AND THROMBOXANE SYNTHETASE INHIBITORS: EVIDENCE FOR A NEW ANTITUMOR AND ANTIMETASTATIC AGENT (BAY g 6575)

Kenneth V. Honn, Jay Meyer, Gregory Neagos, Thomas Henderson, Christine Westley and Vaneerat Ratanatharathorn.
Departments of Radiation Oncology, Radiology and Biological Sciences
Wayne State University, Detroit, Michigan 48202.

INTRODUCTION

As discussed by Clark (1979) the ideal solution to eradication of cancer is prevention of the initial transformation. The more realistic present need is to augment or facilitate the early destruction of metastatic cells while applying maximally effective therapy to the primary tumor. Metastasis as simply defined is a loss of contiguity between a tumor cell or group of tumor cells and the primary neoplasm. Several excellent reviews have examined the various aspects of this process (Hilgard et al., 1972; Warren, 1973; Salsbury, 1975; Weiss, 1976, 1977; Fidler, 1978; Carter, 1978; Spreafico and Garattini, 1978; Donati et al., 1981). Simply stated, after an initial period of primary tumor growth and invasion into adjacent normal tissue, tumor cells either singly or in clumps are detached and penetrate either lymphatic or blood vessels, whereupon they are disseminated. The separation between the two systems of dissemination, lymphatic and hemotagenous, may be more an oversimplification than real. Hilgard (1972) has noted that tumor cells initially distributed in lymphatics will reach the blood vessels. Similar conclusions have been reached by other workers (Fisher and Fisher, 1967; del Regato, 1977). Once introduced into circulation tumor cells possess a relatively short half life. Fidler (1978) has reported that only 1% of the original tumor cell population introduced via tail vein injection survive to form metastatic colonies. It has been stated that the mere presence of tumor cells in circulation does not always lead to metastasis (Weiss, 1977; Sugarbaker and Ketcham, 1977). Salsbury (1975) concluded that there is no evidence that the mere presence of tumor cells in the circulation indicates a poor prognosis as Fidler's (1978) stated 1% survival might indicate. Nevertheless, the use of the hematogenous route for the spread of malignant disease is an event which does undeniably occur at some point in time.

Detachment of tumor cells from a primary tumor may be a continuous process (Weiss, 1976) which might begin early in the growth of the primary tumor (Franks, 1973). It has been suggested that massage, incision or surgical trauma to the primary tumor can send a shower of tumor cells into circulation (Cameron, 1954; Fisher and Fisher, 1967).

Once in circulation tumor cells can assume various forms of transport. They can travel singly, or in clumps enmeshed in a coating of fibrinlike material (Warren and Vales, 1972). They can exist attached to blood platelets, RBC's, leukocytes, etc. All of the above are possible and have been observed in one tumor type or another. It does not necessarily follow that all of the above modes of transport will allow for successful metastasis; one transport form may enhance chances of survival over another. It is intuitive that for successful mestastasis to occur, the metastatic cell must first arrest and adhere to the vascular endothelium and remain intravascular until such time as it can extravasate either by destruction of the endothelium (Hilgard, 1973; Chew et al., 1976) or diapedesis (Dingemans et al., 1978). The actual mechanism for extravasation is unknown and currently a matter of some debate. The fate of the arrested tumor cell is by no means guaranteed. Some cells loosely adherent to the endothelium may become detached, shed into circulation and die (Wood, 1971). It has also been observed both in vivo (Warren and Vales, 1972) and in vitro (Kramer and Nicolson, 1979) that tumor cells adhere more readily to damaged endothelium than to intact endothelium. Whether this is due to a loss of some protective substance produced by intact endothelium, coaggregation with platelets or some other mechanism is unknown.

The role of platelets in tumor metastasis has been the subject of much debate for several years. Morphologists at the beginning of the century described tumor cells enmeshed within a thrombus and adherent to the capillary wall. In a classic study Wood (1971) visualized via continuous cinemicrography of rabbit ear chambers the fate of intraarterially injected V2 squamous carcinoma cells. Small clumps (6-10) of tumor cells were found to arrest and subsequently become enmeshed in a clot of fibrin and platelets. Within 3-6 hrs the tumor cells had extravasated. Wood concluded that the initial site of arrest was not determined by vessel diameter or rate of blood flow, a concept supported by the observations of Zeidman (1961). Gasic et al. (1968) suggested that tumor cells in circulation could aggregate platelets and possibly enhance tumor cell survival and adhesion. In a series of follow-up papers Gasic and coworkers (Gasic et al., 1972, 1973, 1976, 1977) demonstrated that in a number of tumor types which will aggregate platelets in vitro metastasis was reduced by induced thrombocytopenia or aspirin treatment.

The mechanism of tumor cell induced platelet aggregation is unknown, however, recent evidence (Gasic et al, 1978; Dvorak et al., 1981) has demonstrated that tumor cells spontaneously shed plasma membrane derived vesicles. These shed vesicles carry procoagulant activity that is reported to be responsible for activation of the clotting system and the fibrin deposition associated with many types of malignancies. Karpatkin and Pearlstein (Pearlstein et al., 1979, 1980; Karpatkin et al., 1980) have isolated platelet aggregating material (PAM) from SV-40 transformed mouse 3T3 fibroblasts. PAM does not aggregate washed platelets in the presence or absence of fibrinogen, however, this substance will aggregate washed platelets in the presence of platelet poor plasma. These authors indicate a minimum divalent cation requirement and the requirement of a plasma factor for PAM activity. The platelet aggregating activity of tumor cells has been partially characterized (Pearlstein et al., 1979; Hara et al, 1980) and it is found to contain a protein, lipid, and sialic acid component. The presence of all three components appear to be essential for stimulating platelet aggregation (Hara et al., 1980).

The aforementioned results would suggest that some form of anticoagulant therapy might prove efficacious in the treatment of metastasis. Nevertheless results to date have been ambivalent.

HYPOTHESES

Prostaglandin research in the past several years has uncovered the critical role played by thromboxane A_2 (TXA_2; formed by platelets and lung) and prostacyclin (PGI_2; formed by vascular endothelium and lung) in the platelet aggregation mechanism. Both compounds are products of arachidonic acid metabolism (Fig. 1). It has been suggested that a delicate balance exists between the PGI_2 and TXA_2 systems, a balance which is necessary for normal hemostasis. We propose the hypothesis that the primary tumor, tumor cell shed vesicles and/or circulating tumor cells disrupt this balance in favor of platelet aggregation. The failure and/or divergent results which have been obtained by others attempting anticoagulant therapy to alter metastasis could be due to the fact that the agents tested, with the exception of aspirin, do not impinge directly upon the intravascular PGI_2/TXA_2 balance. If this hypothesis is correct then the following criteria should be substantiated by experimental fact: 1) the exogenous adminstration of PGI_2 should reduce lung colony formation by tail vein injected tumor cells, 2) a therapeutic synergism should result from the use of PGI_2 with a phosphodiesterase inhibitor. [Since the effect of PGI_2 is mediated by increasing concentrations of adenosine 3',5'-monophosphate (cAMP) in platelets it follows that phosphodiesterase inhibitors by slowing the breakdown of cAMP should potentiate the antithrombogenic action of PGI_2 and thus the anti-metastatic effect.], 3) an inhibitor of endogenous PGI_2 synthesis should enhance metastasis, 4) thromboxane synthetase inhibitors should also function as antimetastatic agents, and 5) agents that augment in vivo PGI_2 synthesis or activity should function as antimetastatic agents.

A possible criticism of the development of antimetastatic agents for use in human malignancies is the fact that many patients present with occult metastatic disease. This does not obviate the necessity for prevention of further metastasis but suggests that ideally an agent and/or agents should be developed which possess both antimetastatic and antitumor activities. Consideration of the mechanism of action of PGI_2 and TXA_2 in the regulation of platelet cAMP levels has led us to propose the following hypothesis. Prostacyclin increases cAMP levels in the platelet while TXA_2 prevents a rise in cAMP levels in response to external stimuli (Gorman et al, 1977). Considering that increased cAMP levels are believed to inhibit cell division and promote differentiation (Kreider et al, 1975; Friedman et al, 1976), we therefore propose that thromboxane synthetase inhibitors, PGI_2 and agents which may increase PGI_2 production could have an effect on inhibition of tumor growth and possibly tumor cell differentiation. If this hypothesis is correct, then the following criteria should be substantiated by experimental fact: 1) thromboxane synthetase inhibitors should decrease tumor cell replication, 2) TXA_2 or its stable metabolite TXB_2 should enhance tumor cell proliferation, 3) the TXA_2 mimicking agent (Smith, 1980) (15S)-Hydroxy-11α,9α-(epoxymethano) prosta-5Z,13E-dienoic acid, should enhance tumor cell proliferation, 4) exogenous PGI_2 should decrease tumor cell proliferation, and 5) agents that stimulate PGI_2 production should decrease tumor cell proliferation.

MATERIALS AND METHODS

In vivo maintenance of the B16 amelanotic melanoma and Lewis Lung carcinoma tumor lines: The B16 amelanotic melanoma (B16a) and Lewis Lung carcinoma (3LL) were originally obtained from the DCT Human and Animal Tumor Bank. Routine tumor transplantation was performed every three weeks into male, syngeneic host mice (C57BL/6J; Jackson Laboratories). All mice used for transplantation were between 17-22g and were housed under identical conditions of photoperiod, feeding regime, temperature, etc.

Isolation of tumor cells: All of the studies discussed below utilized monodispersed cells obtained from a primary subcutaneous tumor. Cells were obtained by a modification (Sloane et al., 1981) of the procedure described by Honn et al. (1979). The subcutaneous tumors used for these isolations were of similar age post-transplantation and weight (1.5-2.5 g). Briefly, aseptically removed tumors were diced and dispersed using sequential collagenase digestion. The collagenase used was type CLS III (Worthington Biochem). This type of

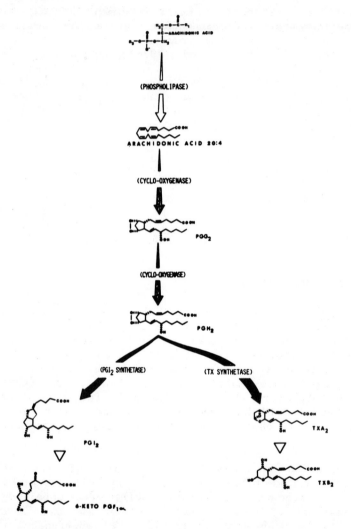

Figure 1. Biosynthesis of PGI_2 and TXA_2 by the oxidative cyclization of the polyunsaturated fatty acid, arachadonic acid (20:4). The endoperoxide intermediate PGH_2, serves as a pivotal point in the biosynthesis of prostaglandins and related compounds. PGH_2 can serve as substrate for prostacyclin synthetase and thromboxane synthetase. PGI_2 and TXA_2 can be nonenzymatically converted to their stable metabolites 6 keto $PGF_{1\alpha}$ and TXB_2 respectively.

collagenase is unusually low in clostripain and all other proteolytic activities normally found in crude collagenase preparations (Mitchell, 1968). The final dispersions were obtained by subjecting the cells to centrifugal elutriation (Meistrich et al, 1977; Sanderson and Bird, 1977) using a Beckman JE-6 Elutriator rotor in a J2-21 centrifuge, as previously described (Sloane et al, 1981). Final cell dispersion characteristics: cell debris, 0; viability > 90%; contaminating host stromal elements, < 1%; monodispersed cells, 100%.

For in vitro work tumor cells were adapted for growth in tissue culture medium [Eagle's MEM with Hank's salts (GIBCO, Grand Island, NY), supplemented with sodium pyruvate, MEM non-essential amino acids (GIBCO), 150 U penicillin-G/ml (Sigma, St. Louis, MO), 100 μ g/ml/neomycin sulfate (Sigma), 25mM HEPES (Sigma) and 10% fetal calf serum (M.A. Bioproducts, Walkersville, MD)]. Growth medium in stock cultures was changed twice per week and the cells were subcultured once per week. For enumeration the cells were detached by a one min trypsinization [0.25% trypsin (Worthington, Freehold, NJ); 0.02% EDTA in MEM], pelleted at 100 X g, resuspended in MEM and counted in a hemocytometer. ^{3}H-thymidine incorporation into DNA was determined as previously described (Honn et al., 1979).

All data were analyzed by an analysis of variance and a two-tailed Student's t test and were considered significantly different from controls if $P \leq 0.05$.

Role of PGI_2 in the prevention of metastasis

Prostacyclin is the most potent antithrombogenic agent (Moncada and Vane, 1979) known both in vitro (Moncada and Vane, 1977) and in vivo (Bayer et al, 1979). In addition, unlike most prostaglandins, PGI_2 is not inactivated by the lungs (Gryglewski et al., 1978) and may function as a circulating hormone (Moncada et al., 1978). In fact, the biological $t_{\frac{1}{2}}$ of PGI_2 in vivo has been demonstrated to be considerably lengthened by binding to serum albumin (Wynalda and Fitzpatrick, 1980). Furthermore, PGI_2 is a potent vasodilator (Moncada and Vane, 1979), even in the microvasculature (Higgs et al., 1979), and inhibits the release of serotonin from platelets (Cazencave et al., 1979). The latter effect is of particular interest since serotonin is a potent vasoconstrictor whose release from platelets can be stimulated by a variety of animal and human tumor cells (Gasic et al., 1976). For the above reasons we chose to investigate the antimetastatic effects of prostacyclin in vivo.

Prostacyclin and its major degradatory product, 6 keto-$PGF_{1\alpha}$ were prepared and injected as previously described (Honn et al., 1981).

Prostacyclin produces a dose related decrease in experimental metastasis of the B16a melanoma (Table 1). A significant decrease was observed with as little as 25 µg/animal. At the highest dose tested (200 µg) lung metastatic colonies were reduced to 14% of controls. Metastasis to the liver and spleen were also reduced with PCI_2 suggesting that the decrease in lung metastasis does not merely reflect a redistribution of tumor cells (Table 1). The byproduct of PCI_2 hydrolysis, 6-keto $PGF_{1\alpha}$ was ineffective at a dose of 50 µg/animal; however, at 200 µg/animal a slight reduction in lung metastasis was noted, probably due to the pulmonary vasodilator effect of this compound (Hyman and Kadowitz, 1979). However, it should be noted that although PCI_2 is a potent pulmonary vasodilator, such vasodilation is not principally responsible for the antimetastatic effects which have been observed. Honn et al. (1981) have demonstrated that prostaglandin D_2 (PGD_2), which is an approximately equipotent vasodilator but does not inhibit platelet aggregation, was ineffective in reducing B16a melanoma tumor cell metastasis. Theophylline, a platelet phosphodiesterase inhibitor (Moncada and Vane, 1979), at 100 µg/animal produced a slight but insignificant decrease in pulmonary metastasis. However, the combination of theophylline plus PCI_2 (100 µg/animal) resulted in a four fold reduction in pulmonary metastasis over PCI_2 alone at the same dose (Table 1).

Heparin and warfarin anticoagulation have been demonstrated to decrease the incidence of metastasis from small numbers ($<10^5$) of i.v. injected tumor cells. However, metastasis from large numbers ($>10^6$) of injected tumor cells was not prevented (Gralnick, 1981).

In order to assess the effect of a single PCI_2 injection on a large circulating tumor burden, mice were pretreated with theophylline (100 µg, i.p.) + PCI_2 (150 µg, i.v.) before the injection of 3×10^6 B16a melanoma cells. A tumor burden of 3×10^6 cells resulted in >500 tumor colonies in the control, while theophylline + PGI_2 treated mice exhibited 37±10 colonies (Fig. 2).

If PGI_2 is efficacious in preventing metastatic tumor colony formation, then an inhibitor of endogenous PGI_2 synthesis should enhance metastasis. To test this hypothesis we synthesized 15-hydroperoxy arachidonic acid (15-HPETE) from arachidonic acid utilizing soybean lipoxygenase. Hydroperoxy fatty acids are in general a very potent and specific inhibitors of prostacyclin synthetase (Salmon et al., 1978; Siegel et al., 1979).

Compared to controls, 100 μg of 15-HPETE increased lung tumor colony formation almost 200% (Table 2). A dose of 200 μg nearly quadrupled that number (Table 2). Liver metastasis increased 667% and 944% with 100 μg and 200 μg of 15-HPETE respectively (Table 2). There was also an increase in metastasis to the spleen (Table 2). In order to ascertain if exogenous PGI_2 could reverse the effects of 15-HPETE, animals were injected with 100 μg 15-HPETE followed by an i.p. injection of 100 μg theophylline. Thirty minutes later 100 μg PGI_2 was administered. The results in Table 2 indicate that PGI_2 significantly reversed the effect of 15-HPETE in lung, liver and spleen, reducing metastatic colony formation to below control levels.

Collectively these results indicate that a vital role may be played by in vivo PGI_2 synthesis in preventing the spread of metastatic disease. Furthermore, exogenous PGI_2 may be efficacious as an adjuvant chemotherapeutic agent to reduce the total tumor burden prior to and immediately following therapies aimed at the the removal or destruction (surgery/radiotherapy) of the primary neoplasm.

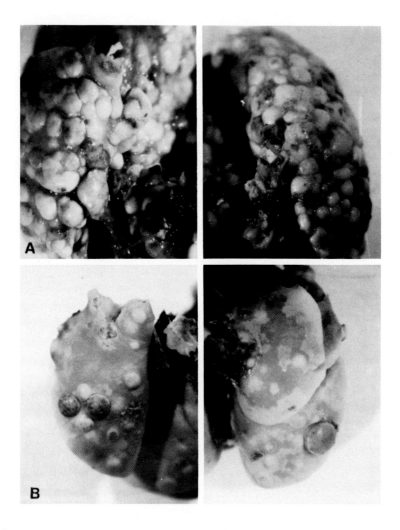

Figure 2. Inhibition of lung colony formation from tail vein injected B16a cells (3×10^6) by PGI_2 pretreatment. A. Representative lungs from untreated controls. B. Representative lungs from animals pretreated with PGI_2 prior to tumor cell injection.

Table 1: Effects of PGI_2 on pulmonary and extrapulmonary metastasis following injection of 3×10^5 viable b16a melanoma cells

Treatment	Lungs	Liver	Kidney	Spleen	Brain
MEM control	144±18[a]	13±5	0	1±0.5	0
TRIS control	130± 9	8±3	0	3±1	0
PGI_2 25 µg	95±10	2±0.5	0	0	0
PGI_2 50 µg	77± 5	0	0	0	0
PGI_2 100 µg	43±13	0	0	0	0
PGI_2 200 µg	15± 7	0	0	0	0
6-keto $PGF_{1\alpha}$ 50 µg	138±15	0	0	0	0
6-keto $PGF_{1\alpha}$ 200 µg	100±14	3±1	0	0	0
Theopnylline 100 µg+PGI_2 100 µg	10± 5	0	0	0	0
Theophylline 100 µg	119± 8	7±3	0	0	0

[a] Mean ± SEM; n = 7

Table 2: Effect of 15-hydroperoxyarachidonic acid (15-HPETE) and PGI_2 on pulmonary and extrapulmonary metastasis following injection of 3×10^5 viable B16a melanoma cells[a].

Treatment	Lung	Liver	Kidney	Spleen	Brain
MEM control	144±18[c]	13± 5	0	1±0.5	0
TRIS control	130± 9	8± 3	0	3±1	0
Ethanol control	127±14	9± 5	0	0	0
15-HPETE 100 μg[b]	380±40	60±21	0	17±6	0
15-HPETE 200 μg	>500	85±33	0	22±8	0
15-HPETE 100 μg + theophylline 100μg + PGI_2 100 μg	39±16	3±2	0	0	0

[a] 25 μl i.v. (tail vein)
[b] Injected in 25 μl ethanol i.v. (tail vein)
[c] Mean ± SEM; n = 7

Thromboxane Synthetase Inhibitors and Metastasis

It is possible that metastasis may result from an imbalance in the PGI_2/TXA_2 system. These two compounds, TXA_2 produced by platelets and PGI_2 produced by vascular endothelium, are central to the control of the thrombogenic mechanism (Marcus, 1978). It is well known that there is a close clinical correlation between microangiopathic haemolytic anemia and disseminated carcinoma (Brain et al., 1970). It is also known that vascular PGI_2 production is decreased in several clinical and experimental conditions associated with thrombotic or microangiopathic conditions (de Gaetano et al, 1979).

We have presented data in the preceding section to indicate that PGI_2 may function as a potent antimetastatic agent. According to the hypothesis which we presented earlier, it follows that thromboxane synthetase inhibitors should also function as antimetastatic agents. We have screened three compounds which are endoperoxide analogues. These compounds and their actions are as follows: 1) 9,11-diazo-prosta-5,13-dienoic acid (U51605), a thromboxane synthetase and prostacyclin synthetase inhibitor (Fitzpatrick and Gorman, 1978); 2) 9,11-iminoepoxy-prosta-5,13-dienoic acid (U54701), an inhibitor of thromboxane synthetase and stimulator of prostacyclin synthetase (Fitzpatrick et al., 1978); 3) 9,11-epoxy imino-prosta-5,13-dienoic acid (U54874), a thromboxane A_2 receptor antagonist (61). The structures of these compounds are shown in Figure 3. It is evident from the data presented in Table 3 that all of these compounds significantly reduced pulmonary metastasis from tail vein injected B16a melanoma cells.

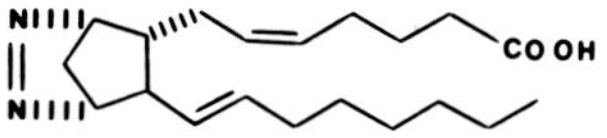

9,11-Diazo-Prosta-5,13-Dienoic Acid
(U51605, Upjohn Company)

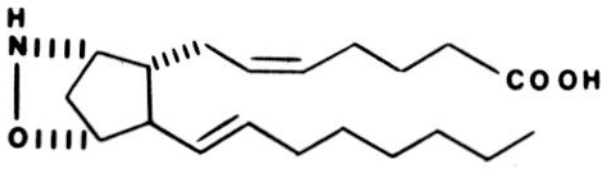

9,11-Imminoepoxy-Prosta-5,13-Dienoic Acid
(U54701, Upjohn Company)

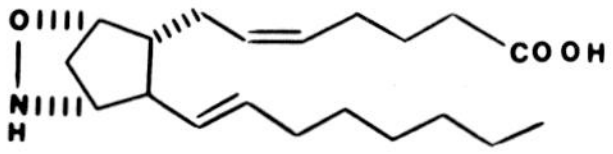

9,11-Epoxyimino-Prosta-5,13-Dienoic Acid
(U54874, Upjohn Company)

Figure 3. Chemical structures of endoperoxide analogues, U51605, U54701 and U54874.

However, the most effective drug tested was 9,11-iminoepoxy-prosta-5,13-dienoic acid, the agent which is reported to inhibit thromboxane synthetase and stimulate

Table 3: Effect of thromboxane synthetase inhibitors and receptor antagonist on pulmonary metastasis from B16a melanoma cells.

Treatment	Tumor Colonies
Control[a]	102 ± 26[b]
U51605	
100 µg	75 ± 23
200 µg	56 ± 13
U54701	
100 µg	21 ± 7
200 µg	6 ± 1.3
U54874	
100 µg	47 ± 17
200 µg	35 ± 9

[a] 3×10^5 B16a cells injected in 50 µl.
[b] Number of metastatic colonies on lung surface (bilateral), mean ± SEM; n = 10

prostacyclin synthetase. At an injected dose of 200 µg this compound reduced tumor colony formation from 102±26 in the controls to 6±1.3 in the treated group. Representative lungs from control and treated animals are shown in Figure 4.

Considering the complexity of events leading to the eventual establishment and growth of a secondary metastatic lesion, there is little doubt that experimental metastasis via tail vein injection of dissociated tumor cells is an artificial and partial model. Nevertheless, it has its usefulness when standardization of experimental conditions is critical. However, in order to assess the effects of thromboxane synthetase inhibitors on spontaneous metastasis, animals bearing subcutaneous B16a melanoma and Lewis lung carcinoma were tumors treated with the thromboxane synthetase inhibitor 1-(7-carboxyheptyl) imidazole (Yoshimoto, 1978). Treatment (2mg/animal/day/i.p.) began with the appearance of a palpable tumor. The data in Table 4 clearly indicates the antimetastatic effects of this thromboxane synthetase inhibitor in the complete metastasis model. Spontaneous metastasis was inhibited 60% for the B16a tumor and 74% for the 3LL tumor.

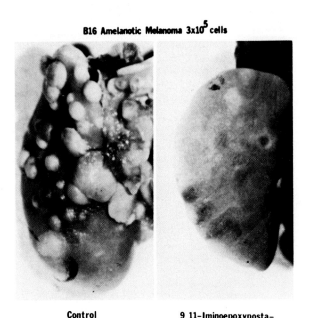

Figure 4. Inhibition of lung colony formation from tail vein injected B16a cells by pretreatment with a thromboxane synthetase inhibitor.

Table 4: Effect of 1-(7-carboxyheptyl) imidazole on pulmonary metastasis from subcutaneous B16a melanoma and Lewis lung carcinoma tumors

	Control	1-(7-carboxyheptyl)imidazole
B16a melanoma	48 ± 10[a]	19 ± 9
Lewis lung carcinoma	89 ± 17	30 ± 11

[a]No. of metastatic colonies on lung surface (bilateral), mean ± SEM; n = 12.

Nafazatrom (Bay g 6575), a New, Potent Antimetastatic Agent.

Nafazatrom (Fig. 5) has been reported to possess significant antithrombotic activity in model systems of experimental thrombosis (Seuter et al., 1979). Thrombus formation in the femoral arteries of rabbits was inhibited at a minimal effective dose of 1 mg/kg p.o. Nafazatrom also possesses significant thrombolytic properties similar to urokinase (Seuter et al., 1979). The mechanism of action for these antithrombotic effects appears related to the ability of the drug to stimulate PGI_2 production by the vascular wall (Vermylen et al., 1979). Nafazatrom significantly increased bioassayable PGI_2 release from aortic rings obtained from normal and diabetic rats (Carreras et al., 1980). In addition, plasma, obtained from human volunteers after ingestion of a single dose (1.2 g) of nafazatrom, stimulated PGI_2 release from slices of rat aorta (Vermylen et al., 1979). The reported action of this new drug on PGI_2 production and its low toxicity in vivo (Seuter et al., 1979; Vermylen et al., 1979) suggested its possible use as an antimetastatic agent.

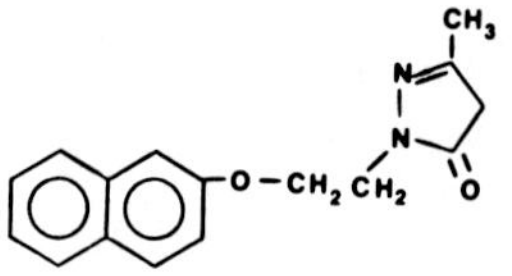

1-[2-(β-napthyloxy)ethyl]-3-methyl-2-pyrazolin-5-one (Bay g 6575, Bayer Pharmaceutical)

Figure 5. Chemical structure of nafazatrom (BAY g 6575).

Mice were pretreated (3 days) with 0.02 and 0.08 mg (s.c.) nafazatrom prior to tail vein injection of 5×10^4 B16a melanoma cells. Both doses produced significant reduction (90% and 99% respectively) in lung tumor colony formation (Table 5). Representative lungs from control and 0.08 mg nafazatrom treated animals are shown in Fig. 6. This compound was also effective when injected 1 hr prior to tumor cell injection, an effect which was potentiated by theophylline (Table 5). Theophylline alone (200 μg i.p.) was ineffective.

Table 5: Effect of nafazatrom on metastasis from tail vein injected B16a melanoma cells[a]

Treatment	Lung Tumor Colonies
Control	181 ± 45[b]
0.02 mg Nafazatrom[c]	19.3 ± 7.5
0.08 mg Nafazatrom[c]	2.7 ± 1.3
Theophylline 200 µg[d]	165 ± 38
Theophylline 200 µg + 0.08 mg Nafazatrom[d]	33.6 ± 18
0.08 mg Nafazatrom[d]	65 ± 36

a. 5×10^4 cells injected intravenously in 50 µl.
b. $\bar{X}$ ± SEM; n = 6.
c. Animals pretreated daily (3 days) before tumor cell injection.
d. Injected 1 hour prior to tumor cells.

Nafazatrom was also evaluated in spontaneous metastasis models. Mice bearing subcutaneous B16a melanoma and Lewis lung carcinoma tumors were injected (s.c.) daily with 0.01 to 0.08 mg nafazatrom. All doses tested significantly reduced spontaneous metastasis to the lungs of tumor bearing mice (Table 6). In addition, the number of mice positive for metastasis was also reduced from 12/12 in the controls to 5/12 in B16a and 2/12 in Lewis lung with 0.08 mg of nafazatrom (Table 6). Collectively, these results point to significant antimetastatic properties of nafazatrom, an effect which may be mediated by the ability of this drug to stimulate endogenous PGI_2 production.

Table 6: Effect of nafazatrom on spontaneous metastasis from subcutaneous B16a melanoma[a] and Lewis lung carcinoma[b] tumors.

Treatment	B16a	3LL
Control	14.1 ± 3.1[c] (12/12)[d]	34.5 ± 6.4[c] (12/12)[d]
Nafazatrom[e]		
0.01 mg	1.7 ± 0.7 (7/12)	--
0.02 mg	2.5 ± 0.9 (7/12)	--
0.04 mg	3.1 ± 0.8 (8/12)	--
0.08 mg	1.4 ± 0.8 (5/12)	2 ± 0.5 (2/12)

a. 1.8×10^5 cells injected subcutaneously.
b. 1×10^5 cells injected subcutaneously.
c. No. of metastatic tumor colonies on bilateral lung surface; $\bar{X}$ ± SEM; n = 12.
d. No. of animals positive for metastasis
e. Injected daily subcutaneously in 0.2 ml.

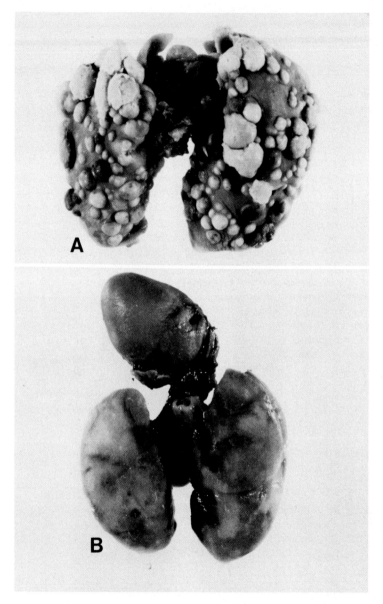

Figure 6. Inhibition of lung colony formation from tail vein injected B16a cells by pretreatment (0.08 mg x 3 days, s.c.) with nafazatrom. A. Representative lungs from untreated control animal. B. Representative lungs from animals receiving nafazatrom pretreatment.

Thromboxane A_2 and Prostacyclin Control of Tumor Cell Growth and Differentiation.

Parker and colleagues (Kelly and Parker, 1979; Kelly et al., 1979; Parker et al., 1979) have demonstrated that arachidonic acid enhances DNA synthesis in mitogen stimulated lymphocytes. Furthermore, selected inhibitors of arachidonic acid metabolism inhibit this response. The causative agent has not been identified, however, it appears likely that it is TXA_2 (Parker et al., 1979). This is not unreasonable considering that the mechanism of action of TXA_2 in the platelet is to prevent a rise in cAMP levels (Moncada and Vane, 1979). Increased cAMP levels are believed to inhibit cell division and promote differentiation. Examination of the literature related to thromboxanes and tumor growth reveals a paucity of information, however, two papers are worth noting. Chang et al. (1977) working with a carrageenin induced granuloma identified the major products of arachidonic acid metabolism in homogenates of this tissue. Classical prostaglandins (PGE_2, $PGF_{2\alpha}$) were found in small amounts, however, the major product was TXB_2, the stable metabolite of TXA_2. Murota et al. (1977) tested the effects of TXB_2 upon a fibroblast cell line derived from a carrageenin granuloma and found it to produce a 9 fold stimulation of DNA synthesis compared to controls. RNA synthesis increased 50%. These combined results suggested to us a possible role for thromboxane synthetase products in tumor cell proliferation and have led to the hypothesis that TXA_2 and PGI_2 possess diametrically opposed actions on cell proliferation. To test this hypothesis we examined the effects of the endoperoxide analogues (U51605, U54701, U54874) on DNA synthesis by B16a and 3LL cells *in vitro*. All three compounds significantly reduced DNA synthesis by these tumor cells at non-cytotoxic doses (Table 7). In the B16a cells the order of effectiveness was: U51605=U54874 >U54701. In 3LL cells U54701 was slightly more efficacious than U51605 or U54874 (Table 7). Similar inhibition of DNA synthesis by B16a and 3LL cells has been obtained with two additional, structurally unrelated, thromboxane synthetase inhibitors (data not shown) indicating that such inhibition was not peculiar to the endoperoxide structure.

Cultured B16a cells exposed to U51605 and U54701 (5-25 µg/ml) for 7 days demonstrated a dose dependent inhibition of proliferation. At 25 µg/ml inhibition of proliferation by U51605 and U54701 was 85% and 70% respectively.

Table 7: Inhibition of B16a melanoma and Lewis lung carcinoma cell DNA synthesis with thromboxane synthetase inhibitors and a thromboxane A_2 receptor antagonist.

Test Compound	Dose (µg/ml)	B16a 4 hr	B16a 18 hr	3LL 4 hr	3LL 18 hr
U51605	1	73.2±3.8[a]	67.0±6.9	88.8±7.4	77.3±11.8
	10	33.7±0.6	37.2±1.5	35.5±2.9	44.3± 2.3
	25	13.1±1.0	3.5±0.9	18.6±1.9	30.2±2.4
U54701	1	86.7±8.6	115.2±9.6	88.0±7.4	50.9±3.4
	10	50.0±0.8	76.9±2.6	53.9±1.7	44.7±8.1
	25	23.3±1.0	23.6±1.8	28.8±2.1	36.1±3.7
U54874	1	108.0±7.0	115.4±8.9	107.2±1.8	84.8±22.4
	10	70.9±4.0	56.2±1.4	66.8±1.3	75.8± 6.2
	25	27.3±1.3	4.3±0.3	38.5±1.6	33.0± 1.5

[a] % control ± SEM

The above results would suggest that endogenous TXA_2 production by the tumor cells is a necessary positive signal for proliferation. In order to test the ability of tumor cells to respond to exogenous thromboxanes, B16a cells were cultured (7 days) in the presence of either TXB_2 (the stable metabolite of TXA_2) or a TXA_2 mimicking agent (Smith, 1980), (15S)-Hydroxy-11α,9α-(epoxymethano) prosta-5Z,13E-dienoic acid. Both compounds produced a dose dependent stimulation of B16a tumor cell proliferation (Figs. 7, 8) with the TXA_2 mimicking agent effective at a dose of 1 µg/ml. The mechanism of action of thromboxanes in the stimulation of tumor cell proliferation is unknown. However, we have observed that B16a melanoma cells pretreated (48 hrs) with TXB_2 (20 µg/ml) and subsequently challenged with PGE_1 (2.5 µg/ml) have diminished release (90% decrease) of cAMP into the media when compared to PGE_1 stimulation of control cultures (unpublished observation). Whether this reflects decreased cyclic nucleotide synthesis or increased catabolism is unknown at present. However, if the mechanism of action of these compounds centers around the downward modulation of cAMP levels, agents such as

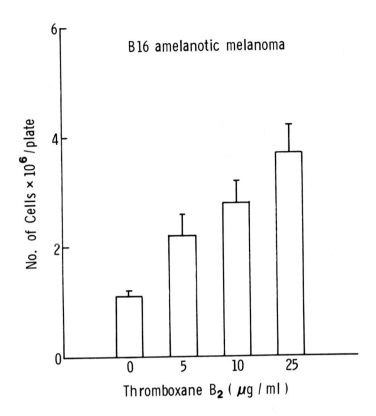

Figure 7. Dose dependent increase in B16a melanoma cells in culture by treatment with TXB_2. Media was changed daily and TXB_2 was added daily at the doses indicated for 7 days. Cells were detached from the plates by brief trypsinization and counted in a hemocytometer.

PGI_2, known to increase cAMP levels, should inhibit proliferation. Treatment of B16a melanoma cells in culture for 7 days with PGI_2 produced a dose dependent decrease in cell proliferation (Fig. 9). Collectively these results suggest the possibility of bidirectional control of tumor cell proliferation by the balance between endogenous and/or exogenous PGI_2/TXA_2.

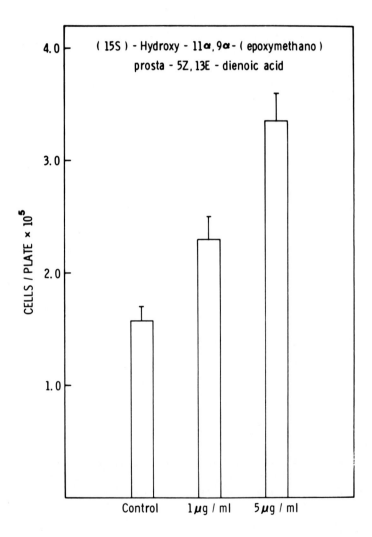

Figure 8. Dose dependent increase in B16a melanoma cell proliferation by treatment with a TXA_2 mimicking agent. Drug was added as described for TXB_2 in Figure 7.

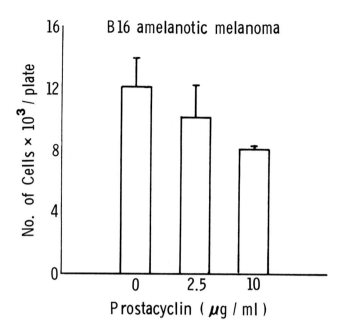

Figure 9. Dose dependent decrease in proliferation of B16a melanoma cells with PCI_2.

As stated earlier, the antithrombogenic and antimetastatic effects of nafazatrom are believed to be mediated in part by an increase in PCI_2 levels. Considering the effects of exogenous PCI_2 on tumor cell replication (Fig. 9) we examined the effects of nafazatrom on DNA synthesis and cell proliferation by B16a melanoma cells in culture. Exposure of B16a melanoma cells to nafazatrom for 4 hr produced a dose dependent decrease in DNA synthesis with an ED_{50} of 0.6 μg/ml (Fig. 10). Similar inhibition by Nafazatrom has been observed with the Lewis lung carcinoma, Friend erythroleukemia and neuroblastoma N-2 cells (unpublished observations). Exposure of B16a cells in culture to nafazatrom (14 days) produces a dose dependent inhibition of proliferation which is not due to drug related cytotoxicity (Fig. 11).

The effects of nafazatrom on cell differentiation were also evaluated. Mouse 3T3L1 fibroblasts differentiate into mature adipocytes (Hopkins and Gorman, 1981) in response to various stimuli (e.g., insulin). Differentiation can be determined by the staining of the mature adipocytes with the lipid stain, oil red O, or the determination of cellular triglyceride levels. Incubation of confluent 3T3L1 cells with nafazatrom over a dose range of 0.1 to 20 µg/ml produced a dose dependent increase in oil red O positive material (Figure 12). A concurrent increase in triglyceride levels were also observed (data not shown). In addition, nafazatrom has been found to increase the differentiation of mouse neuroblastoma N-2 cells (Prasad, 1972) as determine by increased axon formation (unpublished observation), and the differentiation of Friend erythroleukemia cells (Santoro et al., 1979) into mature erythroblasts (unpublished observation).

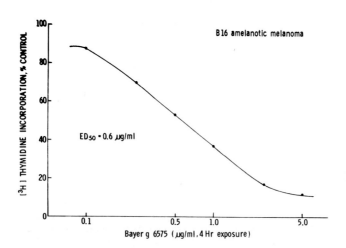

Figure 10. Effects of nafazatrom on ^{3}H-thymidine incorporation into DNA of B16a melanoma cells.

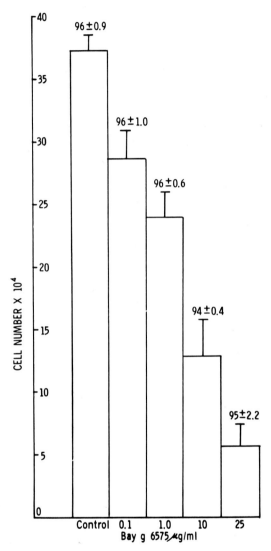

Figure 11. Effects of nafazatrom (BAY g 6575) on replication of B16a melanoma cells in culture. Media was changed on alternate days at which time fresh compound was added at the doses indicated. Cells were counted and cell number per plate was determined after 14 days exposure. Numbers appearing above each bar represent viabilities as determined by vital dye exclusion (mean ± SEM; n = 5).

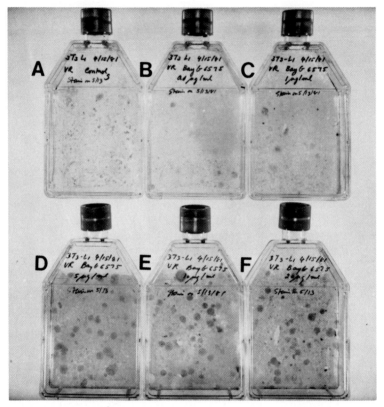

Figure 12. Effects of nafazatrom on differentiation of 3T3L1 fibroblast into adipocytes. Flasks were stained with oil red O for lipid identification. A. Control cultures receiving diluent alone. B. 0.01; C. 1; D. 5; E. 10; and F. 20 μg nafazatrom/ml.

SUMMARY

Honn et al. (1981) have proposed the hypothesis that tumor cells can alter the critical intravascular PGI_2/TXA_2 balance in favor of platelet aggregation. Such alterations could activate the coagulation mechanism not only at the site of the primary tumor, but also systemically, leading to disseminated intravascular coagulation (DIC). The association of disseminated neoplasia with thromboembolic disease is well recognized (Sack et al., 1977). The above hypothesis suggests that specific TXA_2 synthetase inhibitors, PGI_2, or agents which augment in vivo PGI_2 synthesis or activity could function as antimetastatic agents. PGI_2 has been demonstrated to significantly reduce pulmonary tumor colony formation from B16a melanoma cells (Honn et al., 1981). This inhibition did not appear to be related to the effect of PGI_2 on pulmonary vasodilation and was potentiated by the phosphodiesterase inhibitor theophylline (Honn et al., 1981). An inhibitor of PGI_2 synthetase (15-HPETE) when injected in vivo increased the number of pulmonary and extrapulmonary metastatic lesions resulting from tail vein injected B16a melanoma cells (Honn et al., 1981). In addition, the daily injection of a TXA_2 synthetase inhibitor has been reported to reduce spontaneous pulmonary metastasis from subcutaneous Lewis lung carcinoma and B16a melanoma tumors (Honn et al., 1980). Interestingly, the most consistent anticoagulant therapy for the control of metastasis has been reported with warfarin (Lione and Bosmann, 1978), which has been demonstrated to decrease the half-life of TXA_2, presumably by interfering with its binding to albumin (Folco et al., 1977). Such an effect, in vivo, would be expected to shift the PGI_2/TXA_2 balance in favor of PGI_2. These results demonstrate that selective manipulation of the arachidonic acid cascade might be used to therapeutic advantage in control of the hematogenous dissemination of tumor cells.

We have also demonstrated that an agent (nafazatrom) which is believed to stimulate production of endogenous PGI_2 is also a potent antimetastatic agent. We have also suggested the hypothesis that the manipulation of tumor intracellular PGI_2/TXA_2 ratios can regulate tumor cell growth and differentiation. Evidence has been presented demonstrating that selective thromboxane synthetase inhibitors inhibit tumor cell growth. Thromboxane B_2 and a TXA_2 mimicking agent have been shown to stimulate tumor cell proliferation, and exogenous prostacyclin has been demonstrated to inhibit tumor cell proliferation. Nafazatrom, the agent believed to stimulate endogenous PGI_2 production inhibits tumor cell DNA synthesis and proliferation and has been demonstrated to induce differentiation in several models.

ACKNOWLEDGEMENT

Work in the authors' laboratory has been supported by research grants from the National Institutes of Health, CA29405, CA29997, the American Cancer Society, BC-356, the Milheim Foundation for Cancer Research and the Comprehensive Cancer Center of Metropolitan Detroit. The work with nafazatrom has been supported by the Miles Institute for PreClinical Pharmacology. The authors also wish to thank Drs. T.T. Tchen and J.D. Taylor for advice and encouragement. The use of Dept. of Chemistry laboratory facilities is gratefully acknowledged. We thank Dr. John Pike for the generous supply of eicosanoids.
The expert typing of Donna Laciak and Susan Lyman is gratefully appreciated.

REFERENCES

Bayer BL, Blass KE, Forster W (1979). Anti-aggregatory effect of prostacyclin (PGI_2) in vivo. Br J Pharmac 66:10.
Brain MC, Azzopardi JG, Baker LRI, Pineo GF, Roberts PD, Dacie JV (1970). Microangiopathic haemolytic anemia: the possible role of vascular lesions in pathogenesis. Brit J Haemotology 18:183.
Cameron GZ (1954). The liver as a site and source of cancer. Brit Med J 1:347.
Carreras LO, Chamone DAF, Klercky P, Vermÿlen J (1980). Decreased vascular prostacyclin (PGI_2) in diabetic rats. Stimulation of PGI_2 release in normal and diabetic rats by the antithrombotic compound Bay g 6575. Thromb Res 19:663.
Carter RL (1978). Some lymphoreticular reactions and the metastatic process. In Baldwin RW (ed): "Secondary Spread of Cancer", New York: Academic Press, p 67.
Cazencave JP, Dejana E, Kinlough-Rathbone R, Packman MA, Fraser Mustard J (1979). Platelet interactions with the endothelium and the subendothelium: the role of thrombin and prostacyclin. Haemostasis 8:183.
Chang WC, Muroto S, Tsurufuji S (1977). Thromboxane B_2 transformed from arachidonic acid in carrageenin-induced granuloma. Prostaglandins 13:17.

Chew EC, Josephson RL, Wallace AC (1976). Morphologic aspects of the arrest of circulating cancer cells. In Weiss L (ed): "Fundamental Aspects of Metastasis", Amsterdam: North-Holland Publ. Co. p 121.

Clark R (1979). Systemic cancer and the metastatic process. Cancer 43:790.

deGaetano G, Remuzzi G, Mysiliwiec M, Donati MB (1979). Vascular prostacyclin and plasminogen activator activity in experimental and clinical conditions of disturbed haemostasis or thrombosis. Haemostasis 8:300.

delRegato JA (1977). Pathways of metastatic spread of malignant tumors. Seminars Oncol 4:33.

Dingemans KP, Roos E, Vanden Eorgh Weerman MA, VandePavert IV (1978). Invasion of liver tissue by tumor comparative ultrastructure. J Natl Cancer Inst 60:583.

Donati MB, Davidson JF, Garattini S (1981). "Malignancy and the Hemostatic System." New York: Raven Press, p 138.

Dvorak HF, Quay SC, Orenstein NS, Dvorak AM, Hahn P, Bitzer AM, Carvalho AC (1981). Tumor shedding and coagulation. Science 212:923.

Fidler IJ (1978). General considerations for the studies of experimental cancer metastases. Methods Cancer Res 25:399.

Fisher ER, Fisher B (1967). Recent observations on concepts of metastasis. Arch Pathol 83:321.

Fitzpatrick FA, Gorman RR (1978). A comparison of imidazole and 9,11-azoprosta-5,13-deinoic acid. Two selective thromboxane synthetase inihibitors. Biochemica et Biophysica Acta 539:162.

Fitzpatrick FA, Bundy GL, Gorman RR, Honohan T (1978). 9,11-Epoxyiminoprosta-5,13-dienoic acid is a thromboxane A2 antagonist in human platelets. Nature 275:764.

Fitzpatrick FA, Bundy GL, Gorman RR, Honohan T, McGuire J, Sun F (1979). 9,11-iminoepoxyprosta5,13-dienoic acid is a selective thromboxane A2 synthetase inhibitor. Biochimica et Biophysica Acta 573:238.

Folco G, Granstrom E, Kindahl H (1977). Albumin stablizes thromboxane A2. Fed Eur Biochem Soc Letts 82:321.

Franks LM (1973). Structure and biological malignancy of tumors. In Garattini S, Franchi G (eds): "Chemotherapy of Cancer Dissemination and Metastasis", New York: Raven Press.

Friedman DL, Johnson RA, Zeilig LE (1976). The role of cyclic nucleotides in the cell cycle. Adv Cyclic Nucleotide Res 7:69.

Gasic GJ, Gasic TB, Stewart CC (1968). Antimetastatic effects associated with platelet reduction. Proc Natl Acad Sci 61:46.

Gasic GJ, Gasic TB, Murphy S (1972). Anti-metastic effect of aspirin. Lancet 932.

Gasic GJ, Gasic TB, Galanti N, Johnson T, Murphy S (1973). Platelet-tumor cell interactions in mice. The role of platelets in the spread of malignant disease. Int J Cancer 11:704.

Gasic GJ, Gasic TB, Jiminez SA (1977). Effects of trypsin of the platelet-aggregating activity of mouse tumor cells. Thrombosis Res 10:33.

Gasic GJ, Kock PAG, Hsu B, Gasic TB, Niewiarowski S (1976). Thrombogenic activity of mouse and human tumors: Effects on platelets, coagulation, and fibrinolysis, and possible significance for metastasis. Z Krebsforsch 263.

Gasic GJ, Boettiger D, Catalfamo JL, Gasic TB, Stewart GJ (1978). Aggregation of platelets and cell membrane vesiculation by rat cells transformed in vitro by Rous Sarcoma virus. Cancer Res 38:2950.

Gorman RR, Bunting S, Miller OV (1977). Modulation of human platelet adenylate cyclase by prostacyclin (PGX). Prostaglandins 13:377.

Gralnick HR (1981). Cancer cell pro-coagulant activity. In Donati MB, Davidson JF, Garattini S (eds): "Malignancy and the Hemostatic System", New York: Raven Press, p 57.

Gryglewski RJ, Korbut R, Ocetkiewicz A (1978). Generation of prostacyclin by lungs in vivo and its release into arterial circulation. Nature 273:765.

Hara Y, Steiner M, Baldini MG (1980). Characterization of the platelet-aggregating activity of tumor cells. Cancer Res 40:1217.

Higgs GA, Cardinal D, Moncada S, Vane JR (1979). Microcirculatory effects of prostacyclin in the hamster cheek pouch. Microvascular Res 18:245.

Hilgard P (1973). The role of blood platelets in experimental metastases. Brit J Cancer 28:429.

Hilgard P, Beyerle L, Hohage R, Hiemeyci V, Kubler M (1972). The effect of heparin on the initial phase of metastasis formation. Eur J Cancer 8:347.

Honn KV, Cicone B, Skoff A (1981). Prostacyclin: A potent antimetastatic agent. Science 212:1270.

Honn KV, Cicone B, Skoff A, Romine M, Gossett D, Thompson K, Hodge D (1980). Thromboxane synthetase inhibitors and prostacyclin can control tumor cell metastasis. J Cell Biol 87:64.

Honn KV, Dunn JR, Morgan LR, Bienkowski M, Marnett LJ (1979). Inhibition of DNA synthesis in Harding-Passey melanoma cells by prostaglandins A_1 and A_2: Comparison with chemotherapeutic agents. Biochem Biophys Res Commun 87:795.

Hopkins NK, Gorman RR (1981). Regulation of 3T3-L1 fibroblast differentiation by prostacyclin (Prostaglandin I_2). Biochimica et Biophysica Acta 663:457.

Hyman AL, Kadowitz PJ (1979). Pulmonary vasodilator activity of prostacyclin (PGI_2) in the cat. Circ Res 45:404.

Karpatkin S, Smerling A, Pearlstein E (1980). Plasma requirement for the aggregation of rabbit platelets by an aggregating material derived from SV40-transformed 3T3 fibroblasts. J Lab Clin Med 96:994.

Kelly JP, Parker CW (1979). Effects of arachidonic acid and other unsaturated fatty acids on mitogenesis in human lympohcytes. J Immunol 122:1556.

Kelly JP, Johnson HC, Parker CW (1979). Effect of inhibitors of arachidonic acid metabolism on mitogenesis in human lymphocytes: Possible role of thromboxanes and products of the lipoxygenase pathway. J Immunol 122:1563.

Kramer RH, Nicolson GL (1979). Interactions of tumor cells with vascular endothelial cell monolayers: A model for metastatic invasion. Proc Natl Acad Sci 76:5704.

Kreider JW, Wade DR, Rosenthal M, Densley T (1975). Maturation and differentiation of B16 melanoma cells induced by theophylline treatment. J Natl Cancer Inst 54:1457.

Lione A, Bossman HB (1978). The inhibitory effect of heparin and warfarin treatments on the intravascular survival of B-16 melanoma cells in syngeneic C57 mice. Cell Biol Int Rep 2:81.

Marcus AJ (1978). The role of lipids in platelet function with particular reference to the arachodonic acid pathway. J Lipid Res 19:793.

Meistrich ML, Gardina DJ, Meyn DJ, Barlogie RE (1977). Separation of cells from mouse solid tumors by centrifugal elutriation. Cancer Res 37:4291.

Mitchell WM (1968). Pseudocollagenase: A protease from clostridium histolyticum. Biochem Biophys Acta 159:554.

Moncada S, Vane JR (1979). The role of prostacyclin in vascular tissue. Fed Proc 38:66-71.

Moncada S, Vane JR (1977). In Kharasch N, Fried J (eds):"Biochemical Aspects of Prostaglandins and Thromboxanes", New York: Academic Press, p 155.

Moncada S. Korbut R, Bunting S, Vane JR (1978). Prostacyclin is a circulating hormone. Nature 273:767.

Muroto S, Morita I, Abe M (1977). The effects of thromboxane B_2 and 6-ketoprostaglandin F_1 on cultured fibroblasts. Biochimica et Biophysica Acta 479:122.

Parker CW, Stenson WF, Huber MC, Kelly JP (1979). Formation of thromboxane B_2 and hydroxyarachidonic acids in purified human lymphocytes in the presence and abscence of PHA. J Immunol 122:1572.

Pearlstein E, Cooper LB, Karpatkin S (1979). Correlation between spontaneous metastatic potential, platelet-aggregating activity of cell surface extracts, and cell surface sialylation in 10 metastatic-variant derivatives of a rat renal sarcoma cell line. J Lab Clin Med 93:332.

Pearlstein E, Salk PL, Yogees Waran G, Karpatkin S (1980). Proc Natl Acad Sci USA 77:4336.

Prasad KN (1972). Morphological differentiation induced by prostaglandin in mouse neuroblastoma cells in culture. Nature 236:49.

Sack GH, Levin J, Bell WR (1977). Trousseau's syndrome and other manifestations of chronic disseminated coagulopathy in patients with neoplasms: Clinical, pathophysiologic, and therapeutic features. Medicine 56:1.

Salmon JA, Smith DR, Flower RJ, Moncada S, Vane JR (1978). Further studies on the enzymatic conversion of prostaglandin endoperoxide into prostacyclin by porcine aorta microsomes. Biochim et Biophysica Acta 523:250.

Salsbury AJ (1975). The significance of the circulating cancer cell. Cancer Treat Rev 2:55.

Sanderson RJ, Bird KE (1977). Cell separations by counter-flow centrifugation. In "Methods in Cell Biology", Vol 15, New York: Academic Press, p 1.

Santoro MG, Benedetto A, Jaffe BM (1979). Inhibition of friend erthroleukaemia-cell tumours in vivo by a synthetic analogues of prostaglandin E_2. Br J Cancer 39:259.

Seuter F, Busse WD, Meng K, Hoffmeister F, Moller E, Horstmann H (1979). The antithrombotic activity of Bay g 6575. Arzneim-Forsch/Drug Res 29:54.

Siegel MI, McConnell RT, Abrahams L, Porter NA, Cuatreeasa P (1979). Regulation of arachidonate metabolism via lipoxygenase and cyclo-oxygense by 12-hpete, the product of human platelet lipoxygenase. Biochem Biophys ResCommun 89:1273.

Sloane BF, Dunn JR, Honn KV (1981). Lysosomal cathepsin B: Correlation with metastatic potential. Science 212:1151.

Smith JB (1980). The prostanoids in hemostasis and thrombosis. Amer J Pathol 99:743.

Spreafico F, Garattini S (1978). Chemotherapy of experimental metastasis. In Baldwin RW (ed): "Secondary Spread of Cancer", New York: Academic Press, p 101.

Sugarbaker EV, Ketcham AS (1977). Mechanisms and prevention of cancer dissemination: An overview. Seminars Oncol 4:19.

Vermylen J, Chamone DA, Verstraete M (1979). Stimulation of prostacyclin release from vessel wall by Bay g 6575. Lancet 1:518.

Warren BA (1973). Environment of the blood-borne tumor embolus adherent to vessel wall. J Med 4:150.

Warren BA, Vales O (1972). The adhesion of thromboplastic tumour emboli to vessel walls in vivo. J Exp Path 53:301.

Weiss L (1977). A pathobiologic overview of metastasis. Semn in Oncology 4:5.

Weiss LW (1976). "Fundamental Aspects of Metastasis", Amsterdam: North-Holland Publs. Co., p 443.

Wood S (1971). Mechanisms of establishment of tumor metastases. In Ioachim HL (ed): "Pathology Annual", Vol 1, London: Butterworths, p 281.

Wynalda MA, Fitzpatrick FA (1980). Albumins stablize prostaglandin I_2. Prostaglandins 20:853.

Yoshimoto T, Yamamoto S, Hayaishi D (1978). Selective inhibition of prostaglandin endoperoxide thromboxane isomerase by 1-carboxyalkyl-imidazoles. Prostaglandins 16:529.

Zeidman I (1961). The fate of circulating tumor cells. Cancer Res 21:38.

DISCUSSION

Dr. N. Hanna (Frederick Cancer Research Center, Frederick). Two questions: How were the tumor samples treated and were there pathological differences in the treated tumors?

Dr. K.V. Honn (Wayne State University, Detroit). The cells were dispersed using the collagenase dispersion technique the number of the cells that we put into the elutriator was estimated, cells were recovered, corrected for recovery and then the number of cells in the recovered fractions which were tumor cells was estimated, not macrophages, lymphocytes. red blood cells, etc.

With regard to your second point, if you look at a side by side comparison of a cross section through the center of the tumor, the ring of actively growing tumor cells around the periphery is considerably thinner in 6575-treated tumors and the number of inflammatory cells that have migrated into that area appears to be considerably greater.

Dr. S. Karpatkin (New York University Medical Center, New York. Have you got any survival studies on animals which were given tumors?

Dr. Honn. I have about 800 animals right now and survival studies of all combinations with the 6575. We are now at about 70 days and, in the amputation model, we are down to about 40% survivals in the control and we are at about 95% with the 6575.

Dr. J. C. Hoak (University of Iowa, Iowa City). When you incubated cells with prostacyclin and then injected them into the tail vein did the animals never develop tumors?

Dr. Honn. We have not carried it out that far. All animals were sacrificed.

Dr. Hoak. So they go at least two weeks?

Dr. Honn. They go two weeks without tumors. I cannot

say anything about it for longer than two weeks. I had a pathologist look at it and he said that he could find no evidence of tumor colony formation in the tail vein model pretreated with 6575.

Dr. Hoak. If you get rid of prostacyclin by some mechanism?

Dr. Honn. We had an increase in colony formation if we used 15-hydroperoxyarachidonate, but we have since used other hydroperoxides and with all of them we get an elevation of tumor colony formation. Once again, this is a two-week assay.

Dr. E. Pearlstein (New York University Medical Center). What is known about the in vivo toxicity of the drug?

Dr. Honn. I think that you would have to ask a manufacturer's representative for that. What I know is what is published in the literature, that the drug was not toxic at 10 grams/kilogram in acute toxicity studies in rabbits. We administered it to animals over a 60-70 day period and did not see any overt signs of toxicity. The drug is reputed to be very well tolerated.

Dr. M. J. Stuart (SUNY, Upstate Medical Center, Buffalo). What was the time effect with prostacyclin?

Dr. Honn. We can inject prostacyclin an hour after we have injected the tumor cells via the tail vein and still get a significant reduction in colony formation. I have not tried it beyond one hour. Also, we used ^{125}-labeled cells and then followed that with prostacyclin and we calculated that there was a large increase in the number of tumor cells in circulation over the control.

Dr. R. Greig (Smith, Kline, and French, Inc.,Philadelphia). Is it cytostatic?

Dr. Honn. Yes, it is. The effect on DNA synthesis and on cell proliferation was dose-dependent.

Dr. G. J. Gasic (University of Pennsylvania, Philadelphia). What is the stability of prostacyclin?

Dr. Honn. The thinking of the Upjohn group now is that

the long-lived effect is due to the fact that prostacyclin is stabilized by albumin. I must admit that mouse rodent albumin was better than human albumin in terms of stabilization of prostacyclin. We are really looking at about a one hour half-life and not a five minute half-life.

Dr. M. Steiner (The Memorial Hospital, Pawtucket). Is the bound prostacyclin active?

Dr. Honn. Well, it could be. You could also have an equilibrium between free prostacyclin and albumin, albumin functioning as a carrier protein.

TUMOR CELL INTERACTIONS WITH VASCULAR ENDOTHELIAL CELLS
AND THEIR EXTRACELLULAR MATRIX

Randall H. Kramer,* Kathryn G. Vogel[†] and
Garth L. Nicolson[‡]
*Departments of Anatomy and Oral Medicine,
University of California, San Francisco,
California 94143, [†]Department of Biology,
University of New Mexico, Albuquerque, New
Mexico 87131 and [‡]Department of Tumor Biology,
The University of Texas System Cancer Center,
M.D. Anderson Hospital and Tumor Institute,
Houston, Texas 77030

Blood-borne metastasis occurs when malignant cells invade the circulatory system and are transported to distant sites. There they successfully arrest in the microcirculation and invade the endothelium and its underlying basal lamina and escape from the vascular system (Sugarbaker, 1979; Poste and Fidler, 1980; Fidler and Nicolson, 1981). In man and animals there are numerous examples where malignant tumors tend to metastasize to particular secondary locations (Fidler and Nicolson, 1981; Sugarbaker, 1981). For example, prostatic carcinoma metastasizes to bone at very high frequencies (Prout, 1973, small cell carcinoma in the lung colonizes brain (Hansen and Muggie, 1972), and neuroblastoma preferentially colonizes liver and adrenal glands (Jaffe, 1976). Sugarbaker (1981) has concluded that the most common regional metastases can be explained strictly by anatomical-mechanical considerations such as efferent venous circulation and lymphatic drainage, but in many cases distant organ colonization by metastatic cells did not fit this pattern suggesting that the unique properties of metastatic cells and/or the target organs could determine the sites of distant colonization. Recognition of unique capillary vascular endothelial cell surface determinants by circulating malignant cells could be responsible for metastatic colonization at specific sites (Nicolson and Winkelhake, 1975), or alternatively the subsequent growth of randomly arrested cells could be dependent upon local organ environments (Hart and Fidler, 1980).

During transport in the blood, tumor cells appear to undergo a variety of cellular interactions including homotypic aggregation with other tumor cells (Nicolson et al., 1976; Nicolson, 1978) and heterotypic aggregations with platelets (Gasic et al., 1973; Warren, 1973), lymphocytes (Fidler, 1975; Fidler and Bucana, 1977) and interactions with the vascular endothelium (Kramer and Nicolson, 1979, 1981; Poste et al., 1980; Pearlstein et al., 1980). In addition, some tumor cells are thromboplastic and elicit fibrin formation either during their circulation or soon after their arrest in capillary beds (Baserga and Saffiotta, 1955; Wood, 1964). Various drugs and conditions which modify these cellular interactions also modify metastasis. For example, metastasis can be reduced by inducing thrombocytopenia or by administration of anti-platelet agents (Fisher and Fisher, 1961; Gasic et al., 1972; Gastpar, 1977; Honn et al., 1981). In addition, modification of tumor cell surface glycoconjugates using drugs such as tunicamycin causes inhibition of tumor cell interactions with vascular endothelial cells and blocks blood-borne implantation in vivo (Irimura et al., 1981; Nicolson et al., 1981a).

During blood-borne arrest tumor cells must attach to the vascular endothelium. At this point fibrin clots can form around the arrested tumor cells (Chew et al., 1976), and under certain conditions this could lead to vessel wall damage and subsequent accumulation of neutrophils (Warren, 1973) or platelets (Hilgard, 1973).

Invasion of the vascular endothelium probably proceeds via several different mechanisms depending upon the type of vessel being invaded. These include disruption of and invasion at the intracellular junctions of the endothelial cells (Sindelar et al., 1975; Chew et al., 1976), penetration of endothelial cell cytoplasm by tumor cell pseudopodia (Dingemans, 1974; Roos and Dingemans, 1979) or rupture of the vessel as a consequence of expansive tumor growth (Carr et al., 1976). The final step of extravasation involves penetration of the endothelial basement membrane or basal lamina in a process that appears to involve digestion of basal lamina components by tumor cell degradative enzymes.

TUMOR CELL-ENDOTHELIAL CELL INTERACTIONS

In vitro models have been developed for studying the interactions between metastatic tumor cells and the normal vascular endothelium. Tissue culture techniques using monolayers of vascular endothelial cells (Kramer and Nicolson, 1979, 1981; Nicolson, 1981), monolayers of vascular endothelial cells on smooth muscle cell multilayers (Jones and DeClerq, 1980) or multicell biological membranes such as the chorioallantoic membrane (Nicolson et al., 1977; Hart and Fidler, 1978; Poste et al., 1980), urinary bladder membrane (Hart, 1979; Poste et al., 1980) or amniotic membrane (Liotta et al., 1980) have all been used to study invasion. We have chosen for our invasion studies to use monolayers of cultured vascular endothelial cells which produce a uniform cell layer, endothelial cell-endothelial cell junctions and an underlying extracellular matrix that is similar to a basal lamina (Birdwell et al., 1978; Kramer et al., 1981; Sage et al., 1979). Using this model we have found that metastatic tumor cell attachment and invasion of endothelial cell monolayers and their basal lamina proceeds by the following steps (Fig. 1): (1) tumor cell adhesion to the vascular endothelial cells; (2) retraction of the vascular endothelial cells and exposure of underlying basal lamina; (3) migration of the tumor cells onto the exposed basal lamina; (4) destruction of the endothelial basal lamina; and (5) tumor cell invasion through this extracellular matrix.

Tumor Cell-Endothelial Cell Adhesion

Using the vascular endothelial cell monolayer model, we have found that tumor cell-endothelial cell adhesion is the initial step in the process of tumor cell arrest and extravasation. Tumor cell adhesion appears to proceed by the direct active interaction of tumor cell surface microvilli with the apical endothelial cell surface. In culture as well as in vivo vascular endothelial cells possess flat, smooth and apical surfaces that are devoid of microvilli or other surface structures (Haudenschild et al., 1975; Kramer and Nicolson, 1979, 1981). In the tumor cell-endothelial cell adhesion system the tumor cells remain spherical when adherent to the endothelial cell surface (Fig. 2A). Observation by time-lapse, phase-contrast microscopy indicates that tumor cells are capable of active but slow migration over the endothelial cell monolayer (Kramer and Nicolson, 1979). The kinetics of human and mouse tumor cell attachment to bovine aortic endothelial cell monolayers indicates that the adhesion

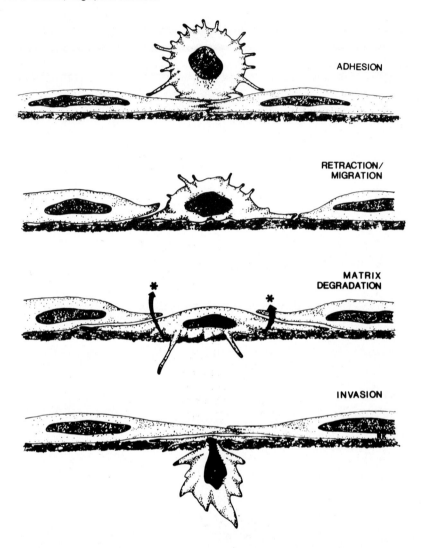

FIGURE 1. The stages of the invasion of vascular endothelium by metastatic tumor cells. Initially blood-borne tumor cell emboli attach to the apical endothelial cell surface. Next, adherent tumor cells induce endothelial cell retraction exposing the underlying subendothelial matrix, and the tumor cell then migrates to and spreads on the matrix. The subendothelial matrix is subsequently digested by the adherent tumor cell, and finally, the tumor cell migrates through the matrix and out of the vascular compartment.

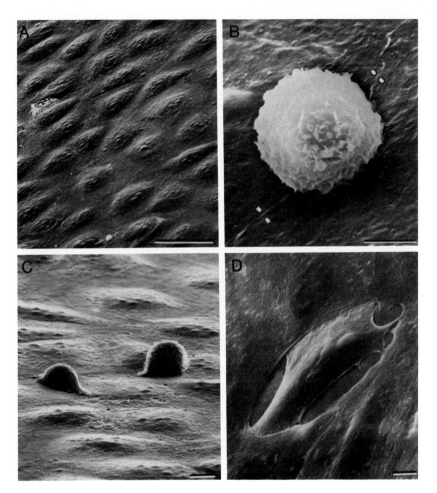

FIGURE 2. Metastatic tumor cell invasion of vascular endothelial cell monolayers as shown by scanning electron microscopy. (A) Confluent bovine aortic endothelial cell monolayer; bar equals 50 µm. (B) Mouse B16 melanoma cell adherent to endothelial cell layer at site of endothelial cell-endothelial cell junction (arrows) after 15 min incubation; bar equals 5 µm. (C) B16 cell invading between adjacent endothelial cells after 30 min incubation; bar equals 5 µm. (D) Extensive retraction of endothelial cells near the site of invading B16 tumor cell; bar equals 5 µm. The tumor cell is highly spread on the subendothelial matrix substratum and has underlapped neighboring endothelial cells at several sites after 1.5 hr incubation.

proceeds with very slow kinetics such that several hours are necessary before a majority of the tumor cells attach firmly to the monolayer (Kramer et al., 1980; Nicolson et al., 1981a). Scanning electron microscopy has revealed that the sites of attachment of metastatic tumor cells to the vascular endothelial cell monolayers are at endothelial cell-endothelial cell junctions (Fig. 2B). These sites may be composed of specific membrane domains or surface specializations that contain a higher than average density of cell adhesion receptors.

Endothelial Cell Retraction and Tumor Cell Migration

Tumor cells capable of adhering to the endothelial monolayer stimulate endothelial cell retraction. Retraction is defined as the breaking of intercellular junctions between the endothelial cells and pulling back of endothelial cell edges with exposure of underlying basal lamina. Retraction is a rapid process which occurs within minutes after adherence of the tumor cells to the endothelial cell monolayer; it is also localized to the area in direct contact with the adherent tumor cells, and portions of the endothelial cells distant to the attached tumor cells remained unaffected (Figs. 2C and 2D). After retraction, the motile tumor cells move and bind tightly to the exposed endothelial cell matrix.

Analysis of the endothelial cell matrix of bovine aortic endothelium (BAE) cells has demonstrated that it is composed primarily of glycoproteins such as fibronectin (Birdwell et al., 1978; Kramer et al., 1980) and laminin (Gospodarowicz et al., 1981), collagens (Sage et al., 1979, 1980) and heparan sulfate proteoglycans (Kramer et al., 1981). The molecular arrangement of these constituents in the matrix is unknown, but the structure appears to be composed of a meshwork of microfibrils when observed by scanning electron microscopy (Kramer et al., 1980).

Arrested metastatic tumor cells display a net movement from the apical endothelial cell surface to the underlying extracellular matrix. We have found that tumor cells attach relatively slowly to the apical surfaces of endothelial cells, while they attach rapidly and strongly to endothelial cell-free extracellular matrix (Kramer et al., 1980; Nicolson et al., 1981a). The difference in adhesive capacities between the endothelial cell surface and its basal lamina is probably important in directing net migration of metastatic tumor cells to the endothelial cell matrix where they rapidly spread. Because fibronectin is a major component of the BAE extracellular matrix but is not present on the apical endothelial cell surface (Birdwell et al., 1978), we tested the ability of tumor cells to adhere to immobilized fibronectin. Tumor cells attached to the immobilized fibronectin or to cell-free endothelial matrix with essentially the same kinetics of adhesion. On the fibronectin surface the adherent tumor cells rapidly spread and develop a highly unusual morphology that mimicks the morphology obtained when tumor cells adhered and spread on endothelial matrix suggesting that fibronectin may be involved in tumor cell adherence to the basal lamina. However, fibronectin is probably not the only component in the basal lamina involved in tumor cell adhesion. When saturating concentrations of affinity purified anti-fibronectin are present in the adhesion assay, tumor cell attachment to basal lamina is reduced only 20-30%, while adhesion to immobilized fibronectin is blocked completely (Nicolson et al., 1981a). Thus, it is likely that other basal lamina molecules, besides fibronectin, are involved in tumor cell adhesion. Murray et al. (1980) have found that metastatic tumor cells will bind to type IV collagen in relation to their metastatic potential indicating that several classes of different basal lamina molecules may be involved collectively in tumor cell adhesion. Tumor cells which bound to the endothelial extracellular matrix were motile and were capable of extensive migration on the matrix surface (Kramer and Nicolson, 1979).

Solubilization and Invasion of the Endothelial Basal Lamina

The final step in extravasation is the penetration of the basal lamina or basement membrane by invading cells. We have observed that this probably occurs by local destruction of the basal lamina (Fig. 3), presumably by enzymatic degradation. Solubilization of BAE endothelial extracellular matrix by metastatic tumor cells has been measured using metabolically-labeled matrix. Kramer et al. (1981) used $[^3H]$leucine and $[^{35}S]O_4^=$ to label the proteins and glycosaminoglycans, respectively, in the basal lamina and measured the ability of B16 melanoma cells to solubilize and degrade components of the isolated matrix. Highly metastatic tumor cells solubilize basal lamina components at higher rates compared to cells of lower metastatic potential (Nicolson, 1981). In these studies we have found that metastatic tumor cells solubilize endothelial cell matrix components and release

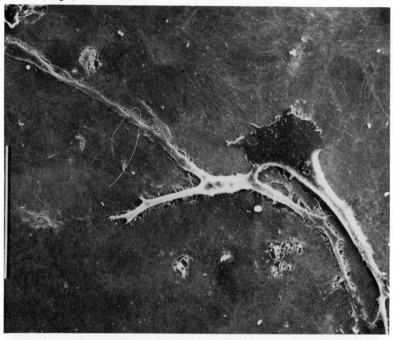
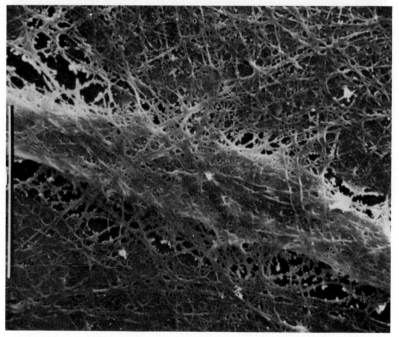

them in macromolecular form (90-95% of [^{3}H]leucine material was precipitable with 10% trichloroacetic acid; similar proportions of [^{35}S]O$_4^=$-labeled material were precipitable with 1% cetylpyridinium chloride). Analysis of the labeled proteins solubilized from endothelial basal lamina by SDS-polyacrylamide gel electrophoresis demonstrates the release of fibronectin in a lower molecular weight form. Fibronectin isolated from the BAE matrix normally migrates as a single molecular species of approximately 220,000 molecular weight under reducing conditions. Analysis of medium from tumor cell-treated endothelial cell matrix shows the presence of a smaller species of fibronectin with an apparent molecular weight of approximately 200,000. The appearance of this lower molecular weight form of fibronectin occurs presumably because of the action of a degradative enzyme (either a protease or a glycosidase) derived from the metastatic tumor cells. The possibility that plasmin mediates the degradation and solubilization of fibronectin was eliminated by removing the plasminogen from the serum used in the assays (Kramer et al., 1981).

Metastatic tumor cells are also capable of solubilizing endothelial cell matrix proteoglycans (Kramer et al., 1981). Matrix proteoglycans produced by BAE cell monolayers are about 1x10^6 daltons in size as determined by Sepharose-CL chromatography of labeled proteoglycans, although it is difficult to accurately determine the size of proteoglycans by this procedure. Analysis of the glycosaminoglycans present on these proteoglycans indicates that they have a molecular weight of ∿25,000 and are about 70% heparan sulfate. Incubation of metastatic tumor cells with [^{35}S]O$_4^=$-labeled endothelial cell matrix results in the appearance in the media of a smaller glycosaminoglycan fragment which is approximately one-third the size of the intact glycosaminoglycan chains and is produced presumably by a tumor cell endoglycosidase. Compositional analysis of this smaller fragment indicates that it is almost entirely heparan sulfate suggesting that tumor cells

FIGURE 3. Subendothelial matrix invasion by B16 melanoma cells. Matrix was isolated by the lytic removal of the endothelial cell monolayer (Kramer et al., 1980), and B16 tumor cells were then added and the culture incubated for 20 hr at 37°C. In the upper panel two tumor cells are seen invading the matrix; the B16 cells have an unusual and highly elongated morphology. Bar equals 50 μm. In the lower panel an enlargement of matrix penetration by a tumor cell pseudopodia shows invasion of matrix microfibrils. Bar equals 5 μm.

elaborate an endoglycosidase capable of cleaving the heparan sulfate glycosaminoglycans at intrachain sites.

DISCUSSION

The endothelial cell monolayer model for the study of arrest and invasion of metastatic tumor cells has proven to be valuable in studying the interactions of metastatic cells with their environment. Using metastatic murine, rat and human tumor cells and normal invasive host cells such as peripheral blood leukocytes, we have examined their ability to attach to and invade endothelial monolayers (Kramer and Nicolson, 1979, 1981). Metastatic tumor cells which attach successfully to endothelial cells do so predominantly at or near the sites of endothelial intercellular junctions suggesting that these regions may have specific cell surface components involved in adhesion. These observations are similar to those made in vivo (Fasske et al., 1975; Sindelar et al., 1975; Carr et al., 1976; Chew et al., 1976). In some cases tumor cells in vivo have been observed to migrate directly through the endothelial cell cytoplasm (Marchesi and Florey, 1960; Dingemans, 1974). This trans-cellular migration may be particular to certain highly specialized endothelium such as in the liver where fenestrated vascular endothelial cells line the liver sinusoids. Roos and Dingemans (1979) have found that several different types of tumor cells, as well as blood leukocytes, are able to penetrate the liver sinusoid endothelium by passing through pre-existing pores or by trans-cellular passage. Recently Roos and Dingemans (1981) have studied the arrest and extravasation of B16 melanoma, MB6A lymphosarcoma and TA3 mammary carcinoma in the lung. In contrast to what they found in the liver, tumor cell immigration in the lung did not occur by trans-cellular passage. Instead, the invading tumor cells migrated between adjacent endothelial cells similar to what we have found in the endothelial cell model presented here. They also found that the MB6A lymphosarcoma which was highly invasive in the liver was non-invasive in the lung and failed to penetrate the endothelial cell layer correlating to what we have found with metastatic RAW117 murine lymphosarcoma (Nicolson et al., 1980). In vitro RAW117 lymphosarcoma attached poorly to BAE endothelial cell monolayers and rarely induced endothelial cell retraction. Recently we have been able to select RAW117 lymphosarcoma sublines that are highly adhesive to murine lung vascular endothelial cells. These variant sublines are now metastatic to lung in vivo and are capable of attaching to and invading lung-derived vascular endothelial cells in vitro (Nicolson and Belloni, in preparation). Thus, it appears that the properties of the endothelium comprising the microvasculature of various organs may be unique for these sites, and this could explain, in part, why certain tumors exhibit unique patterns of organ colonization.

We have found considerable variability in the time required for individual cells to bind to the vascular endothelial monolayer, stimulate retraction and undergo invasion (Kramer and Nicolson, 1979, 1981; Nicolson, 1981). Cell variability in attachment and invasion could be due to the heterogeneous nature of metastatic tumor cell populations. It is well known that metastatic cells are extremely heterogeneous with respect to metastatic (reviewed in Fidler and Nicolson, 1981), invasive (Hart and Fidler, 1978; Poste et al., 1980) potentials, immunologic properties (Pimm and Baldwin, 1977; Olssen and Ebbesen, 1979) and sensitivities to chemotherapeutic drugs (Heppner et al., 1978; Nicolson et al., 1981b), heat (Tomasovic et al., 1981) and radiation (Hill et al., 1979).

Finally, metastasizing tumor cells must elaborate degradative enzymes to penetrate the basement membrane or basal lamina. In order for tumor cells to penetrate the basal lamina it is probably necessary that they possess different classes of degradative enzymes that will break down the major structural components of this matrix. Recently some of these degradative enzymes have been described. Liotta et al. (1980) have shown that highly metastatic tumor cells possess greater activities of type IV collagenase compared to tumor cells of low metastatic potential, and Nakajima et al., (1981) have found that highly metastatic melanoma cells degrade heparan sulfate at rates based on their metastatic potential. There are probably other enzymes which may also be important in the degradation and solubilization of basal lamina and invasion of metastatic tumor cells. The usefulness of the model presented here and other invasion systems based on endothelial cell monolayers is that these types of interactions and others can be studied and characterized in detail.

ACKNOWLEDGMENTS

This investigation was supported by research grants RO1-AG01214 (to K.G.V.), RO1-CA28867 and RO1-CA28844 (to G.L.N.) from the National Institutes of Health, U.S.P.H.S.

REFERENCES

Baserga R, Saffiotta U (1955). Experimental studies on histogenesis of blood-borne metastases. Arch Pathol 59:26.

Birdwell CR, Gospodarowicz D, Nicolson GL (1978). Identification, localization and the role of fibronectin in cultured bovine endothelial cells. Proc Natl Acad Sci USA 75:3273.

Carr I, McGinty F, Norris, P (1976). The fine structure of neoplastic invasion: invasion of liver, skeletal muscle and lymphatic vessels by the Rd/3 tumour. J Pathol 118:91.

Chew EC, Josephson RK, Wallace AC (1976). Morphologic aspects of the arrest of circulating cancer cells. In Weiss L (ed): "Fundamental Aspects of Metastasis," Amsterdam: North-Holland Publishing Co., p 121.

Dingemans KP (1974) Invasion of liver tissue by blood-borne mammary carcinoma cells. J Natl Cancer Inst 53:1813.

Fasske E, Fetting R, Ruhland D, Schubert T, Themann H (1975). Colonization of the mouse liver by transplanted virogenic leukemia cells. Electron microscopic investigation. Z Krebsforch 84:257.

Fidler IJ (1975). Biological behavior of malignant melanoma cells correlated to their survival in vivo. Cancer Res 35:218.

Fidler IJ, Bucana C (1977). Resistance of tumor cells to lysis by syngeneic lymphocytes: a possible mechanism. Cancer Res 37:3945.

Fidler IJ, Nicolson GL (1981). Immunobiology of experimental metastatic melanoma. Cancer Biol Rev 2:171.

Fisher B, Fisher ER (1961). Experimental studies of factors which influence hepatic metastases. VIII. Effect of anticoagulants. Surgery 50:240.

Gasic GJ, Gasic TB, Murphy S (1972) Antimetastatic effects of aspirin. Lancet 2:932.

Gasic GJ, Gasic TB, Galanti N, Johnson T, Murphy S (1973). Platelet-tumor cell interaction in mice. The role of platelets in the spread of malignant disease. Int J Cancer 11:704.

Gastpar H (1977). Platelet-cancer cell interaction in metastasis formation: a possible therapeutic approach to metastasis prophylaxis. J Med 8:103.

Gospodarowicz D, Greenburg JM, Foidart JM, Savion N (1981). The production and localization of laminin in cultured vascular and corneal endothelial cells. J Cell Physiol (in press).

Hansen HH, Muggia FM (1972). Staging of inoperable patients with bronchogenic carcinoma with special reference to bone marrow examination and peritoneoscopy. Cancer 30:1395.

Hart IR (1979). The selection and characterization of an invasive variant of B16 melanoma. Am J Pathol 97:587.

Hart IR, Fidler IJ (1978). An in vitro quantitative assay for tumor cell invasion. Cancer Res 38:3218.

Hart IR, Fidler IJ (1980). Role of organ selectivity in the determination of metastatic patterns of B16 melanoma. Cancer Res 40:2281.

Haudenschild CC, Cotran RS, Gimbrone MA, Jr, Folkman J (1975). Fine structure of the vascular endothelium in culture. J Ultrastruct Res 51:22.

Heppner GH, Dexter DL, DeNucci T, Miller FR, Calabresi P (1978). Heterogeneity in drug sensitivity among tumor cell subpopulations of a single mammary tumor. Cancer Res 38:3758.

Hilgard P (1973). The role of blood platelets in experimental metastases. Br J Cancer 28:429.

Hill HZ, Hill GJ, Miller CF, Kwong F, Purdy J (1979). Radiation and melanoma: Response of B16 mouse tumor cells and clonal lines to in vitro irradiation. Radiation Res 80:259.

Honn KV, Cicone B, Skoff A (1981) Prostacyclin: a potent antimetastatic agent. Science 212:1270.

Jaffee N (1976) Neuroblastoma: review of the literature and an examination of factors contributing to its enigmatic character. Cancer Treatment Rev 3:61.

Jones PA, DeClercq YA (1980) Destruction of extracellular matrices containing glycoproteins, elastin, and collagen by metastatic human tumor cells. Cancer Res 40:3222.

Kramer RH, Nicolson GL (1979). Interactions of tumor cells with vascular endothelial cell monolayers: a model for metastatic invasion. Proc Natl Acad Sci USA 76:5704.

Kramer RH, Nicolson GL (1981). Invasion of vascular endothelial cell monolayers and underlying matrix by metastatic human cancer cells. In Schweiger S (ed): "International Cell Biology 1980-1981," Heidelberg: Springer-Verlag, p 794.

Kramer RH, Gonzalez R, Nicolson GL (1980). Metastatic tumor cells adhere preferentially to the extracellular matrix underlying vascular endothelial cells. Int J Cancer 26:639.

Kramer RH, Vogel K, Nicolson GL (1981). Solubilization and degradation of subendothelial matrix glycoproteins and proteoglyancs by metastatic tumor cells. J Biol Chem (in press).

Liotta LA, Tryggvason S, Garbisa S, Hart I, Foltz CM, Shafie S (1980). Metastatic propensity correlates with tumor cell degradation of basement membrane collagen. Nature 284:67.

Marchesi VT, Flotey HW (1960). Electron micrographic observations of the imigration of leucocytes. Quart J. Exp Physiol 45:343.

Murray CJ, Liotta LA, Rennard SI, Martin GR (1980). Adhesion characteristics of murine metastatic and non-metastatic tumor cells in vitro. Cancer Res 40:347.

Nakajima M, Irimura T, DiFerrante DT, DiFerrante N, Nicolson GL (1981). Rates of heparan sulfate degradation correlate with invasive and metastatic activities of B16 melanoma sublines. J Cell Biol (abst) (in press).

Nicolson GL (1978). Cell and tissue interactions leading to malignant tumor spread (metastasis). Amer Zool 18:77.

Nicolson GL (1981). Metastatic tumor cell attachment and invasion assay utilizing vascular endothelial cell monolayers. J Histochem Cytochem (in press).

Nicolson GL, Winkelhake JL (1975). Organ specificity of blood-borne tumour metastasis determined by cell adhesion? Nature 255:230.

Nicolson GL, Winkelhake JL, Nussey AC (1976). An approach to studying the cellular properties associated with metastasis: some *in vitro* properties of tumor variants selected *in vivo* for enhanced metastasis. In Weiss L (ed): "Fundamental Aspects of Metastasis," Amsterdam: North-Holland Publishing Co., p 291.

Nicolson GL, Birdwell CR, Brunson KW, Robbins JC, Beattie G, Fidler IJ (1977). Cell interactions in the metastatic process: some cell surface properties associated with successful blood-borne tumor spread. In Lash J, Burger MM (eds): "Cell and Tissue Interactions," New York: Raven Press, p 225.

Nicolson GL, Reading CL, Brunson KW (1980). Blood-borne tumor metastasis: some properties of selected tumor cell variants of differing malignancies. In Crispen RG (ed): "Tumor Progression," Amsterdam: North-Holland Publishing Co, p 31.

Nicolson GL, Irimura T, Gonzalez R, Ruoslahti E (1981a). The role of fibronectin in adhesion of metastatic melanoma cells to endothelial cells and their basal lamina. Exp Cell Res (in press).

Nicolson GL, Lotan R, Rios A (1981b). Tumor cell heterogeneity and the *in vitro* sensitivities of metastatic B16 melanoma sublines and clones to retinoic acid or BCNU. Cancer Treatment Rep (in press).

Olsson L, Ebbesen P (1979). Natural polyclonality of spontaneous AKR leukemia and its consequences for so-called specific immunotherapy. J Natl Cancer Inst 62:623.

Pearlstein E, Gold LI, Garcia-Pardo A (1980). Fibronectin: a review of its structure and biological activity. Mol Cell Biochem 29:103.

Pimm MV, Baldwin RW (1977). Antigenic differences between primary methylcholanthrene-induced rat sarcomas and postsurgical recurrences. Int J Cancer 20:37.

Poste G, Fidler IJ (1980). The pathogenesis of cancer metastasis. Nature 283:139.

Poste G, Doll J, Hart IR, Fidler IJ (1980). In vitro selection of murine B16 melanoma variants with enhanced tissue invasive properties. Cancer Res 40:1636.

Prout GR, Jr (1973). Prostate gland In Holland JF, Frei E (eds): "Cancer Medicine," Philadelphia: Lea & Febiger, p 1680.

Roos E, Dingemans KP (1979). Mechanisms of metastasis. Biochim Biophys Acta 560:135.

Roos E, Dingemans KP (1981). Infiltration of metastasizing tumor cells into liver and lungs. In Schweiger HS (ed): "International Cell Biology 1980-1981," Berlin: Springer-Verlag, p 774.

Sage H, Crouch E, Bornstein P (1979). Collagen synthesis of bovine aortic endothelial cells in culture. Biochemistry 18:5433.

Sage H, Pritzl P, Bornstein P (1980). A unique, pepsin-sensitive collagen synthesized by aortic endothelial cells. Biochemistry 19:5747.

Sindelar WF, Tralka TS, Ketcham AS (1975). Electron microscopic observations on formation of pulmonary metastases. J Surg Res 18:137.

Sugarbaker EV (1979). Cancer metastasis: A produce of tumor-host interactions. Curr Prob Cancer 3:3.

Sugarbaker EV (1981). Patterns of metastasis in human malignancies. Cancer Biol Rev 2:235.

Tomasovic SP, Thames HD Jr, Nicolson GL (1981) Heterogeneity in hyperthermic sensitivities of rat 13762NF mammary adenocarcinoma cell clones of differing metastatic potentials. Radiation Res (submitted).

Warren BA (1973) Environment of the blood-borne tumor embolus adherent to vessel wall. J Med 4:150.

Wood S Jr (1964) Experimental studies of the intravascular dissemination of ascitic V2 carcinoma cells in the

rabbit, with special reference to fibrinogen and fibrinolytic agents. Bull Schweiz Akad Med Wiss 20:92.

DISCUSSION

Dr. J. C. Hoak (The University of Iowa, Iowa City). You used several cross-species reactors; what effect does this have?

Dr. R. H. Kramer (University of California, San Francisco). As I tried to indicate, we also tested a system with human tumor cells and endothelium. Essentially the effects are the same in adhesion and invasion and growth. Changing the species does not have an effect.

Dr. K. V. Honn (Wayne State University, Detroit). Did you try to inhibit the invasion process?

Dr. Kramer. No, all I can say is there are protease inhibitors in serum and inactivating these inhibitors with acid treatment facilitates the degradation of matrix components. We believe these proteases and endoglycosidases are present at the cell surface.

Dr. N. Hanna (Frederick Cancer Research Center, Frederick). Do you find any quantitative or qualitative differences?

Dr. Kramer. Yes, indeed. Preliminary data indicate that the F-10 variant has a much greater capacity for degradation of the heparin-sulphate proteoglycan. In fact, the rate of solubilization is possibly eight times greater; there does seem to be a parallel there in terms of the more malignant cells possessing a more radical degradation of the matrix.

Dr. W. Eisert (University of Hannover, West Germany). What else causes retraction of the human cells?

Dr. Kramer. Cyclase and local anesthetics will also do it.

Dr. Eisert. Have you tried the other way around - get something in there that actually prevents the retraction?

Dr. Kramer. Just lowering the temperature of the pH in the medium will reduce the opening of gaps between cells. It is a relatively easy thing to do when the calls are put in an adverse situation.

Dr. E. Pearlstein (New York University Medical Center, New York). What are your criteria that both of the bands in the matrix are derived from fibronectin?

Dr. Kramer. We have labeled the matrix by iodination. We see no significant labeling in the components of fibronectin but we still see the split which would suggest that this doubling is coming from fibronectin.

Dr. K. Brunson (Indiana University, Gary). Have you tried grouping these cells?

Dr. Kramer. In other words, trying to select a metastatic, malignant cell? No, we have not tried that. I think one could do it but the problem is that most cells invade the monolayer very quickly. As I said, normal cells will also invade so it really is not that easy pulling out specific cell variants that have these properties.

Dr. K. Brunson. Do you have any ideas about the portion of the endothelium which causes adhesion?

Dr. Kramer. We really have no idea as to what mediates the initial cell adhesion. If, in the vasculature, you have small gaps the tumor cells have a much better chance of anchoring there. I think one has to look at the various systems involved. For instance, tumor invasion of the liver is quite different from the lung. We have demonstrated that B16 cells will penetrate and migrate underneath endothelial cells.

Dr. M. Steiner (The Memorial Hospital, Pawtucket). Have you tried to study degradative enzymes in the tumor cells?

Dr. Kramer. No, that has not been done. I think one possibility here is to isolate plasma membrane from the tumor cells and incubate that with matrix and see if one can obtain degradation responses to try to localize the actual process. We have isolated B16 cell membranes that are essentially ghosts somewhat like RBC ghosts and there is some attachment of the isolated membrane. I think one has to try localizing the degradative enzymes.

Dr. Mustafa A. Hay (El Safat, Kuwait). I wonder if there is some factor that inhibited the reaction of lymphosarcoma.

Dr. Kramer. I have not tried modifying the medium. One group in Holland has looked at lymphosarcoma invasion in the liver and lung and they do not see actual penetration of the endothelium. This is essentially what I see - there is some adherence but no breaking of the endothelial cells or migration into the endothelium.

INTERACTIONS OF TUMOR CELLS WITH WHOLE BASEMENT MEMBRANE IN THE PRESENCE OR ABSENCE OF ENDOTHELIUM

Calvin M. Foltz, Raimondo G. Russo, Gene P. Siegal, Victor P. Terranova, and Lance A. Liotta

National Institutes of Health, Bethesda, Maryland 20205

Malignant tumors are characterized by their invasion of surrounding tissues. This invasiveness and the metastasis which often accompanies it constitute the primary basis for many deaths from cancer. Metastasis is believed to occur by a series of more or less discrete steps, such as, invasion, intravasation, tumor cell detachment, transport in vascular or other body fluids, arrest, extravasation, proliferation, and angiogenesis. In order to form metastases tumor cells must also resist host defense mechanisms and accomplish the process within a certain time frame (Liotta et al., 1978). The vulnerability of circulating tumor cells and differences in their capacities to accomplish metastasis are attested to by the fact that less than one percent of them are successful in forming metastases (for review see Fidler et al., 1978). Study of the biology and biochemistry of the steps of metastasis will increase our understanding of the process and may reveal points for which it is possible to develop effective measures for disrupting the process.

Basement membranes are extracellular matrices produced by parenchymal cells and separating them from underlying connective tissue stroma. They present a barrier to macromolecules and cells, serve as a scaffold for tissue architecture, and promote cell differentiation (Vracko, 1974). In the formation of metastases tumor cells must penetrate basement membrane several times. For this reason the interaction of tumor cells with basement membrane and its components has been one of the principal focuses of work in this laboratory.

Over the years the participation of enzymes in the invasiveness of tumor cells has been postulated. Among investigators who have performed histologic and electron microscopic studies of tumor cell invasion of basement membranes are the following: Ozello (1959, 1970), Vlaeminck et al. (1972), and Babai (1976). Such studies indicate that the process occurs in three steps: 1) contact of tumor cell surface with basement membrane, 2) local dissolution of the basement membrane at the points of contact with the tumor cell, and 3) passage of the tumor cell through the basement membrane. The principal components of basement membrane are collagen, glycoproteins, and glycosaminoglycans (Kefalides, and Denduchis, 1969; Kefalides, 1975). Type IV collagen is the principal collagen of basement membrane. Using radioactively labeled type IV collagen as a substrate and media from the culture of tumor explants, it was discovered that a highly metastatic murine tumor derived from the T241 fibrosarcoma produces a neutral metalloprotease which degrades type IV collagen and requires trypsin activation for maximal activity. The enzyme fails to degrade other collagens or fibronectin under conditions in which reaction products specific for type IV collagen are produced from type IV collagen (Liotta et al., 1979a, 1981). Previously described animal collagenases fail to degrade type IV collagen. Type IV collagen degrading activity has been measured in a series of metastatic tumors and the level of degradative activity correlates with the metastatic potential of the tumors. The activity has also been detected in cultured endothelial cells, in leukocytes, and in involuting mammary epithelium (Liotta et al., 1979b, 1980b). In order to study the mechanism of tumor cell invasion of basement membrane a method has been developed to quantify type IV collagen degradation by living tumor cells. Biosynthetically [^{14}C]proline-labeled type IV collagen is used to coat 16 mm tissue culture wells. With such coated wells the adherence of cells to the substrate and the production of soluble labeled degradation products can be measured. Using this method also, a correlation of type IV collagen degradation with metastatic potency was found and the results suggest that the mechanism of degradation is local activation of a latent type IV collagen-specific enzyme at the cell-substrate interface with no significant cell phagocytosis of substrate (Garbisa et al., 1980). In the case of aggressive tumor cells an increased level of the enzyme could contribute significantly to the invasiveness of the cells.

It has long been realized that the malignant behavior of tumor cells is related to abnormalities in their adhesive properties (Coman, 1954). During the metastatic process attachment of tumor cells to basement membrane must precede penetration. Differences in the abilities of tumor cells to attach to basement membrane may be a factor in differences in their metastatic potencies. Certain metastatic cells attach preferentially to type IV collagen over other types of collagen or plastic substrates (Murray et al., 1980). Attachment of fibroblasts to collagen is mediated by fibronectin while attachment of epithelial cells to type IV collagen is mediated by laminin, a basement membrane glycoprotein (Terranova et al., 1980). Two metastatic murine tumor cell lines tested, BL6 derived from the B16 melanoma and PMT derived from the T241 fibrosarcoma, attached preferentially to type IV collagen and laminin increased both the rate and extent of attachment, while fibronectin had no effect. Preincubation of the metastatic cells with cycloheximide reduced the attachment to type IV collagen by three-fourths and the inhibition was overcome by the addition of laminin, indicating that laminin is an attachment protein synthesized by the metastatic cell lines. Antibodies to laminin prevented the attachment of the metastatic cells to type IV collagen, and incubation of the cells with antibody to laminin before intravenous injection markedly reduced their metastatic activity as measured by the number of pulmonary metastases produced. Cells selected by attachment to laminin had five times the potency of unbound cells as measured by the number of lung metastases produced after intravenous injection. The foregoing results suggest that laminin participates in attachment and arrest of cells during metastasis and that laminin-mediated attachment of tumor cells to type IV collagen in vitro can select highly metastatic variants in a tumor cell population (Terranova et al., 1981).

Invasiveness is the principal distinguishing feature of malignant cells, and understanding invasiveness is basic to understanding malignancy and metastasis. Because of the complexity of the process and the number of interactions which may contribute to it, in vivo studies of invasion are difficult to conduct and interpret. On the other hand in vitro methods can permit flexibility in design of experiments, have the potential for close examination of one or more aspects of the process, and may yield results more amenable to clear interpretation. Studies of invasion in vitro were begun by Fischer (1925). In Table 1 are listed some recent studies in which malignant cells have been allowed to react with aggregates of normal cells and embryonic or adult tissues. These systems are largely heterologous, do not allow much freedom in design of experiments, and in most cases do not permit recovery of viable penetrating tumor cells. Evaluations are usually done by microscopic exam-

TABLE 1

Examples of in vitro invasion assay systems

Chick chorioallantoic membrane (Easty and Easty, 1974)
Chick chorioallantoic membrane (Scher et al., 1976)
Chick chorioallantoic membrane (Hart and Fidler, 1978)
Chick blastoderm (Mareel et al., 1975)
Chick embryonic mesonephros (Pourreau-Schneider et al., 1977)
Chick embryonic skin (Noguchi et al., 1978)
Chick wing bud (Tickle et al., 1978)
Chick embryonic heart (Mareel et al., 1979b)
Normal cell aggregates (Liebrich and Paweletz, 1976)
Human decidual tissue (Schleich et al., 1976)
Extracted bovine articular cartilage (Pauli et al., 1981)
Mouse lung (Schirrmacher et al., 1979)
Rat embryo yolk sac (Maignan, 1979)
Canine vein, mouse bladder, and chick (Poste et al., 1980)
  chorioallantoic membrane

ination of invaded tissue or by determining the distribution of labeled invading cells, and quantitation often is difficult and unsatisfactory. In the systems described by Poste et al. (1980) penetrating tumor cells can be collected for further study but complex nonhuman organs are used, and the mouse bladder and canine vein systems, which show correlation of invasiveness with metastatic potency, are not readily adaptable to large numbers of invasion assays.

An in vitro invasion assay system has been developed employing the human amnion (Russo et al., 1981). In the term placenta the amnion and chorion layers comprise the fetal membranes with the amnion facing the fetus and joining with the umbilical cord. The typical human amnion is less than 0.5 mm in thickness and contains no blood vessels or nerves. Morphologically the human amnion consists of a single layer of cuboidal to low columnar epithelial cells and the basement membrane resting on nonvascular connective tissue stroma (Fig. 1a). The interepithelial cell junctions are tight and the epithelium is bound to the continuous basement membrane by numerous hemidesmosomes (Van Herendael et al., 1978; King, 1978). The basement membrane can be identified using periodic acid Schiff staining or by immunohistology using antibodies to basement membrane components such as type IV collagen or laminin. The underlying stroma contains banded collagen fibers and fibroblasts. By immunohistology the stroma contains

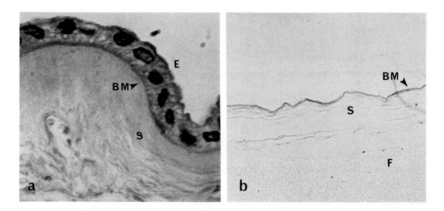

Figure 1a. Human amnion. Epithelium (E), basement membrane (BM), connective tissue stroma (S). (Hematoxylin-eosin, PAS; x630). Figure 1b. Human amnion denuded with 0.1% ammonium hydroxide. Basement membrane (BM), connective tissue stroma (S), Millipore filter (F). (Hematoxylin-eosin, PAS; x400).

type I collagen, type III collagen, type V collagen and fibronectin (Alitalo et al., 1980). The stroma appears fibrillar but it is a dense barrier impermeable to colloidal carbon particles and therefore does not contain preformed channels through which cells can passively migrate (Russo et al., 1982). Supported in the kind of chemotaxis chamber depicted in Figure 2 the amnion was used in a study of the migration

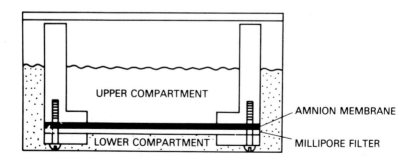

Figure 2. Amnion chemotaxis chamber.

of human polymorphonuclear leukocytes (PMN). The rate of PMN traversal of the amnion was dependent on the concentration of the chemoattractant (N-formyl-methionyl-leucyl-phenylalanine, FMLP) as well as the slope of the gradient. The PMNs passed through the full thickness of the amnion by disrupting tight interepithelial junctions and penetrating the basement membrane and stroma. The amnion supported in the holder shown in Figure 2 can be denuded by treatment with 0.1% ammonium hydroxide to produce a basement membrane-stroma system which can be used in invasion and chemotaxis studies. With this system and serum-free media containing 0.1% fetuin a number of human and murine tumor cells were tested. All tumor cells tested adhered poorly to the epithelium of whole amnion but bound avidly to the basement membrane denuded of epithelial cells. Within 24 h highly invasive tumor cells penetrated the basement membrane and stroma and entered the collecting Millipore filter. The collected tumor cells were viable and formed colonies on the filter. In this system with the cells employed, the participation of host cells was not required for invasion (Russo et al., 1982).

By growing monolayers of human endothelial cells on the amnion denuded of epithelial cells with detergent or alkali and supported in the holder, a homologous human system comprised of endothelium, basement membrane and stroma is produced. It reproduces in vitro the vascular wall of venules and capillaries. A-431 and MCF-7 human tumor cells penetrate this preparation and viable single cells and colonies are collected on the filter. A monolayer of bovine endothelial cells can also be grown on the denuded human amnion. These methods will be described.

MATERIALS AND METHODS

Materials

Hank's balanced salt solution without Ca and Mg (HBSS), solution of penicillin, streptomycin and Fungizone (Squibb) (PSF), medium 199 in Earle's BSS with 25 mM Hepes buffer, newborn calf serum (NCS), and human serum (HS) were purchased from M.A.Bioproducts, Walkersville, Md. Phosphate buffered saline (PBS), solution of penicillin and streptomycin (PS), Eagle's minimal essential medium with Earle's salts (MEM), Dulbecco's modification of Eagle's medium (DMEM), and fetal bovine serum (FBS) were purchased from Flow Laboratories,

McLean, Va. Type I clostridial collagenase was bought from Worthington Biochemical Corp., Freehold, N.J. Epidermal growth factor (EGF), culture grade, was purchased from Collaborative Research, Inc., Waltham, Mass. Type III fetuin was purchased from Sigma Chemical Co., Saint Louis, Mo. Trypsin solutions (0.25% in PBS without Ca and Mg and 0.25% with 0.1% EDTA in HBSS without Ca and Mg), and glutamine were obtained from the NIH Media Unit. All media were supplemented with L-glutamine (2 mmol/liter). All sera were heat inactivated (30 min at 56°C) before use. The Lucite device for supporting membranes was fabricated in the NIH shops. Tissue culture flasks and 6-well tissue culture cluster dishes were manufactured by Costar, Cambridge, Mass.

Isolation of human amnion

Fresh normal term placentas were obtained as soon as possible after delivery and kept on ice until used. The amnion faces the amniotic fluid compartment and is adherent to the chorion. The amnion was separated from the chorion by blunt dissection as described (Liotta et al., 1980a) and rinsed twice with PBS containing 0.01% sodium hypochlorite. The membrane was immersed in PBS containing PS (100 units/ml and 100 µg/ml). Amnions were finally placed in MEM containing PS and used immediately or stored refrigerated. The amnion is available in large sheets, and one placenta supplies enough material for many quantitative assays.

Preparation of human amnion membrane

A portion of the amnion is smoothed and clamped in the ring-shaped holder depicted in Figure 2. The holder is formed from two concentric Lucite rings of 3.2 cm outside diameter and 1.2 cm inside diameter. The upper ring is 1 cm in thickness and the lower 0.2 cm. The two rings are screwed together as shown with the amnion and a Millipore filter (8 µm pore size, 2.3 cm in diameter) placed between them with the stromal side facing the filter. When placed in a 17.8 mm x 35 mm well of a 6-well tissue culture cluster dish with a membrane and filter in place, a two-compartment chamber is produced which can be used for chemotaxis or invasion studies.

The amnion can be denuded of epithelial cells by treatment with 4% deoxycholate solution for 1 h at 4°C followed by gentle agitation with a rubber policeman in a modification of a method for isolation of capillary basement membrane (Meezan et al., 1975) or by treatment with 0.1% ammonium hydroxide for 15 min at 25°C followed by gentle scraping with a rubber policeman (Fig. 1b). The membranes were checked for leaks with colloidal carbon particles as previously described (Liotta et al., 1980a).

Isolation and culture of cells

Human MCF-7 breast carcinoma cells were obtained from Dr. Samir Shafie and human A-431 squamous carcinoma cells from Dr. David Salomon, both of the Laboratory of Pathophysiology, National Cancer Institute, Bethesda, Md. Both lines were cultured in DMEM with 10% FBS and PSF (100 units/ml, 100 µg/ml and 0.25 µg/ml). They were harvested in log phase of growth with the use of trypsin solution.

Bovine capillary endothelial cells were obtained in established culture from Dr. Giulio Alessandri of the Laboratory of Pathophysiology, National Cancer Institute, Bethesda, Md. The cells were grown in culture flasks coated with gelatin, applied as a 1% solution, in a mixture of DMEM supplemented with 20% FBS and PS (100 units/ml and 100 µg/ml) and ar equal volume of conditioned culture medium from S180 murine sarcoma cells. At confluence the cells were harvested with trypsin solution. After harvesting, the cells were grown in the mixed medium described above. At the first medium change after passage the cells were placed in DMEM with 10% FBS and PS and continued in that medium.

Human endothelial cells were obtained from the human umbilical cord vein by the method described by Jaffe et al. (1973b) with minor modifications. The cells were isolated as soon as possible after delivery and within 3 h. The vein from sound parts of the cord was cannulated and rinsed with HBSS containing PSF (200 units/ml, 200 µg/ml and 0.5 µg/ml) until free of blood. Ten ml of a 0.2% solution of collagenase in HBSS was introduced into the vein. The cord was then suspended in HBSS with PSF and incubated at 37°C with slow shaking for 20 min. After the digestion the cord was kneaded gently briefly, and the suspension of endothelial cells was flushed from the vein with 30 ml of medium 199 with PSF and 10% NCS. The suspension was centrifuged at 200 x g for 10 min. The pelleted cells were resuspended by gentle vortex mixing in medium 199 with Earle's salts, 25 mM Hepes buffer, PSF (100 units/ml, 100 µg/ml and 0.25 µg/ml), 10% HS, and 100 ng/ml of EGF. The suspension was placed in one or more

25 cm² tissue culture flasks and placed in the incubator. The medium was replaced with fresh medium after 24 h and thereafter every two or three days. Such primary cultures became confluent in about 7 days. As the cells approached confluency the amount of EGF in the medium was reduced from 100 to 50 to 10 or 0 ng/ml. Confluent cells were released for harvest or passage at a low ratio with trypsin-EDTA solution. In agreement with previous reports (Gospodarowicz et al., 1978) we have found medium 199 with 10% HS and 100 ng/ml EGF to be superior to medium 199 with 20% FBS and 100 ng/ml EGF for culturing human endothelial cells. The cells were used in invasion studies after 1 to 4 passages. They were verified to be endothelial cells by positive immunofluorescent staining for Factor VIII antigen (Jaffe et al., 1973a).

All cell cultures and preparations containing cells were incubated at 37°C in a humid atmosphere of 5% carbon dioxide and 95% air. The viability of cells was measured by trypan blue exclusion.

Invasion studies

The area of the membrane exposed in the holder is 1.13 cm². In the case of bovine capillary endothelial cells 10^5 to 2×10^5 cells were applied to each membrane in the mixed medium described above with change to DMEM with 10% FBS after 2 days. The amnion basement membrane had been coated previously by treatment with a 1% solution of gelatin. The cells were allowed to grow to confluency in 4 to 5 days. The state of the monolayer was followed by examination with the inverted microscope and by microscopic examination of control monolayers after staining with alizarin red (0.1% solution) and counterstaining with methylene blue (0.14% solution) (Fig. 3a). The same medium was placed in both compartments of the invasion chamber.

Human endothelial cells, 10^5 to 2×10^5/membrane, were applied to the membrane (Fig. 3b) and grown for four days in medium 199 with 25 mM Hepes buffer, 10% HS, 100 ng/ml EGF and PSF (100 units/ml, 100 µg/ml and 0.25 µg/ml). The monolayers were monitored as described for bovine endothelial cell monolayers.

The two human tumor cell lines, A-431 squamous carcinoma and MCF-7 breast carcinoma, were grown in DMEM with 25 mM Hepes buffer, 10% FBS and PSF. They were harvested in log phase of growth with trypsin solution and rinsed with plain medium before suspension in DMEM with 25 mM Hepes buffer,

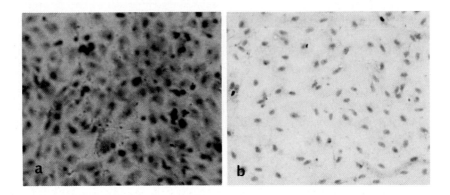

Figure 3a. Bovine capillary endothelial cells growing on human amnion basement membrane four days after seeding, x250. Figure 3b. Human endothelial cells growing on human amnion basement membrane one day after seeding, x100. Both stained with alizarin red (0.1% solution) with methylene blue (0.14% solution) counterstain.

0.1% fetuin and PSF (100 units/ml, 100 µg/ml and 0.25 µg/ml) for application to basement membranes (Figs. 4 and 5) or endothelial monolayers (Fig. 6), 4×10^5 cells/membrane. The membranes and endothelial layers were also rinsed with plain medium before application of the cells. The same medium was placed in both compartments of the invasion chamber during assays.

After 2 or 4 days the cells which penetrated the membrane or the endothelial cell layer and membrane and were trapped in or adherent to the filter were stained with hematoxylin-eosin. The whole filter was scanned (x400 magnification) and the total number of cells and colonies on each filter counted. Cells adherent to the basement membrane or the endothelial layer were stained with hematoxylin-eosin and counted in 10 to 15 random areas at x400 magnification and the total number calculated with a constant correction factor for area. The experiments were done in triplicate and the means and ranges recorded. The ratios of the numbers of the cells trapped on the filters to the numbers of nonpenetrating cells attached to the basement membranes or endothelial cell monolayers were calculated for studies with A-431 cells (Fig. 6b).

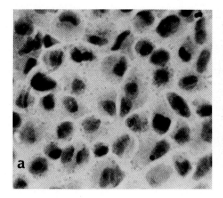 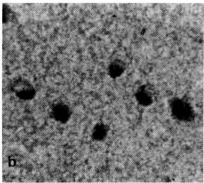

Figure 4a. A-431 cells, human squamous carcinoma, confluent on human amnion basement membrane four days after seeding. (Hematoxylin-eosin, x400). Figure 4b. A-431 cells on Millipore filter after penetrating human basement membrane-connective tissue stroma. (Hematoxylin-eosin, x400).

RESULTS AND DISCUSSION

We have found that endothelial cells can be grown on human amnion denuded of epithelial cells with detergent or alkali to produce a three-component system, endothelium, basement membrane and stroma, which reproduces in vitro the walls of venules and capillaries. Two species of endothelial cells have been applied to the membrane supported in a suitable holder and both have grown to confluent monolayers. When human tumor cells and endothelial cells are employed, a homologous human system is produced.

Tumor cells applied to the human amnion basement membrane attach, pass through the membrane and are collected on a filter (Fig. 4). Figure 5 shows the time course of invasion of the basement membrane-stroma by MCF-7 cells; between days 2 and 7 the number of penetrating cells increased threefold. The penetrating cells were viable, formed colonies on the filter, and could be counted or recovered and used in further studies. The fact that the tumor cells penetrate the basement membrane indicates that in this in vitro system host cells are not required for invasion of the basement membrane and connective tissue stroma.

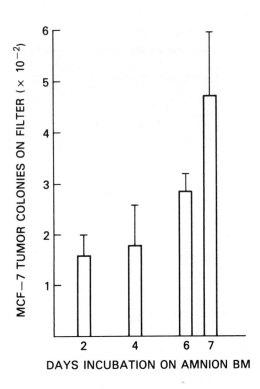

Figure 5. Time course of migration of MCF-7 cells through the human basement membrane-stroma. To each membrane 5 x 10^5 cells in 1.7 ml of DMEM with 0.2% fetuin were applied. The whole filters in triplicate were stained with hematoxylin-eosin and the number of tumor cell colonies counted at x400 at the intervals shown after application of cells. The means and ranges are indicated.

A four-day monolayer of human endothelial cells growing on the basement membrane-stroma produced a significant reduction in the rate of invasion by MCF-7 cells relative to the rate of invasion of the basement membrane-stroma (Fig. 6a). Present results do not show a significant difference in the migration of human A-431 cells through bovine or human endothelial monolayers (Fig. 6b). As four-day monolayers, both produced significant reductions in the rates of invasion by A-431 cells relative to the rates of penetration of basement membrane-stroma. The ratios of the numbers of cells penetrating to the numbers of nonpenetrating cells adhering to the basement membranes or endothelial monolayers indicate that the differences in the numbers of penetrating cells are not due simply to differences in the numbers of cells attaching to the basement membranes or endothelial monolayers (Fig. 6b). The fact that the penetration of the endothelial monolayer-basement membrane-stroma composite takes place in vitro without host cells other than endothelial cells may indicate the occurrence of an interaction of the tumor and endothelial cells leading to exfoliation or retraction of endothelial cells and to exposure and penetration of the basement membrane.

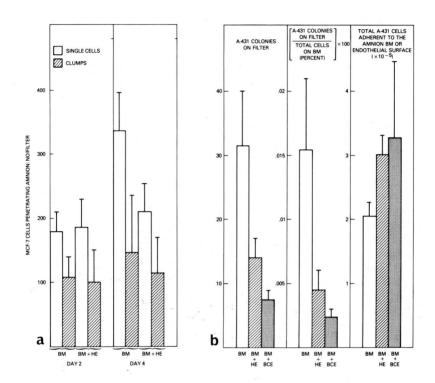

Figure 6a. Migration of MCF-7 cells through human amnion basement membrane-stroma without or with a four-day monolayer of human endothelial cells. The whole filters in triplicate were stained with hematoxylin-eosin and the number of tumor cell colonies counted at x400 on days 2 (left panel) and 4 (right panel) after application of the tumor cells, 4 x 10^5/ membrane. The means and ranges are indicated. By day 4 the rate of invasion was reduced by the endothelial cells.

Figure 6b. Migration of A-431 cells, 4 x 10^5/membrane, through human amnion basement membrane-stroma without or with a four-day monolayer of human or bovine capillary endothelial cells in the same type of experiment. Tumor cells on the filter and the basement membrane or endothelium were counted on day 4 in triplicate and the ratios of penetrating to attached nonpenetrating cells calculated.

Only a small proportion of the human tumor cells applied to the human amnion basement membrane or the membrane bearing a monolayer of bovine or human endothelial cells penetrates the whole thickness of these barriers even after 7 days (Figs. 5 and 6). This may indicate that the systems are exercising a rigorous selection process which could make the results of invasion assays reliable measures of malignancy and metastatic potencies (Poste et al., 1980). Three-dimensional in vitro invasion assays are more reliable measures of in vivo invasiveness and malignancy (Mareel, 1979a) than are two-dimensional in vitro assays. The systems which have been reported here are three-dimensional and quantitation is facile. Penetrating cells collected on the filter and nonpenetrating cells attached to the basement membrane or the endothelial layer can be stained and counted or used for further study. Another advantage of the method is the fact that many assays can be performed using the membrane prepared from one placenta; this reduces the importance of the tissue source as a basis for variation. Performing the assays with serum-free medium is also an advantage since it eliminates the potential variability from serum itself and from interactions of serum components with biochemical agents or drugs which might be used experimentally to modulate the invasion process.

Endothelial cells of two species, human and bovine, have been applied to the human amnion basement membrane and both have grown to form confluent monolayers. Endothelial cells of other species may be expected to do the same, thus providing opportunities for chemotaxis and invasion studies with a wide range of systems homologous in tumor and endothelial cells and supported on a natural basement membrane-stroma matrix. While present results do not show significant differences between human and bovine endothelial cells, future studies will surely reveal some such differences, which may be exploitable to increase our understanding of chemotaxis, invasion and tumor cell-endothelial cell interactions. The use of endothelial cells from different organs of a given species will make it possible to do similar studies which may increase our understanding of the phenomenon whereby circulating tumor cells from most tumors produce metastases in a limited number of organs.

The extravasation of circulating tumor cells that is essential for hematogenous metastasis requires attachment and arrest of appropriate duration and character to permit survival of the tumor cells, breaching of the endothelium, and penetration of the basement membrane and connective tissue stroma. The endothelial lining provides blood vessels with a nonthrombogenic surface and also plays a role in regulating vascular permeability (Mason et al., 1979). If the endothelium is intact, it is the first barrier to extravasation confronted by circulating tumor cells. At the present time it is thought that the principal route of egress of cells from blood vessels to surrounding tissues is through interendothelial junctions (Mason and Balis, 1980). In some cases tumor cells may pass through interendothelial gaps produced by migrating leukocytes (Ludatscher et al., 1967; Sindelar et al., 1975). In others, injuries of the endothelium can lead to separation of cells at junctions or necrosis and exfoliation of cells exposing subendothelium and thus facilitating extravasation by removal of the endothelial barrier and permitting a kind of interendothelial passage (Mason et al., 1979; Mason and Balis, 1980). Tumor cells adhere avidly to basement membrane and more readily to damaged than to intact endothelial cells. Interaction of tumor cells with damaged endothelial cells or with endothelial cells to produce damage can lead to passage through endothelial cells (Warren and Vales, 1972; Warren, 1973; Chew et al., 1976; Warren, 1981). Breaching of the endothelium thus can take place intracellularly or intercellularly depending apparently on the kind of tumor cell and the circumstances. In some cases platelet aggregates appear to play an important role in the arrest of tumor cells and the interaction of tumor cells and tumor cell emboli with endothelium (for review see Karpatkin and Pearlstein, 1981).

Extravasation appears to be a very complex process. At the present time there is little understanding of the interactions of tumor cells with endothelial cells, subendothelium, blood cells and soluble blood components which are involved in the required arrest and invasion. The process will be understood more readily if the stages can be studied in vitro as separate phenomena. The versatile systems which we have described should be particularly useful in studies of this process and in other studies of endothelium and endothelial cells. Studies of and with these systems are continuing.

REFERENCES

Alitalo K, Kurkinen M, Vaheri A, Krieg T, Timpl R (1980). Extracellular matrix components synthesized by human amniotic epithelial cells in culture. Cell 19:1053.
Babai F (1976). Etude ultrastructurale sur la pathogenie de l'invasion du muscle strie par des tumeurs transplantables.

J Ultrastr Res 56:287.

Chew EC, Josephson RL, Wallace AC (1976). Morphologic aspects of the arrest of circulating cancer cells. In Weiss L (ed): "Fundamental Aspects of Metastasis," Amsterdam: North-Holland Publishing Co, p 121.

Coman DR (1954). Cellular adhesiveness in relation to the invasiveness of cancer. Cancer Res 14:519.

Easty DM, Easty GC (1974). Measurement of the ability of cells to infiltrate normal tissues in vitro. Brit J Cancer 29:36.

Fidler IJ, Gersten DM, Hart IR (1978). The biology of cancer invasion and metastasis. Adv Cancer Res 28:149.

Fischer A (1925). Beitrag zur Biologie der Gewebezellen. Eine vergleichend biologische Studie der normalen und malignen Gewebezellen in vitro. Wilhelm Roux Archiv fur die Entwicklungsmechanik der Organismen 104:210.

Garbisa S, Kniska K, Tryggvason K, Foltz C, Liotta LA (1980). Quantitation of basement membrane degradation by living tumor cells in vitro. Cancer Lett 9:359.

Gospodarowicz D, Brown KD, Birdwell CR, Zetter BR (1978). Control of proliferation of human vascular endothelial cells. J Cell Biol 77:774.

Hart IR, Fidler IJ (1978). An in vitro quantitative assay for tumor cell invasion. Cancer Res 38:3218.

Jaffe EA, Hoyer LW, Nachman RL (1973a). Synthesis of antihemophilic factor antigen by cultured human endothelial cells. J Clin Invest 52:2757.

Jaffe EA, Nachman RL, Becker CG, Minick CR (1973b). Culture of human endothelial cells derived from umbilical veins. J Clin Invest 52:2745.

Karpatkin S, Pearlstein E (1981). Role of platelets in tumor cell metastases. Ann Intern Med 95:636.

Kefalides NA (1975). Basement membranes: structural and biosynthetic considerations. J Invest Dermatol 65:85.

Kefalides NA, Denduchis B (1969). Structural components of epithelial and endothelial basement membranes. Biochemistry 8:4613.

King BF (1978). A cytological study of plasma membrane modifications, intercellular junctions and endocytic activity of amniotic epithelium. Anat Rec 190:113.

Liebrich W, Paweletz N (1976). Ultrastructure of aggregates of tumor and normal cells and of mixed aggregates. Virch Arch B Cell Pathol 22:39.

Liotta LA, Abe S, Robey PG, Martin GR (1979a). Preferential digestion of basement membrane collagen by an enzyme derived from a metastatic murine tumor. Proc Natl Acad Sci USA

76:2268.
Liotta LA, Lee CW, Morakis DJ (1980a). New method for preparing large surfaces of intact human basement membrane for tumor invasion studies. Cancer Lett 11:141.
Liotta LA, Tryggvason K, Garbisa S, Hart I, Foltz CM, Shafie S (1980b). Metastatic potential correlates with enzymatic degradation of basement membrane collagen. Nature 284:67.
Liotta LA, Tryggvason K, Garbisa S, Robey PG, Abe S (1981). Partial purification and characterization of a neutral protease which cleaves type IV collagen. Biochemistry 20:100.
Liotta LA, Vembu D, Saini RK, Boone C (1978). In vivo monitoring of the death rate of artificial murine pulmonary micrometastases. Cancer Res 38:123.
Liotta LA, Wicha MS, Foidart JM, Rennard SI, Garbisa S, Kidwell WR (1979b). Hormonal requirements for basement membrane collagen deposition by cultured rat mammary epithelium. Lab Invest 41:511.
Ludatscher RM, Luse SA, Suntzeff V (1967). An electron microscopic study of pulmonary tumor emboli from transplantable Morris hepatoma 5123. Cancer Res 27:1939.
Maignan MF (1979). Etude ultrastructurale des interactions entre des cellules normales ou malignes et le sac vitellin de rat, explante in vitro. Biologie Cellulaire 35:229.
Mareel M (1979a). Is invasiveness in vitro characteristic of malignant cells? Cell Biol Int Rep 3:627.
Mareel M, De Ridder L, De Brabander M, Vakaet L (1975). Characterization of spontaneous, chemical, and viral transformants of a C3H/3T3 type mouse cell line by transplantation into young chick blastoderms. J Natl Cancer Inst 54:923.
Mareel M, Kint J, Meyvisch C (1979b). Methods of study of the invasion of malignant C3H-mouse fibroblasts into embryonic chick heart in vitro. Virch Arch B Cell Pathol 30:95.
Mason RG, Balis JU (1980). Pathology of the endothelium. In Trump BF, Arstila AU (eds): "Pathobiology of Cell Membranes," Vol II, New York: Academic Press, p 425.
Mason RG, Mohammad SF, Saba HI, Chuang HYK, Lee EL, Balis JU (1979). Functions of endothelium. In Ioachim HL (ed): "Pathobiology Annual 1979," Vol 9, New York: Raven Press, p 1.
Meezan E, Hjelle JT, Brendel K, Carlson EC (1975). A simple, versatile, nondisruptive method for the isolation of morphologically and chemically pure basement membranes from several tissues. Life Sci 17:1721.
Murray JC, Liotta LA, Rennard SI, Martin GR (1980). Adhesion characteristics of murine metastatic and nonmetastatic tumor cells in vitro. Cancer Res 40:347.

Noguchi PD, Johnson JB, O'Donnell R, Petricciani JC (1978). Chick embryonic skin as a rapid organ culture assay for cellular neoplasia. Science 199:980.

Ozello L (1959). The behavior of basement membranes in intraductal carcinoma of the breast. Am J Pathol 35:887.

Ozello L, Sanpitak P (1970). Epithelial-stromal junction of intraductal carcinoma of the breast. Cancer 26:1186.

Pauli BU, Memoli VA, Kuettner KE (1981). In vitro determination of tumor invasiveness using extracted hyaline cartilage. Cancer Res 41:2084.

Poste G, Doll J, Hart IR, Fidler IJ (1980). In vitro selection of murine B16 melanoma variants with enhanced tissue-invasive properties. Cancer Res 40:1636.

Pourreau-Schneider N, Felix H, Haemmerli G, Strauli P (1977). The role of cellular locomotion in leukemic infiltration. An organ culture study on penetration of L 5222 rat leukemia cells into the chick embryo mesonephros. Virch B Cell Pathol 23:257.

Russo RG, Liotta LA, Thorgeirsson U, Brundage R, Schiffmann E (1981). Polymorphonuclear leukocyte migration through human amnion membrane. J Cell Biol 91:459.

Russo RG, Thorgeirrson U, Liotta LA (1982). In vitro quantitative assay of invasion using human amnion. In Liotta LA, Hart IR (eds): "Cancer Invasion and Metastases," Boston: Martinus Nijhoff, in press.

Scher CD, Handenschild C, Klagsbrun M (1976). The chick chorioallantoic membrane as a model system for the study of tissue invasion by viral transformed cells. Cell 8:373.

Schirrmacher V, Shantz G, Clauer K, Komitowski D, Zimmermann HP, Lohmann-Matthes ML (1979). Tumor metastases and cell-mediated immunity in a model system in DBA/2 mice. I. Tumor invasiveness in vitro and metastasis formation in vivo. Int J Cancer 23:233.

Schleich AB, Frick M, Mayer A (1976). Patterns of invasive growth in vitro. Human decidua graviditatis confronted with established human cell lines and primary human explants. J Natl Cancer Inst 56:221.

Sindelar WF, Tralka TS, Ketcham AS (1975). Electron microscopic observations on formation of pulmonary metastases. J Surg Res 18:137.

Terranova VP, Foltz CM, Murray JC, Liotta LA, Martin GR (1981). Laminin as an attachment factor for metastatic cells. Am Assn Cancer Res Proc 72nd Ann Mtg 22:14.

Terranova VP, Rohrbach DH, Martin GR (1980). Role of laminin in the attachment of PAM 212 (epithelial) cells to basement membrane collagen. Cell 22:719.

Tickle C, Crawley A, Goodman M (1978). Cell movement and the mechanism of invasiveness: a survey of the behaviour of some normal and malignant cells implanted into the developing chick wing bud. J Cell Sci 31:293.

Van Herendael BJ, Oberti C, Brosens I (1978). Microanatomy of the human amniotic membranes. Am J Obstst Gyn 131:872.

Vlaeminck MN, Adenis L, Mouton Y, Demaille A (1972). Etude experimentale de la diffusion metastatique chez l'oeuf de poule embryonne. Int J Cancer 10:619.

Vracko R (1974). Basal lamina scaffold--anatomy and significance for maintenance of orderly tissue structure. Am J Pathol 77:314.

Warren BA (1973). Environment of the blood-borne tumor embolus adherent to the vessel wall. J Med 4:150.

Warren BA (1981). Cancer cell-endothelial reactions: the microinjury hypothesis and localized thrombosis in the formation of micrometastases. In Donati MB, Davidson JF, Garattini S (eds): "Malignancy and the Hemostatic System," New York: Raven Press, p 5.

Warren BA, Vales O (1972). The adhesion of thromboplastic tumor emboli to vessel walls in vivo. Brit J Exp Pathol 53:301.

THE USE OF A PERFUSION MODEL FOR STUDYING AGGREGATION AND
ATTACHMENT OF PLATELETS AND TUMOR CELLS AT SUBENDOTHELIAL
SURFACES.

Antonio Ordinas, J. Michael Marcum, Manley McGill,
Eva Bastida, and G. A. Jamieson
American Red Cross Blood Services
Blood Research Laboratories
9312 Old Georgetown Road
Bethesda, Maryland 20814

ABSTRACT

The Baumgartner perfusion apparatus has been applied to the study of the interaction of platelets and tumor cells and their attachment to subendothelial structures. Cells derived from an anaplastic murine tumor (Hut 20 line) induced platelet aggregation and were included in platelet thrombi that deposited on vascular subendothelium in perfusion experiments with heparinized human blood. In contrast, perfusion of blood samples containing cells from a line derived from a human epithelial carcinoma of the lung (A549 line), which did not interact with platelets, resulted in the deposition of platelets alone, with no tumor cells or blood cells other than platelets being observed in the thrombus. Extremely large platelet-tumor cell thrombi were found at the vascular surface in Hut 20 perfusions using vessel segments which had been treated with α-chymotrypsin. These large heterogeneous thrombi perturbed blood flow through the system and entrapped both erythrocytes and white cells. In order to quantitate the deposition of tumor cells, Hut 20 cells were labeled with ^{125}I-deoxyuridine and perfused in whole blood at a concentration of 3.7×10^5/ml. Tumor cell incorporation into platelet-tumor cell thrombi on chymotrypsinized segments yielded about 30,000 cpm/mg of vascular tissue but this value was reduced some 2 orders of magnitude by the inclusion of PGE_1 (1 ng/ml of perfusing blood; 2.8 μM) in parallel samples. Aspirin at 100 μM reduced tumor cell-dependent platelet aggregation but did not decrease the platelet-dependent deposition of radiolabeled Hut 20 cells on vascular subendothelium, suggesting that the release reaction may not be of major significance in this interaction. Tumor cell-induced platelet aggregation was not observed in a perfusion experiment using blood from a patient with severe von Willebrand's disease. However addition of 0.1 vol of ABO-compatible, heterologous plasma as a source of factor VIII to the von Willebrand blood sample restored the platelet-dependent deposition of radiolabeled tumor cells to control values.

INTRODUCTION

Gasic and coworkers (Gasic et al, 1973; Gasic et al, 1976) have demonstrated that cells from a variety of tumors can induce platelet aggregation in vitro and can cause thrombocytopenia when injected into animal models. Additionally, the number of metastases produced by intravenous injection of tumor cells was significantly reduced in animals previously rendered thrombocytopenic or treated with anti-platelet drugs (Gasic et al, 1973). The ability of tumor cells to be incorporated into platelet emboli may provide selective advantages to the tumor cells by enabling them to escape immune surveillance during circulation and by increasing the size and adhesiveness of the cell emboli. Platelet-containing emboli would be especially apt to adhere to damaged vasculature as in the formation of metastases at sites of surgical intervention after removal of the primary tumor (Fidler et al, 1978; Sugarbaker and Ketcham, 1977). In fact, the in vivo study by Warren and Vales (1972) indicated that the attachment of tumor cells to intact endothelium was reduced compared with their attachment at sites of damage to vascular surfaces.

We have recently described the adaptation of the Baumgartner perfusion system, previously used for the study of platelet function (Baumgartner, 1973; Turritto, 1974, Weiss et al, 1975; Baumgartner, 1977; Baumgartner et al, 1977; Tschopp et al, 1979; McGill and Brindley, 1979) to demonstrate the formation of platelet-tumor cell thrombi on vascular subendothelium in flowing blood in vitro (Marcum et al, 1981). This system, although more complex than aggregation, is a closer approximation to the in vivo situation because it takes into account rheological parameters and the contribution of the vessel wall as well as the presence of other blood cells.

MATERIALS AND METHODS

The Hut 20 cell line was kindly provided by Dr. Adi Gazdar, Veterans Administration Hospital, Washington, D.C. Although originally reported as being of human origin (Marcum et al, 1981; Gazdar et al, 1980), subsequent karyotyping and isozyme phenotyping has shown it to be an anaplastic tumor of murine origin derived from the nude mouse through which the original human tumor was passed. The A549 line from a defined human epithelial lung carcinoma was also provided by Dr. Gazdar. Cell culture conditions were as described (Bastida et al, 1981).

When required, cells were labeled by 24 hr incubation with culture medium containing 0.8 µCi/ml ^{125}I-deoxyuridine (72 Ci/µmol; New England Nuclear, Boston, Mass.), followed by a 2 hr chase with nonradioactive medium. Cells were harvested by mechanical scraping and washed.

Heparinized blood was used in all studies since the presence of citrate destroys the ability of platelets to interact with tumor cells (Gasic et al, 1973; Gasic et al, 1976).

These experiments were performed with the standard Baumgartner perfusion chamber, which has been described in detail elsewhere (Baumgartner, 1973; Turritto and Baumgartner, 1974; Baumgartner et al, 1977; Tschopp et al, 1979; Weiss et al, 1975; Baumgartner, 1977; McGill and Brindley, 1979). Rabbit aortas above the iliac arteries were de-endothelialized in situ with a balloon catheter or by digestion overnight with 20 U/ml α-chymotrypsin (Sigma) to remove cellular material and expose fibrillar collagen. Segments 10 to 14 mm in length were used in the perfusion apparatus. Rabbit aorta was used, rather than human material, because of its ready availability and the fact that it appears to cause no significant differences in platelet interaction (Tschopp et al, 1979). Fresh samples of whole blood were obtained from laboratory staff and anticoagulated by addition of 5 U/ml Na heparin. These samples were incubated at 37°, and concentrated aliquots of tumor cells with or without drugs were added to a 30 ml volume of blood with gentle mixing, to a final concentration of 4 to 5×10^5 cells/ml, since this concentration was found to give optimal results in the aggregation studies. The samples were then pumped through the perfusion chamber at flow rates of either 10 or 40 ml/min in order to simulate the low shear of capillary and venous flow. After 30 min of perfusion at 37° the vessel segments were rinsed by perfusion with phosphate-buffered saline and fixed with glutaraldehyde. The fixed segments were sliced from the supporting rod, dehydrated through a graded series of ethanol concentrations, embedded in Epon or in JB-4 embedding material (Polysciences, Inc., Warrington, Pa.), sectioned for light microscopy, and stained by conventional histological procedures.

Blood from a patient with severe von Willebrand's disease (F VII R: Ag 10%, F VIII R: vWF 0%) was obtained with permission. Whole blood samples containing an additional 0.1 vol of either autologous or ABO-compatible heterologous plasma as a source of factor VIII were perfused with ^{125}I-labeled Hut 20 cells in the Baumgartner chamber.

RESULTS

Aggregation Studies

Incubation of a heparinized PRP with 1.25×10^5 cells/ml from the tumor cell line Hut 20 resulted in platelet aggregation equal to 90% of the ADP control after a prolonged lag period. However, the lag time was dependent on tumor cell concentration. For example, at twice the Hut 20 cell concentration (5×10^5 cell/ml) there was equal platelet aggregation within 5 to 6 min. This concentration was used in all subsequent aggregation and perfusion studies with the Hut 20 cell line. Exposure of identical PRP aliquots to the epithelial human lung tumor line, A549, did not result in platelet aggregation after more than 20 min of incubation, even at cell concentrations as high as 1.3×10^7/ml. Aggregation induced by Hut 20 cells was inhibited by PGE_1 (2.8 µM), a potent inhibitor of platelet aggregation.

Perfusion Studies

When heparinized whole blood was perfused through chambers in the absence of added tumor cells or in the presence of non-platelet-reactive cells, only platelets were observed in association with the vascular subendothelium. No tumor cells or blood cells other than platelets were observed at the vascular surface after extensive examination of numerous, noncontiguous sections. In contrast, the inclusion of approximately the same concentration of Hut 20 cells (5×10^5/ml) in an identical blood sample in a parallel perfusion showed very large platelet thrombi at the vascular surface containing many tumor cells. The arrows indicate several tumor cells that were observed both within the mass of aggregated platelets and at its periphery. The size of the coaggregates ranged from single tumor cells in association with several platelets to very large platelet-tumor cell thrombi similar to that shown in Fig. 1, b. Perfusion of Hut 20 cells in the presence of 2.8 µM PGE_1, a concentration which totally inhibited Hut 20 cell-induced platelet aggregation, similarly inhibited deposition of both tumor cells and platelets at the vascular surface (Fig. 1, c). The corresponding control perfusion in the absence of PGE_1 yielded results comparable to that shown in Fig. 1, b.

In subsequent perfusion experiments, vessel segments were used that had been digested overnight with α-chymotrypsin to expose fibrillar collagen and increase their thrombogenicity. Additionally, flow rates were reduced to 10 ml/min to enhance the probability of larger elements (tumor cells and cell-platelet emboli) contacting the vascular surface. The results of perfusing blood samples containing Hut 20 cells under these conditions are shown in Fig. 1, d, which includes a small portion of a thrombus containing many nucleated cells embedded within the thrombus as well as along the edges. In addition, several erythrocyte profiles can be seen in this micrograph, suggesting that red cells were physically entrapped within the large thrombi.

In the above studies, it was difficult to differentiate the added Hut 20 tumor cells from nucleated blood cells trapped non-specifically in the thrombus. For quantitative studies the Hut 20 cells grown in the presence of ^{125}I-deoxyuridine were used in the perfusion experiments and small cylinders of tissue (diameter, 2mm) were cut from the vessel segments and counted. Approximately 20,000 - 30,000 cpm/mg vascular tissue (dry weight) were found in control samples but this fell to about 1% (∼300 cpm/mg) in the presence of PGE_1. In contrast, in the presence of aspirin (100 µM) the number of radioactive tumor cells deposited on the vessel (18,500 ± 9,800 cpm/mg) was the same as control values.

Fig. 1. Light micrographs of de-endothelialized vessel segments after perfusion with whole blood samples containing tumor cells. a, Perfusion 4.3×10^5 A549 cells/ml of blood. b, Parallel perfusion of 4.3×10^5 Hut 20 cells/ml of blood. c, Perfusion of 5.1×10^5 Hut 20 cells/ml of blood containing 2.8 μM PGE1. d, Perfusion of whole blood samples containing Hut 20 cells (1.5×10^5/ml) over digested vessel segment (x2000 for all). (Reprinted with permission, from Marcum et al. 1980).

The radioactive tumor cell perfusion technique was utilized to investigate the role of factor VIII/von Willebrand factor in the deposition of platelet-tumor cell thrombi on subendothelium. Whole blood from a patient with severe von Willebrand's disease (<10% factor VIII antigen levels) was mixed with radioactive Hut 20 cells and 0.1 vol of ABO-compatible plasma as a source of normal factor VIII. In the absence of added factor VIII, a very low amount of radioactivity was observed at the vessel wall (approximately 200 cpm/mg). However, addition of 0.1 volume of heterologous plasma a a source of factor VIII resulted in incorporation of normal levels of ^{125}I-labeled tumor cells (∼ 40,000 cpm/mg) at vessel surface.

DISCUSSION

In the present work, the Hut 20 line consistently 30). This aggregation was preceded by a lag time that was inversely related to cell concentration and was totally inhibited by PGE_1. Cells from this line were also incorporated into the thrombi that formed the basement membrane of de-endothelialized vessel segments during perfusion experiments and, to a much greater extent, on the intima exposed by prior treatment of the vessel segments with chymotrypsin. In the latter case, the thrombi were so large that they entrapped both erythrocytes and leukocytes from the circulating blood and physically occluded the flow of blood through the annular spaces of the perfusion chamber. In contrast, the A549 line, which served as a negative control, showed no interaction with platelets in the aggregation studies or in the perfusion apparatus. Aggregation was not affected by treatment of the Hut 20 cultures with collagenase, indicating that aggregation induced by the cells was not an artifact of contaminating fibroblasts.

The perfusion system has also been used to evaluate the effect of various antiplatelet drugs on the Hut 20-induced platelet deposition. Under the normal conditions of the Baumgartner perfusion system, PGE_1 inhibits platelet interaction with subendothelium (Baumgartner et al, 1976), whereas aspirin permits adhesion but blocks the platelet release reaction and thrombus formation (Sakariassen et al, 1979). In the presence of Hut 20 cells, PGE_1 continued to completely block the interaction with the vessel wall. However, 100 µM aspirin was without effect on the deposition of ^{125}I-labeled Hut 20 cells at the vascular surface, whereas the Hut 20-induced aggregation of platelets was reduced about 50% whether PRP from aspirinated donors was used or aspirinated in vitro. These results suggest that the platelet release reaction may not be of critical importance for the interaction of Hut 20 cells and platelets with subendothelium.

The finding that factor VIII/von Willebrand factor is necessary for the formation of platelet-tumor cell thrombi on the vascular surface is consistent with the demonstrated role of factor VIII/von Willebrand factor in platelet adhesion to subendothelium (Weiss et al, 1975; Sakariassen et al, 1979).

The previous applications of the Baumgartner perfusion device have involved the interaction of platelets with thrombogenic surfaces in normal blood (Baumgartner, 1973; Turrito and Baumgartner, 1974; Tschopp et al, 1979), in a variety of pathological conditions (Baumgartner et al, 1977; Weiss et al, 1975) and, in our laboratories, in an assessment of platelet function after storage (McGill and Brindley, 1979). The present results suggest that this system may also have utility in studying the interaction of subendothelial elements with platelets and tumor cells as an ex vivo model of the early stages of blood-borne metastasis.

ACKNOWLEDGEMENTS

This work was supported, in part, by USPHS Grants HL 14697, HL 20971 and Biomedical Research Support Grant RR 05737. Publication No. 548 from the American Red Cross.

REFERENCES

Agostino D, Clifton EE (1965). Trauma as a cause of localization of blood-borne metastases: preventive effect of heparin and fibrinolysin. Ann Surg 161:97.
Bastida E, Ordinas A, Jamieson GA (1981). Differing platelet aggregating effects by two tumor cell lines: Absence of role for platelet-derived ADP. A J Hematol (In press).
Baumgartner HR (1973). The role of blood flow in platelet adhesion, fibrin deposition and formation of mural thrombi. Microvasc Res 5:167.

Baumgartner HR, Tschopp TB, Weiss HG (1977). Platelet interaction with collagen fibrils in flowing blood. II. Impaired adhesion-aggregation in bleeding disorders. A comparison with subendothelium. Thromb Haemost 37:17.

Baumgartner HR (1977). Platelet interaction with collagen fibrils in flowing blood. I. Reaction of human platelets with chymotrypsin digested subendothelium. Thromb Haemost 37:1.

Baumgartner HR (1979). Effects of acetylsalicylic acid, sulfinpyrazone and dipyridamole on platelet adhesion and aggregation in flowing native anticoagulated blood. Haemostasis 8:340.

Baumgartner HR, Muggli R (1976). Adhesion and aggregation: morphological demonstration and quantitation in vivo and in vitro in platelets. In Gordon JL (ed): "Biology and Pathology," Amsterdam:Elsevier-North Holland Biomedical Press, p 23.

Chew EC, Wallace AC (1976). Demonstration of fibrin in early stages of experimental metastases. Cancer Res 36:1904.

Donati MB, Poggi A (1980). Malignancy and haemostasis. Br J Haematol 44:182.

Fidler IJ, Gersten DM, Hart IR (1978). The biology of cancer invasion and metastasis. Adv Cancer Res 28:149.

Fidler IJ (1974). Immune stimulation-inhibition of experimental cancer metastasis. Cancer Res 34:491.

Gasic GJ, Gasic TB, Galanti N, Johnson T, Murphy S (1973). Platelet-tumor cell interaction in mice. The role of platelets in the spread of malignant disease. Int J Cancer 11:704.

Gasic GJ, Koch PAG, Hsu B, Gasic TB, Niewiarowski S (1976). Thrombogenic activity of mouse and human tumors: effects on platelets, coagulation, and fibrinolysis and possible significance for metastases. Z Krebsforsch 86:263.

Gazdar AF, Carney DN, Russell EK, Sims HL, Baylin SB, Bunn PA, Jr, Guccion JG, Minna JD (1980). Establishment of contious clonable cultures of small cell carcinoma of the lung which have amine precursor uptake and decarboxylation cell properties. Cancer Res 40:3502-3507.

Hara Y, Steiner M, Baldini MG (1980). Characterization of the platelet aggregating activity of tumor cells. Cancer Res 40:1217.

Lieber M, Smith B, Seakal A, Nelson-Rees W, Todaro G (1976). A continuous tumor-cell line from a human lung carcinoma with properties of type II alveolar epithelial cells. Int J Cancer 17:62.

Marcum JM, McGill M, Bastida E, Ordinas A, Jamieson GA (1980). The interaction of platelets, tumor cells and vascular subendothelium. J Lab Clin Med 96:1046-1053.

McGill M, Brindley DC (1979). Effects of storage on platelet reactivity to arterial subendothelium during blood flow. J Lab Clin Med 94:370.

Pearlstein E, Cooper LB, Karpatkin S (1979). Extraction and characterization of a platelet-aggregating material from SV40-transformed mouse 3T3 fibroblasts. J Lab Clin Med 93:332.

Sakariassen KS, Bolhuis PA, Sixma JJ (1979). Human blood platelet adhesion to artery subendothelium is mediated by factor VIII-von Willebrand factor bound to the subendothelium. Nature 279:636.

Sugarbaker EV, Ketcham AS (1977). Mechanisms and prevention of cancer dissemination: an overview. Semin Oncol 4:19.

Tschopp TB, Baumgartner HR, Silberbauer K, Sinzinger H (1979). Platelet adhesion and platelet thrombus formation on subendothelium of human arteries and veins exposed to flowing blood in vitro. A comparison with rabbit aorta. Haemosstasis 8:19.

Turritto VT, Baumgartner HR (1974). Effect of physical factors on platelet adherence to subendothelium. Thromb Diath Haemorrh (Suppl) 60:17.

Warren BA (1973). Environment of the blood-borne tumor embolus adherent to vessel wall. J Med 4:150.

Warren BA (1974). Tumor metastasis and thrombosis. Thromb Diath Haemorrh (Suppl) 59:139.

Weiss HJ, Tschopp TB, Baumgartner HT (1975). Impaired interaction of platelets with subendothelium in bleeding disorders. N Engl J Med 293:619.

Warren BA, Vales O (1972). The adhesion of thromboplastic tumor emboli to vessel walls in vivo. Br J Exp Pathol 53:301.

Wood S (1964). Experimental studies of the intravascular dissemination of ascitic V2 carcinoma cells in the rabbit with special reference to fibrinogen and fibrinolytic agents. Bull Schweiz Akad Med Wiss 20:92.

INTERACTION OF PLATELETS AND TUMOR CELLS

Manfred Steiner

Division of Hematologic Research, The Memorial
Hospital, Pawtucket, and Brown University,
Providence, Rhode Island 02860

It has been suggested that an interaction between tumor cells and platelets may play a role in initiating the arrest of circulating malignant cells (Weiss, 1976) and in producing coagulopathies (Goodnight, 1974; Slichter and Harker, 1974). Walker 256 tumor cell emboli arrested in lung capillaries of the rat were associated with large platelet aggregates and platelets were sometimes found to form a tail at one pole of the tumor cells (Jones et al., 1971; Chew et al., 1976). Fibrin was also associated with platelet clumps in early stages of metastasis (Chew and Wallace, 1976). Platelets with dendrite pseudopodia were seen to be adherent to the spiky surface of Walker 256 tumor cells when the latter were agitated with rat platelet rich plasma (Warren, 1976). Degranulation of platelets was recognized within 30 min to a few hours followed by their disappearance presumably by phagocytosis (Gasic et al., 1968). Endothelium experimentally damaged by air emboli appeared to provide a more secure adhesional site for tumor cell emboli (Warren, 1976). An even more direct link between platelets and tumor cells has been provided by the studies of Gasic and coworkers (1968) who reduced tumor cell metastases in mice by decreasing their circulating platelet count. Conversely, infusion of tumor cells with platelet aggregating activity led to a higher incidence of lung metastases than that of tumor cells lacking this activity (Gasic et al., 1973).

These early studies have made a very convincing case for the involvement of platelets in the development of blood-borne metastases. This process, according to Sugarbaker and

Ketcham (1977), requires at least 4 steps; the separation of cancer cells from the primary tumor with subsequent transfer into the blood stream, their transport to distant sites, followed by arrest at a vessel wall as tumor cell emboli and finally their survival and growth. Studies in our laboratory have confirmed and further elucidated some of the early steps in this scheme in which platelets play an important role (Hara et al., 1980a). In addition, we have found convincing evidence for their involvement in the final sequence of blood-borne metastatic development, i.e. the survival and growth of the tumor cells at the site of their arrest during embolization (Hara et al., 1980b).

As the rapidly growing volume of studies in this area of research has clearly shown, only certain tumor cells have the ability to aggregate platelets (Gasic et al., 1976 and 1978) and platelets, in turn, have variable growth-promoting effects for the relatively few cancer cell lines that were tested for this phenomenon (Eastment and Sibrasku, 1978; Hara et al., 1980b; Kohler and Lipton, 1977). Most of the studies reported here have been carried out with two well-defined animal tumor cell lines which exhibit both aspects of the tumor cell-platelet interaction, i.e. the platelet aggregating activity of the former and the growth stimulation exerted by the latter.

MATERIALS AND METHODS

Cell Lines and Cultures

Four tumor cell lines including three from the mouse, i.e. neuroblastoma (Neuro-2a), renal adenocarcinoma (RAG), and mammary tumor 060562 and one from the rat, Leydig cell testicular tumor were obtained from the American Type Culture Collection (Rockville, MD). These cells were maintained in culture as described previously (Hara et al., 1980b).

Platelet Aggregation Studies

For aggregation studies, tumor cells grown to confluency were washed with Ca^{2+}-Mg^{2+}-free Hanks' balanced salt solution (HBSS), were harvested, resuspended in HBSS, dialyzed overnight, and counted electronically. Platelets for these studies were obtained either from rabbits or humans using heparin (10 U/ml), sodium citrate (0.38%, final concentration) or ACD as anticoagulant. Platelet rich plasma (PRP) was prepared as previously described (Hara et al., 1980a). In general, platelet counts were adjusted to 300,000-350,000/µl. Aggregations were monitored in a regular or a Lumi-aggregometer according to conventional techniques.

Preparation of Platelet Lysate and Tumor Cell Membrane Fragments

Platelet lysate was prepared from human platelets either isolated from freshly drawn blood or from outdated concentrates (>36 but <72 hours old). Platelets were washed, freed of contaminating red and white cells, and lysed by sonication (Hara et al., 1980b). The 100,000 x g supernatant of the platelet sonicate was used either without further purification or was subjected to isoelectric focusing on a pH gradient of ampholytes (LKB Produkter AB) from 3 to 10. Membrane fragments were prepared by exposing confluent monolayer cultures to an isotonic low ionic strength medium (Hara et al., 1980a) and removing undisrupted cells as well as large cellular debris by centrifugation.

Assay of Growth-Promoting Activity

Growth promoting activity was measured by the ability of a particular additive to the basic Dulbecco's modified Eagle's culture medium to increase target cell number over a designated incubation period. Cells were plated at about 1-2 x 10^5/30 ml plastic flask containing the above culture medium with 1% fetal calf serum. After 24 hours at 37°C, the cell monolayers were washed twice with Ca^{2+}-Mg^{2+}-free HBSS and the medium changed to one containing the desired additive in Dulbecco's modified Eagle's medium. Day 0 cell counts taken after the 24 hour conditioning period were compared to cell counts after 2 or 3 days of incubation (rat Leydig cell testicular tumor line was measured after 3 days, all other cell lines after 2 days). All experiments were performed in triplicate.

RESULTS

Tumor Cell-Induced Platelet Aggregation

The platelet aggregating activity of tumor cells is characterized by certain distinctive features which clearly set it apart from most other aggregation stimuli. A fairly high species specificity has been recognized (Gasic et al., 1976). According to our (Hara et al., 1980a) and other investigators' (Gasic et al., 1976) previous studies, mouse and rat tumor cell lines were thought to have no platelet aggregating activity for human platelets. This concept, however, has to be revised in the light of some of the experiments described below. In view of the greater tumor cell-induced aggregation of rabbit platelets compared to that of other rodents, most of the aggregation experiments were performed with rabbit PRP.

It has been the general experience of investigators in this field that heparinized PRP was best suited to demonstrate the aggregating activity of tumor cells. This peculiarity was further studied. A clear dependence of tumor cell-induced aggregation on Mg^{2+} could be demonstrated (Figure 1). Even relatively strong chelation of divalent cations by sodium citrate and especially by ACD could be overcome by the addition of Mg^{2+} ions. It is interesting to note that Ca^{2+}, up to 120 mM, failed to restore platelet aggregation in a citrate- or ACD-inhibited system.

Tumor cell-induced platelet aggregation always showed a lag time, the length of which was found to be inversely related to the number of tumor cells added. Aggregation of platelets was accompanied by a secretory release (Figure 2) which could be inhibited by an ADP-clearing system and by pretreatment of platelets with acetylsalicylic acid. The possibility that tumor cells by releasing procoagulant activity activated the coagulation sequence with eventual formation of thrombin, a powerful aggregating agent for platelets, could be ruled out by experiments using hirudin. This specific thrombin inhibitor had absolutely no effect on platelet aggregation promoted by RAG or Neuro-2a cells.

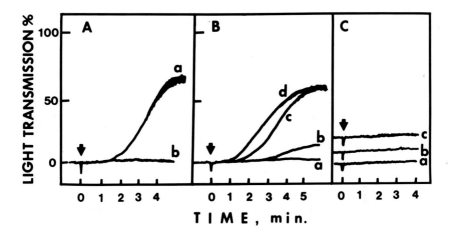

Reprinted with permission of Cancer Research.

Figure 1: Effect of Divalent Cations on Tumor Cell-Induced Platelet Aggregation. In A, to heparinized rabbit PRP was added either 0.05 ml of Ca^{2+}-Mg^{2+}-free HBSS (Curve a) or 0.05 ml of sodium citrate at a final concentration of 0.19% (Curve b). In B, to heparinized rabbit PRP containing sodium citrate in a final concentration of 0.19% were added Mg^{2+} in the following concentrations: Curve a, no Mg^{2+} (control); Curve b, 0.8 mM Mg^{2+}; Curve c, 1.6 mM Mg^{2+}; Curve d, 3.2 mM Mg^{2+}. In C, to heparinized rabbit PRP containing sodium citrate in a final concentration of 0.19% were added the following concentrations of Ca^{2+}: Curve a, 10 mM Ca^{2+}; Curve b, 60 mM Ca^{2+}; Curve c, 120 mM Ca^{2+}. For all these experiments the aggregating agent which was added at the time of the arrow was RAG cell membrane fragments (40 μg/ml).

Platelet Aggregation Induced by Tumor Cell Membrane Fragments

The aggregating activity of the malignant cells we examined appeared to be an integral part of the cells. The concentrated supernatant of tumor cells stirred either by themselves or together with platelets showed no aggregating activity when tested on heparin-anticoagulated rabbit PRP. However, isolation of membrane fragments by isotonic low ionic strength treatment of intact RAG cells yielded a preparation of membranous structures and vesicles which had potent aggregating activity (Figures 3 and 4). At concentrations $\geq$ 10 µg protein per ml such cell fragment preparations were able to induce a maximal aggregation response in heparinized mouse or rabbit PRP. The characteristics of this aggregation were identical to those described above for intact tumor cells which included the relation of lag time to potency of aggregation stimulus, the behavior with respect to divalent cations and the stimulation of a secretory release that could be inhibited by aspirin or by an ADP-clearing system of creatine phosphate:creatine phosphokinase.

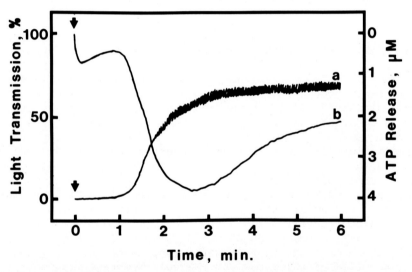

Figure 2: Tumor Cell-Induced Aggregation of Rabbit Platelets. To heparinized rabbit PRP were added 2 x 10^5 RAG cells. Aggregation (Curve a) and ATP release (Curve b) were measured in a Lumi-Aggregometer. An ATP standard was added at the termination of the experiment which was used to calibrate the right ordinate. The arrow indicates the time of addition of the tumor cells.

Sensitization of Platelets to Tumor Cell-Induced Aggregation

Maximal platelet aggregation induced either by intact tumor cells or by cell fragments was found to be dependent on the presence of plasma which could be reduced to $\leq$ 5%.

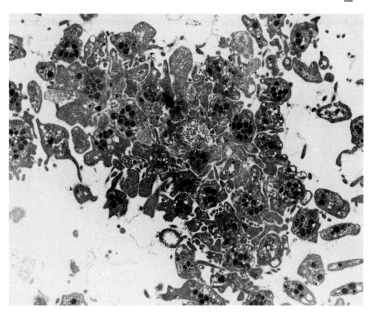

Reprinted with permission of Cancer Research.

Figure 3: Electronmicrograph of RAG Cell Membrane Fragments-Induced Aggregation of Rabbit Platelets. A large platelet aggregate is seen surrounding a central core of compressed membrane fragments. Platelets near the center show complete degranulation but in the periphery intact electron dense granules can still be recognized in several platelets. Magnification x4,500.

Fibrinogen in concentrations of 50-100 mg/dl could restore 40-50% of the maximal aggregability of platelets in a plasma-containing medium. Although platelets suspended in a plasma-free buffer medium were unable to respond to tumor cells with a secretory release, they showed gradually increasing adherence to the tumor cells and underwent shape

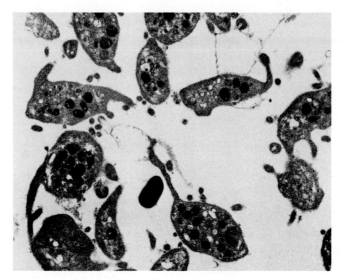

Figure 4: Close up View of the Contact Area Between Platelets and RAG Cell Membrane Fragments. Most of the platelets have undergone shape change and are partially degranulated. Magnification x10,000.

change in the course of this agglutination. This effect could be markedly enhanced by pretreating platelets with neuraminidase. The enzyme which was purified by the method of Hatton and Regoeczi (1973) was demonstrably free of protease activities and did not cause spontaneous platelet aggregation when added in concentrations up to 0.01 U/ml for 30 min at 37°C. The stimulatory effect on tumor cell-induced agglutinability of platelets was not due to an action of the enzyme on tumor cells. Neuraminidase-treated platelets were washed twice before exposure to RAG or Neuro-2a cells. Similar pretreatment of platelets with fucosidase, β-galactosidase or albumin did not enhance tumor cell-promoted platelet agglutination above that seen in non-enzyme treated platelet suspensions.

Human platelets suspended in plasma were able to adhere to tumor cells in single layers when stirred at 37°C for extended periods of time. Preincubation of human platelets with neuraminidase in concentrations between 0.005-0.01 U/ml for 15-30 min at 37°C sensitized the platelets to subsequent exposure to tumor cells (Figure 5). Aggregation accompanied by release reaction occurred under these conditions after 9-13 min following a lag period of 5-8 min and a slow agglutination phase of approximately 4-5 min during which platelets began to adhere to tumor cells. Sensitization or pretreatment of platelets by other means, e.g. by exposure to low non-aggregating concentrations of ADP (0.1 µM) or by preincubation with β-galactosidase or fucosidase did not produce tumor cell-promoted aggregation of human platelets.

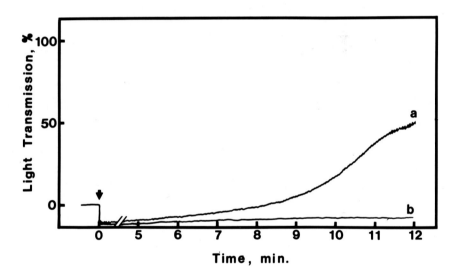

Figure 5: Aggregation of Human Platelets by Neuro-2a Cells. Platelets suspended in Ca^{2+}-free Tyrode's buffer were incubated with purified neuraminidase (Hatton and Regoeczi, 1973) at a concentration of 0.005 U/ml for 30 min at 37°C. Platelets were washed in ACD-containing Ca^{2+}-free Tyrode's buffer and resuspended in heparinized platelet poor plasma. Aggregation was measured after addition of 2×10^5 tumor cells (arrow). Non-enzyme-treated control platelets (Curve b), neuraminidase-treated platelets (Curve a).

The Nature of the Platelet Aggregating Principle of Tumor Cells

The activity of the tumor cell-associated platelet-aggregating principle was readily destroyed by treatment with trypsin, neuraminidase, phospholipase A_2 and sonication and was drastically reduced by nonionic detergents. Heating at 56°C did not reduce this activity but almost completely destroyed it at 85°C. Collagenase, in concentrations capable of abolishing the aggregating activity of calf skin collagen, did not affect the aggregating principle of tumor cells. In parallel experiments identical results were obtained with isolated tumor cell membrane fragments. Treatment of tumor cells with trypsin eliminated their platelet aggregating activity. However, the latter could not be recovered in the supernatant of the incubates. Reculturing trypsin-digested tumor cells for 24 hours fully regenerated their aggregating activity.

Release of Platelet Glycosaminoglycans by Tumor Cells.

In other experiments we investigated the possibility that glycosaminoglycan (GAG)-hydrolyzing enzyme(s), e.g. chondroitin sulfatases, chondroitinase, etc., were responsible for the aggregating activity of RAG or Neuro-2a tumor cells. Platelets are known to contain GAGs which were shown to be primarily chondroitin 6-sulfate and dermatan sulfate (Kerby and Taylor, 1959) or chondroitin 4-sulfate (Olsson and Gardell, 1967). To facilitate demonstration of a GAG-specific enzyme reaction on the part of tumor cells, platelets were harvested from rabbits injected with $[^{35}SO_4]$. Sulfate is primarily incorporated into GAGs (Odell and Andersen, 1957), only to a very minimal extent into sulfur-containing amino acids (Dziewiatkowski, 1954). $[^{35}S]$-labeled platelets were collected 3-4 days after the last of 3 consecutive daily injections of $[^{35}SO_4]$. Tumor cell-induced aggregation of $[^{35}S]$-labeled rabbit platelets led to release of $[^{35}S]$ from platelets. The magnitude of this release was related to the potency of the aggregation stimulus (Table 1). However, this effect is not specific for tumor cells. Other aggregating agents also release GAGs in the course of platelet clumping, especially during the secretory release phase (Riddell and Bier, 1965). The incubation of tumor cells with increasing numbers of $[^{35}S]$-labeled platelets suggested approach of a limiting release value (Table 2).

Table 1

Release of Radioactivity from [^{35}S]-Labeled Platelets by Tumor Cells. Variation of Tumor Cell Number.

Rabbits were injected with 0.5 mCi [^{35}SO$_4$] daily for a total of 3 days. The animals were bled 3-4 days after the last injection and glycosaminoglycans extracted (Olsson and Gardell, 1967) from one aliquot of the platelets. The radioactivity in this fraction constituted > 95% of the total activity associated with platelets. Platelets were incubated with varying concentrations of RAG cells for 30 min at 37°C with continuous stirring. Radioactivity of the solubilized platelet pellet and of the supernatant were measured.

Number of Tumor Cells	[^{35}S] Activity, cpm† Released to Supernatant	Remaining in Platelets
0 x 10^5	76	732
1 x 10^5	118	700
1.5 x 10^5	131	687
2.5 x 10^5	160	648
4.0 x 10^5	181	615
5.0 x 10^5	204	596
7.5 x 10^5	246	550
10.0 x 10^5	295	501

† Mean of 2 experiments

These findings are compatible with the existence of an enzymatic reaction but do not necessarily prove that tumor cell-induced aggregation of platelets is mediated by GAG-hydrolyzing enzyme(s). More direct evidence linking activity of the latter to platelet aggregation is currently being sought in our laboratory.

Growth-Promoting Activity of Platelets for Tumor Cells

The interaction of platelets and tumor cells was not found to be limited to aggregation of the former by certain malignant cell lines but in addition platelets could release a growth promoting factor(s) for tumor cells during aggregation. In previous experiments (Hara et al., 1980b), we were able to characterize some aspects of this activity which are briefly summarized below.

Table 2

Release of Radioactivity from [^{35}S]-Labeled Platelets by Tumor Cells. Variation of Platelet Number.

The experimental conditions were similar to those described in legend to Table 1. RAG cells, 4×10^5/ml, were incubated with a series of platelet suspensions varying in cell number from 1×10^5 to 4×10^5/µl. The radioactivities remaining in platelets and released into the supernatant were measured after an incubation at 37°C for 30 min with constant stirring.

Number of Platelets	[^{35}S] Activity, cpm†	
	Released to Supernatant	Remaining in Platelets
1×10^5/µl	73	200
1.5×10^5/µl	115	320
2.0×10^5/µl	168	400
2.5×10^5/µl	202	570
3.0×10^5/µl	232	638
3.5×10^5/µl	240	672
4.0×10^5/µl	251	684

† Mean of 2 experiments

Lysate of human platelets showed a dose-dependent promotion of cell growth with RAG and Neuro-2a cells (Table 3). Minimal activity was found at a concentration of 40-50 µg/ml while maximal activity plateaued at about 250 µg protein per ml. Compared to equal concentrations of human serum, platelet lysate always exhibited lower growth stimulation (Figure 6). No obvious synergism between platelet lysate and human serum could be detected. The platelet-derived growth-promoting activity for tumor cells was released in the course of aggregation-induced secretory release. Both RAG and Neuro-2a cells were able to release growth-promoting activity from platelets. In fact, the activity was higher when tumor cells induced aggregation than with thrombin, collagen or ADP as the aggregating agent (Table 4).

Table 3

Dose-Dependent Stimulation of Tumor Cell Growth by Platelet Lysate and Serum. Counts of RAG and Neuro-2a cells were taken after 2 days of culture in various concentrations of human serum or platelet lysate. Growth-promoting activity is expressed as the ratio of day 2 over day 0 cell counts.

Conc. of Serum or Platelet Lysate µg/ml	Target Cell Line	Growth-Promoting Activity ($count_2/count_0$) Serum	Platelet Lysate
20	Neuro-2a	1.57	1.04
40		2.30	1.11
80		3.48	1.68
200		3.42	2.30
400		3.73	2.64
50	RAG	1.22	1.12
100		2.05	1.80
200		2.22	1.89
400		2.84	2.05
800		2.89	2.05

Table 4

Release of Growth-Promoting Activity from Platelets. Washed human platelets suspended in Tyrode's buffer were incubated with aggregating agents under constant stirring for 5 min. Growth-promoting activity was measured in the supernatants. Values are the means of 3 experiments.

Aggregating Agent	Concentration	Target cell line	Growth-Promoting Activity ($count_2/count_0$) Platelets	Control
Thrombin	2 U/ml	Neuro-2a	2.31	1.23
		RAG	3.81	1.80
Collagen	200 µg/ml	Neuro-2a	1.66	1.23
ADP	50 µM	Neuro-2a	1.29	1.23
RAG cells	5×10^5/ml	RAG	4.98	1.79

Reprinted with permission of Cancer Research

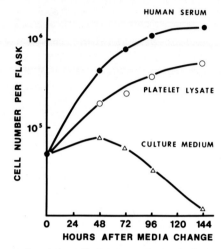

Reprinted with permission of Cancer Research.

Figure 6: Stimulation of Neuro-2a Cell Growth by Unfractionated Platelet Lysate and Serum. Cells were cultured in human serum (200 µg protein per ml), platelet lysate (200 µg protein per ml), or culture medium only. The means of 3 experiments, each performed in triplicate, are plotted.

Isolation of Tumor Cell-Directed Growth Promoting Activity of Platelets

In recent studies we have attempted to purify this platelet-derived growth factor. Isoelectric focusing of crude platelet lysate on a column having a pH gradient of 3 to 10 is shown in figure 7. Individual fractions making up the peaks were pooled, exhaustively dialyzed, and freeze-dried. Reconstituted with HBSS, the isolated fractions were adjusted to an optical absorbance reading of approximately 0.5 at 280 nm. After sterilization each fraction was tested for its growth-promoting activity with RAG and Neuro-2a cells. Activity resided primarily in peaks of high pI, i.e. $\geq$ 9.0. Compared to human serum at a concentration of about 200 µg/ml culture medium, the maximal growth promotion of platelet lysate fractions was either equal or only slightly lower when expressed as day 2 cell count over day 0 cell count. There was no significant activity in any of the isoelectric fractions with pI $\leq$ 8.0. The rate of growth-promotion associated with isolated platelet lysate fractions appeared to drop off after 2-3 days of incubation. It was interesting to note that the addition of human plasma at concentrations of $\leq$ 20 µg/ml culture medium to platelet lysate fractions prevented this reduction in the rate of growth stimulation. By itself, human plasma was unable to maintain an active growth of tumor cells. Denaturation of platelet lysate fractions by heating at $\geq$ 70°C or by treatment with sodium dodecyl sulfate or Triton x-100 destroyed any growth-promoting activity the fraction had.

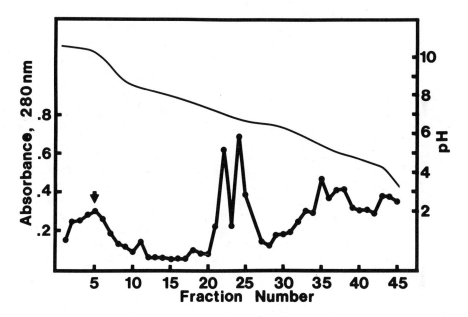

Figure 7: Isoelectric Focusing Profile of Human Platelet Lysate. A total of approximately 10 mg protein were focused in a pH gradient of 3-10. Individual fractions were collected and optical absorbance at 280 nm measured (full circles). The pH gradient is shown by the continuous line in the upper half of the figure. The arrow indicates the fraction of greatest growth-promoting activity.

DISCUSSION

The observations on several animal tumor cell lines,

especially a mouse neuroblastoma and renal adenocarcinoma, clearly demonstrated that platelets have a special relationship to tumor cells. Although in our studies only a small group of cell lines were studied, sufficient information on a considerable number and wide variety of malignant cells is available to conclude that the capability to aggregate platelets is a characteristic shared only by a select group of tumor cells. At this time it is not known whether the aggregating principle is similar or identical in the various tumor cells possessing the ability to clump platelets. The number of cell lines investigated has been too small to draw any conclusion. There are, however, indications that the factor(s) responsible for aggregation are not identical. Our findings are in broad agreement with those of Gasic et al. (1973, 1976, 1977a and b, 1978).

The need to use heparinized PRP in order to demonstrate tumor cell-induced platelet aggregation has been found by all investigators in this field. Our studies demonstrate quite clearly that chelation of Mg^{2+} appears to be responsible for this behavior. This finding sets tumor cell-induced platelet aggregation decidedly apart from that of any other aggregating agent. Nevertheless, there are several features which are common to aggregation induced by tumor cells and especially by collagen. Both share a lag period, inhibition by aspirin and by an ADP-destroying system. Both are accompanied by a secretory release and depend for maximal response on the presence of a minimal concentration of plasma. Fibrinogen appears to be only one of probably a number of factors that are required for optimal aggregation.

The aggregating principle is an integral part of tumor cell membranes which has so far resisted all attempts at solubilization whether by low ionic strength media, proteolytic enzymes, lithium diiodosalicylate or urea. The active principle as demonstrated by electronmicroscopy was free from any subcellular organelles and consisted of a preparation of apparently pure tumor cell membranes to which platelets were adhering in monolayers and clumps. A constellation of lipids, carbohydrates, and protein made up the aggregation-inducing activity. Loss or major modification of any one of these components resulted in complete abolition of the activity.

Especially interesting was the diametrically opposite effect of neuraminidase treatment on tumor cells and platelets. While N-acetylneuraminic acid was needed on the reducing end of the tumor cell-associated aggregating principle, on the platelet the terminal sialic acid residues proved to be a hindrance to the action of that factor in inducing maximal aggregation. The role of N-acetylneuraminic acid appears to be extremely critical as only small reductions in platelets and tumor cells will initiate large changes of their respective functions in the system. These findings bring up a point of possible clinical implication. In unrelated studies examining platelet N-acetylneuraminic acid as a function of platelet age (unpublished observations), we have found a gradual decrease of sialic acid in the course of platelet senescence. Therefore, platelets nearing the end of their life span may be more susceptible to the aggregating action of tumor cells. This contrasts with the widely held belief that young platelets are hemostatically more active than older cells.

The discovery that platelets can provide growth-promoting activity for certain tumor cell lines is not totally unexpected as their stimulation of nontransformed fibroblasts, glial cells and arterial smooth muscle cells by a specific growth factor is well known (Ross et al., 1974; Kohler and Lipton, 1974; Pledger et al., 1977). The existence of other platelet-derived growth promoting factors with specificity for a wider variety of mammalian cells, including transformed (Kohler and Lipton, 1977) and malignant cell lines (Eastment and Sibrasku, 1978) has been suggested. In our studies (Hara et al., 1980b), we were able to demonstrate that human platelets possess a growth-promoting activity for at least 4 tumor cell lines of mice and rats. Whole, i.e. unfractionated, platelet lysate was able to sustain growth of tumor cells without any additives other than those provided in Dulbecco's modified Eagle's medium, indicating that it contained a "survival factor" (Paul et al., 1971) for these malignant cells. This behavior set the growth-stimulating activity for malignant cells clearly apart from the platelet-derived growth factor for non-transformed fibroblasts and smooth muscle cells, which is unable to stimulate cell growth without a plasma factor (Rutherford and Ross, 1976). At least one of these factors in plasma appears to be somatomedin C which allows quiescent fibroblasts, rendered competent to replicate their DNA by exposure to platelet-derived growth factor, to progress into S phase (Stiles et al., 1979). As our studies have shown, there are strong indications that the high pI platelet fraction, showing growth promoting activity, is unable to maintain continued cell growth without a plasma factor. We have not yet isolated this activity from platelet lysate which appears to mimic the effect of the plasma factor. At comparable protein concentrations, serum always had a more potent growth-promoting effect than platelet lysate and always attained higher saturation density. Growth promotion by serum and platelet lysate produced cells which were morphologically identical.

A comparison of the biochemical characteristics of platelet-derived growth factors with specificity for tumor and normal cells reveals similarities but also some significant differences. Among the latter, most important are probably the heat resistance (Heldin et al., 1977) and extreme sensitivity to trypsin digestion of the platelet-derived growth factor for normal cells (Ross and Vogel, 1978), whereas that for tumor cells showed complete destruction by periodate oxidation (Hara et al., 1980b). Notable similarities are the nondialyzable nature of the growth factors and their extremely alkaline isoelectric points.

Based on our findings we conclude that the tumor cell-directed platelet-derived growth factor activity resides in a glycoprotein or -peptide. From the potency of its release in response to specific aggregating agents, we tentatively conclude that the activity is localized in α-granules. Platelet-derived growth factor activity for normal cells is also thought to be distributed in α-granules (Kaplan et al., 1979; Witte et al., 1978).

The pathophysiologic implications of these findings which have been somewhat speculative in the past, are now beginning to take on more concrete form. The survival of only a small number of tumor cells in the course of hematogenous tumor spread (Fidler, 1970) relegates an important role to platelets in arresting tumor cell in the vascular tree and in providing growth-promoting factors for the survival of tumor emboli. Realization of the importance of the platelet release reaction in mediating especially the latter response establishes a secure basis for clinical trials of agents inhibiting platelet secretory release.

This work was supported by DOE Contract 79EV02783.

REFERENCES

Chew EC, Josephson RL, Wallace ACL (1976). Morphological aspects of the arrest of circulating cancer cells. In Weiss L (ed): "Fundamental Aspects of Metastasis," New York: American Elsevier Publishing Co., p 121.
Chew EC, Wallace AC (1976). Demonstrations of fibrin in early stages of experimental metastases. Cancer Res 36: 1904.
Dziewiatkowski DD (1954). Utilization of sulfate sulfur in the rat for the synthesis of cystine. J Biol Chem 207: 181.
Eastment CT, Sibrasku DA (1978). Platelet-derived growth factor(s) for a hormone-response rat mammary tumor cell line. J Cell Physiol 97:17.
Fidler IJ (1970). Metastasis:quantitative analysis of distribution and fate of tumor emboli labeled with ^{125}I-5-iodo-2'-deoxyuride. J Natl Cancer Inst 45:773.
Gasic GJ, Gasic TB, Stewart CC (1968). Antimetastatic effects associated with platelet reduction. Proc Natl Acad Sci USA 61:46.
Gasic GJ, Gasic TB, Galanti N, Johnson T, Murphy S (1973). Platelet-tumor-cell interactions in mice. The role of platelets in the spread of malignant disease. Int J Cancer 11:704.
Gasic GJ, Koch PAG, Hsu B, Gasic TB, Niewiarowski S (1976). Thrombogenic activity of mouse and human tumors: effects on platelets, coagulation and fibrinolysis and possible significance for metastases. Z Krebsforsch 86:263.
Gasic GJ, Gasic TB, Jimenez SA (1977a). Effects of trypsin on the platelet-aggregating activity of mouse tumor cells. Thromb Res 10:33.
Gasic GJ, Gasic TB, Jimenez SA (1977b). Platelet aggregating material in mouse tumor cells, removal and regeneration. Lab Invest 36:413.
Gasic GJ, Boettiger D, Catalfamo JL, Gasic TB, Stewart GJ, (1978). Aggregation of platelets and cell membrane vesiculation by rat cells transformed in vitro by Rous sarcoma virus. Cancer Res 38:2950.
Goodnight SH (1974). Bleeding and intravascular clotting in malignancy: a review. Ann NY Acad Sci 230:271.
Hara Y, Steiner M, Baldini M (1980a). Characterization of platelet-aggregating activity of tumor cells. Cancer Res 40:1217.
Hara Y, Steiner M, Baldini M (1980b). Platelets as a source of growth-promoting factor(s) for tumor cells.

Cancer Res 40:1212.

Hatton MWC, Regoeczi E (1973). A simple method for the purification of commercial neuraminidase preparations free from proteases. Biochim Biophys Acta 327:114.

Heldin CH, Wasteson A, Westermark B (1977). Partial purification and characterization of platelet factors stimulating the multiplication of normal human glial cells. Exp Cell Res 109:429.

Jones DS, Wallace AC, Fraser EE (1971). Sequence of events in experimental metastases of Walker 256 tumor: light, immunofluorescent and electronmicroscopic observations. J Natl Cancer Inst 26:493.

Kaplan DR, Chao FC, Stiles CD, Antoniades HN, Scher CD (1979). Platelet α granules contain a growth factor for fibroblasts. Blood 53:1043.

Kerby GP, Taylor SM (1959). The acid mucopolysaccharide and 5-hydroxytryptamine content of human thrombocytes in rheumatoid arthritic and nonarthritic individuals. J Clin Invest 38:1059.

Kohler N, Lipton A (1974). Platelets as a source of fibroblast growth-promoting activity. Exp Cell Res 87:297.

Kohler N, Lipton A (1977). Release and characterization of a growth factor for SV40 virus-transformed cells from human platelets. Proc Am Assoc Cancer Res 18:244.

Odell TT, Anderson B (1957). Isolation of a sulfated mucopolysaccharide from blood platelets of rats. Proc Soc Exp Biol Med 94:151.

Olsson I, Gardell S (1967). Isolation and characterization of glycosaminoglycans from human leucocytes and platelets. Biochim Biophys Acta 141:348.

Paul D, Lipton A, Klinger I (1971). Serum factor requirements of normal and simian virus 40-transformed 3T3 mouse fibroblasts. Proc Natl Acad Sci USA 68:645.

Pledger WJ, Stiles CD, Antoniades HN, Scher CD (1977). Induction of DNA synthesis in BALB/c3T3 cells by serum components: reevaluation of the commitment process. Proc Natl Acad Sci USA 74:4481.

Riddell PE, Bier AM (1965). Electrophoresis of 35S-labeled material released from clumping platelets. Nature 205:711.

Ross R, Glomset J, Kariya B, Harker L (1974). A platelet-dependent serum factor that stimulates the proliferation of arterial smooth muscle cells in vitro. Proc Natl Acad Sci USA 71:1204.

Ross R, Vogel A (1978). The platelet-derived growth factor. Cell 14:203.

Rutherford RB, Ross RJ (1976). Platelet factors stimulate fibroblasts and smooth muscle cells quiescent in plasma serum to proliferate. J Cell Biol 69:196.

Slichter SJ, Harker LA (1974). Hemostasis in malignancy. Am NY Acad Sci 230:252.

Stiles CD, Capone GT, Scher CD, Antoniades HN, Van Wyk JJ, Pledger WJ (1979). Dual control of cell growth by somatomedins and platelet-derived growth factor. Proc Natl Acad Sci USA 76:1279.

Sugarbaker EV, Ketcham AS (1977). Mechanisms and prevention of cancer dissemination: an overview. Semin Oncol 4:19.

Warren BA (1976). Some aspects of blood borne tumor emboli associated with thrombosis. Z Krebsforsch 87:1.

Weiss L (1976). Biophysical aspects of the metastatic cascade. In Weiss L (ed): "Fundamental Aspects of Metastasis," New York: American Elsevier Publishing Co., p 51.

Witte LD, Kaplan KL, Nossel HL, Lages BA, Weiss JH, Goodman DS (1978). Studies of the release from human platelets of the growth factor for cultured human arterial smooth muscle cells. Circ Res 42:402.

MECHANISMS OF PLATELET AGGREGATION BY HUMAN TUMOR CELL LINES

G. A. Jamieson, E. Bastida, A. Ordinas
American Red Cross Blood Services
Blood Services Laboratories
9312 Old Georgetown Road
Bethesda, Maryland 20814

ABSTRACT

Three distinct mechanisms of platelet aggregation have been observed with tumor cell lines of human origin based on their differential responses to apyrase, hirudin and phospholipase D. Aggregation induced by the first group of cells, SKBR3 (adenocarcinoma), HT 29 (adenocarcinoma) and HT 144 (melanoma) is probably initiated by ADP derived from the tumor cells. Aggregation by the second group of cells, Hut 28 (mesothelioma) and U87MG (glioblastoma), involves activation of the clotting system. However aggregation by U87MG can be differentiated from that by Hut 28 since the latter is inhibited by phospholipase D. Thus, two major categories of tumor cells can be described reflecting ADP-dependent and coagulant-dependent mechanisms of platelet aggregation but, within the latter group, there appear to be two separate mechanisms at work as shown by different sensitivities to phospholipase D.

INTRODUCTION

Several lines of evidence have implicated platelets in the metastatic dissemination of cancer. It has been noted by Warren and Vales (1972) that, in lectures to medical students in 1878, Billroth stated that the spread of tumors could be brought about by the circulation of thrombi containing tumor cells. The adherence of these platelet-tumor cell emboli was oabserved postmortem (Iwasaki, 1915; Saphir, 1947; Warren and Gates, 1963), in a variety of animal model systems

(Takahashi, 1915; Baserga and Saffioti, 1955; Wood et al., 1957; Boeryd, 1966; Hilgard, 1973; Ambrus et al, 1978) and by the use of microcinematography to study the time sequence of the interactions (Wood, 1958; Gastpar, 1970).

Gasic and co-workers made the important observation that the incidence of tumor metastases was reduced in thrombocytopenic mice (Gasic et al, 1968) and, in subsequent work, they utilized aggregation as a means of screening tumor cells for their ability to interact with platelets (Gasic et al, 1973; 1976; 1978). The basis for this aggregaion by tumor cells has not been determined but has generally been ascribed as being due to the release of ADP from platelets (Gasic et al, 1976; Holme et al, 1978), or to the tumor cells thromboplastic activity (Gastpar, 1970) which may be derived from a lipo-glycoprotein complex of the tumor cell surface (Hara et al, 1980). This complex may be shed as microvesicles (Gasic et al, 1976) and may be identical to the platelet aggregating material (PAM) which has been isolated from mouse tumor cells (Pearlstein et al, 1979).

Much of this previous work has utilized animal tumor cell lines, mainly of rat or mouse origin, and human or rabbit platelets. We have recently begun to re-examine the aggregation of platelets and tumor cells (Bastida et al, 1981a), their attachment as mixed thrombi at the vessel wall (Marcum et al, 1980) and individual susceptibilities to tumor cells among different donors (Bastida et al, 1981b).

We have now extended these studies by examining several tumor cell lines, all of human origin. We have been led to the use of a homologous system of human platelets together with cultured tumor cell lines of human origin because of the relevance of the system and because of apparent discrepancies in reports from different laboratories using homologous systems of non-human origin or using heterologous systems. The present studies confirm the complementary roles of ADP and thrombin but show that two different sub-categories can be recognized within the thrombin-mediated systems based on their sensitivities to phospholipase D.

Tumor Cells and Cell Cultures

We are indebted to Dr. Jørgen Fogh, Sloan Kettering Institute, New York, for providing us with the U87 MG (glioblastoma),

HT 29 (adenocarcinoma), SKBR3 (adenocarcinoma) and HT 144 (melanoma) lines, and to Dr. Adi Gazdar, Veterans Administration Hospital, Washington, D.C., for the Hut 23 adenocarcinoma) and Hut 28 (mesothelioma) lines. Cells were cultured under standard conditions as described elsewhere (Bastida et al, 1981b), and were harvested without exposure to proteases.

RESULTS

Aggregation by Tumor Cells

Each of the cell lines examined gave different aggregation patterns and required different amounts of tumor cells to effect aggregation. The U87Mg line aggregated platelets at 5×10^5 cells/ml while the HT 29 line required 10^6 cells/ml. The Hut 28, HT 144 and SKBR3 lines required 5×10^6 cells/ml while the Hut 23 line did not cause platelet aggregation at tumor cell concentrations as high as 10^7/ml. Differences were also observed between the different cell lines in the effect of apyrase, hirudin and phospholipase D on the course of aggregation. These results are described below for each of the cell lines examined and are summarized in the Table.

U87MG cells caused full monophasic platelet aggregation with a lag time which became more prolonged at lower tumor cell concentrations. This lag time was also prolonged in the presence of hirudin with a 4-fold increase at a hirudin concentration of 100 units/ml although the rate of aggregation was only slightly affected. Both apyrase and phospholipase D were without effect on the aggregation patterns.

The aggregation profile of platelets exposed to SKBR3 cells consisted of a brief phase of reversible aggregation followed immediately by a larger irreversible phase. This profile is similar to that observed with the murine Hut 20 line (Bastida et al, 1981a). Aggregation by the SKBR3 line was completely blocked by apyrase (250 ug/ml). The second wave of aggregation, but not the first, was inhibited by phospholipase D while hirudin (100 u/ml) had no effect on the aggregation profile.

The HT 144 line caused monophasic irreversible aggregation with a prolonged lag time. Apyrase further prolonged the lag phase to the onset of irreversible aggregation although the extent of aggregation was unaffected. Both phospholipase D and hirudin completely inhibited the aggregation response.

The HT 29 line also showed biphasic aggregation but with no apparent dissociation between the first and second waves and a brief lag phase to the onset of aggregation. Aggregation by this cell line was completely blocked in the presence of apyrase or phospholipase D. On the other hand, neither the lag phase to the onset of aggregation, nor the biphasic aggregation curve were affected by the presence of hirudin.

Hut 28 cells caused monophasic aggregation but with a slowly increasing base line during the prolonged lag phase to the onset of aggregation. Apyrase had no effect on the aggregation profile but it was completely inhibited by phospholipase D and by hirudin.

The Hut 23 cell line gave no significant aggregation at tumor cell concentrations as high as 10^7/ml.

DISUSSION

The present examination using human tumor cell lines shows that three mechanisms of platelet aggregation by cultured human tumor cells can be recognized based on aggregation responses and the inhibitory effects of apyrase, hirudin and phospholipase D. The results are summarized in the Table. Some groupings among the various lines appear to be possible. In all cases, there was an absolute, irreversible requirement for Ca^{2+} since aggregation occurred only with heparinized PRP.

SKBR3 shows biphasic aggregaton that is inhibited by apyrase but not by hirudin, and phospholipase D eliminates the second wave of aggregation. This pattern is similar to that previously observed with the murine Hut 20 line (Bastida et al, 1981a). Secretion of ADP from the tumor cells appears to initiate the first wave of aggregation, which leads to platelet activation and to a second wave of aggregation. This second wave is associated with release which must be independent of thrombin production since it is not blocked by hirudin. HT 29 and HT 144 also probably belong in this class. There is no detectable first wave of reversible aggregation with these two cell lines, but they show similar effects with apyrase, hirudin and phospholipase D to the other two cell types.

Hut 28 gives a rising baseline during platelet aggregation, rather than a clearly marked first wave. Aggregation is not affected by apyrase and is presumably independent of ADP. However, aggregation is blocked by hirudin suggesting that thrombin production is involved in the initiating step. While the aggregating effect of the U87MG cell line also involves activation of the coagulation system, it falls in a different group since it is not affected by phospholipase D.

The various phospholipases have been of value in characterizing the platelet aggregating effects of the tumor cell lines. Phospholipase A2 has been reported to completely inhibit platelet aggregation induced by two mouse tumor cell lines (Hara et al, 1980) and we have found that it also inhibits the human tumor lines which caused aggregation in the present study.

We have also found that tumor cell-induced platelet aggregation is inhibited by lysolecithin, the hydrolytic product of the action of phospholipase A2 on phosphatidylcholine, which also inhibits platelet aggregation induced by ADP, epinephrine, collagen and thrombin (Joist et al, 1977). Lysolecithin is known to inhibit prostaglandin sysnthesis (Shier et al, 1977) and this may be the basis for its antiaggregating effects although it can also affect membrane fluidity (Utsumi et al, 1978) and the levels of nucleotide cyclases (Aunis et al, 1978; Shier et al, 1976). It may be noted that the metastatic potential of B16 melanoma variants is known to be inversely proportional to their content of prostaglandin D_2 (Fitzpatrick and Stringfellow, 1979; Stringfellow and Fitzpatrick, 1979).

Phospholipase D has been particularly valuable in this regard since it could be used to separate the platelet aggregating effects of Hut 28 from those of U87Mg although activation of the coagulation system appeared to be involved in each case. Little is known about the effects of phospholipase D on membranes but the enzyme can alter calcium translocation in sarcoplasmic reticulum (Fiehn, 1978). This may explain its differential effects on the first and second waves of tumor cell-induced platelet aggregation.

In addition to the usual aggregating agents, platelet aggregation can be induced by the platelet aggregating factor (PAF) elaborated by IgE-sensitized basophils. PAF is lipidic in nature and is destroyed by phospholipases A2, C and D (Benveniste et al, 1977). Since phospholipase C had no effect on platelet aggregation induced by any of the human tumor cell lines examined, and several phases are unaffected by phospholipase D, it is unlikely that PAF-like material is involved.

In summary, our results suggest that there are two major mechanisms by which cultured human tumor cells initiate platelet aggregation. One mechanism involves the secretion of ADP from the tumor cells themselves resulting in platelet stimulation and then irreversible aggregation. The second mechanism involves the initial activation of the coagulation system and the generation of thrombin as the mediator of aggretion. Within this second group, two subgroups can be identified based on whether or not agggregation can be inhibited by phospholipase D. These results suggest that no single mechanism will explain the nature of the interaction between platelets and tumor cells under all circumstances.

ACKNOWLEDGEMENTS

This work was supported, in part, by USPHS Grants HL 14697, HL 20971 and Biomedical Research Support Grant RR 05737. Publication No. 547 from the American Red Cross.

REFERENCES

Ambrus JL, Ambrus CM, Gastpar H (1978). Studies on platelet aggregation and platelet interaction with tumor cells. In Gaetano G and Garratini S (eds): "Platelets: A Multidisciplinary Approach," Raven Press: New York, p. 467.
Aunis D, Pescheloche M, Zwiller J, Mandel P (1978). Effects of lysolecithin on adenylate cyclase and guanylate cyclase in bovine adrenal medullary plasma membranes. J Neurochem 31:355.
Baserga R, Saffioti U (1955). Experimental studies on the histiogenesis of blood-borne metastases. Arch Path 59:26.
Bastida E, Ordinas A, Jamieson GA (1981a). Differing platelets aggregating effects by two tumor cell lines: Absence of role for platelet-derived ADP. Am J. Hematol 11:367.
Bastida E, Ordinas A, Jamieson GA (1981b). Idiosyncratic platelet responses to human tumor cells. Nature 291:661.
Beneveniste J, Le Couedic JP, Polonsky J, Tence M (1977). Structural analysis of purified platelet activating factor by lipases. Nature 269:170.

Billroth T (1878). Lectures on Surgical Pathology and Therapeutics. A Handbook for Students and Practitioners. Vol II (in translation). London: The New Sydenham Society, p. 355.

Boeryd B (1966). Effect of heparin and plasminogen inhibitor (EACA) on intravenously injected ascites tumor cells. Acta Path Microbiol Scand 68:547.

Fiehn W (1978). The effect of phospholipase D on the function of fragmented sarcoplasmic reticulum. Lipids 13:264.

Fitzpatrick FA, Stringfellow DA (1979). Prostaglandin D_2 formation by malignant melanoma cells correlates inversely with cellular metastatic potential. Proc Natl Acad Sci USA 76:176S.

Gasic GJ, Gasic TB, Stewart CC (1968). Antimetastatic effects associated with platelet reduction. Proc Natl Acad Sci USA 61:46.

Gasic GJ, Gasic TB, Galanti N, Johnson T, Murphy S (1973). Platelet-tumor cell interactions in mice. The role of platelets in the spread of malignant disease. Int J Cancer 11:704.

Gasic GJ, Koch PAG, Hsu B, Gasic TB, Niewiarowski S (1976). Thrombogenic activity of mouse and human tumors: Effects on platelets, coagulation and fibrinolysis and possible significance for metastases. Z Krebforsch 86:263.

Gasic GJ, Boettiger D, Catalfamo JL, Gasic TB, Stewart CC (1978). Aggregation of platelets and cell membrane vesiculation by rat cells transformed in vitro by Rous Sarcoma Virus. Cancer Res 38:2950.

Gastpar H (1970). Stickiness of platelets and tumor cells induced by drugs. Thromb Diath Haemorrh Suppl 42:291.

Hara Y, Steiner M, Baldini MC (1980). Characterization of the platelet aggregating activity of tumor cells. Cancer Res 40:1217.

Hilgard P (1973). The role of blood platelets in experimental metastases. Br J Cancer 28:429.

Holme R, Oftebro R, Hovig T (1978). In vitro interaction between cultured cells and human blood platelets. Thrombos Haemostas 40:89.

Iwasaki T (1915). Histological and experimental observations on the destruction of tumor cells in the blood vessels. J Path Bacteriol 20:85.

Joist JH, Dolezel G, Cucuianu MP, Nishizawa EE, Mustard JF (1977). Inhibition of potentiation of platelet function by lysolecithin. Blood 49:101.

Marcum JM, McGill M, Bastida E, Ordinas A, Jamieson GA (1980). The interaction of platelets, tumor cells and vascular subendothelium. J Lab Clin Med 96:1046.

Pearlstein E, Cooper LB, Karpatkin S (1979). Extraction and characterization of a platelet aggregating material from SV-40 transformed mouse 3T3 fibroblasts. J Lab Clin Med 93:332.

Saphir Q (1947). The fate of carcinoma emboli in the lung. Am J Path 23:245.

Shier WT, Baldwin JH, Nilsen-Hamilton M, Hamilton RT, Thanassi NM (1976). Regulation of guanylate and adenylate cyclase activities by lysolecithin. Proc Natl Acad Sci USA 73:1587.

Stringfellow DA, Fitzpatrick FA (1979). Prostaglandin D_2 controls pulmonary metastasis of malignant melanoma cells. Nature 282:76.

Takahashi M (1915). An experimental study of metastasis. J Path Bacteriol 20:1.

Utsumi H, Inoue K, Nojima S, Kwan T (1978). Interaction of spin-labeled lysophosphatidylcholine with rabbit erythrocytes. Biochemistry 17:1990.

Warren BA, Vales O (1972). The adhesion of thromboplastic tumor cell emboli to vessel walls in vivo. Br J Exp Pathol 53:301.

Warren S, Gates O (1963). The fate of intravenously injected tumor cells. Am J Cancer 27:485.

Wood S Jr, Yardley JH, Holyoke ED (1957). The relationship between intravascular coagulation and the formation of pulmonary metastases in mice injected intravenously with tumor cell suspension. Proc Am Cancer Res 2:260. Abstr.

Wood S Jr (1958). Pathogenesis of metastasis formation observed in vivo in rabbit ear chamber. Arch Path 66:550.

Table: Comparison of Platelet Aggregation by Human Tumor Cell Lines

Cell Line	Origin	Conc.	Aggregation Profile	Apyrase	Hirudin	Phospholipase D
HT 29	Adenocarcinoma of colon	10^6		Inhibition	No Inhibition	Inhibition
SKBR 3	Adenocarcinoma of breast	5×10^6		Inhibition	No Inhibition	2nd wave only
Hut 28	Mesothelioma	5×10^6		No Inhibiton	Inhibition	Inhibition
HT 144	Melanoma	5×10^6		Inhibition	No Inhibition	Inhibiton
Hut 23	Poorly differentiated adenocarcinoma	10^7	No Aggregation	—	—	—
A549	Epithelial lung carcinoma	10^7	No Aggregation	—	—	—
U87MG	Glioblastoma	10^5		No Inhibition	Inhibition	No Inhibition

ISOLATION AND PARTIAL PURIFICATION OF AN ACTIVITY FROM HUMAN PULMONARY EPIDERMOID CARCINOMA CELLS IN CULTURE THAT CLOTS HEPARINIZED PLASMA

S.F. Mohammad[1], H. Y. K. Chuang[1] J. Szakacs[2], and R. G. Mason[1]

[1]Department of Pathology, College of Medicine University of Utah, Salt Lake City, Utah and
[2]Department of Pathology, Saint Joseph's Hospital, Tampa, Florida

Human pulmonary epidermoid carcinoma cell lines obtained from surgical biopsies or pneumonectomies and grown in tissue culture were found to contain a soluble fraction capable of clotting heparinized, but not citrated, human plasma. Soluble fractions from a number of normal cell lines and non-pulmonary malignant cell lines did not clot heparinized plasma under similar test conditions. These nonreactive cell lines included adult fibroblasts, fetal endothelial cells, five different melanomas, an adenocarcinoma of the breast, a renal adenocarcinoma, a nasopharyngeal carcinoma, a chondrosarcoma, a Schwannoma, the HeLa cell line, and the Hep-2 cell line. The activity that clotted heparinized plasma eluted from a column of Sephadex G-50 just after the void volume. These data along with information obtained from ultrafiltration studies indicate that the probable molecular weight of the active species is between 50,000 and 100,000 daltons. The activity that clots heparinized plasma is destroyed by treatment with trypsin. Rapid loss of this activity occurs above pH 8.0. Fractions capable of clotting heparinized plasma can be preserved for up to 4 weeks at temperatures below 0°C but are unstable for that interval at higher temperatures, while all activity disappears in 15 minutes at 100°C. The active molecular species has a strong affinity for glass and plastic surfaces and appears to remain active in the adsorbed state.

INTRODUCTION

Patients with certain malignancies are known to have a higher incidence of thrombosis than other cancer patients or patients with most nonmalignant diseases (Harris, 1971; Born, 1981). Despite considerable study, the mechanisms by which neoplasms initiate thrombosis have not been elucidated clearly. There are convincing data to indicate that necrotic neoplasms contribute thromboplastic material that can enhance blood coagulation in vivo (Granlick, 1981). Recent studies have indicated that certain nonhuman neoplasms possess high molecular weight proteins or vesicular structures that produce aggregation of blood platelets, but this has been shown not to be the case with a large number of human neoplastic cell lines studied in our laboratory (unpublished observations). The finding that 10 different pulmonary epidermoid carcinoma cell lines possessed an activity that produced clotting of heparinized plasma was of interest, since similar activity was not found to be present in a number of human normal and nonpulmonary neoplastic cell lines.

MATERIALS AND METHODS

Pulmonary Epidermoid Carcinoma (PEC) Cell Lines

Ten different pulmonary epidermoid carcinoma cell lines were used in these studies, and their growth in tissue culture was initiated in our laboratories. Neoplastic cells were obtained from surgical biopsies or pneumonectomies. The cell type of each tumor was confirmed by the Surgical Pathology Section of the Department of Pathology, V.A. Hospital, Tampa, FL. In each case, the cells in culture appeared morphologically to be of a single type.

Approximately 1 gm of neoplastic tissue was minced and incubated with a solution of collagenase (Type IV, Worthington, Freehold, NJ) for 15 min at 37°C with occasional mixing. Larger fragments of the enzyme-treated tissue were allowed to settle, and suspended cells were removed gently by use of a Pasteur pipette. The cell suspension then was centrifuged at 100 g for 5 min at 23°C. The supernatant was discarded, and the sedimented cells were resuspended in medium M-199 (Grand Island Biological Company, Grand Island, NY, GIBCO) and centrifuged a second time to remove any residual collagenase. The cells were suspended finally in medium RPMI-1640 (GIBCO) containing 15 to 20% fetal calf serum (Reheiss Chemical Company, Phoenix, AZ) and seeded in standard 25 cm^2 tissue culture flasks (Corning Glass Works, Corning, NY). Seeded cells were kept at 37°C in an incubator that was maintained in an atmosphere of 5% CO_2 and 95% air with a humidity of 95%. Cell culture medium was changed twice each week. When cells had achieved confluent growth, they were detached from the support medium either by scraping with a rubber policeman or by treatment with trypsin - EDTA solution (0.05% trypsin 1:250, 0.02% ethylene diaminetetraacetic acid (EDTA), GIBCO). Cells recovered from culture flasks were subcultured by a previously described method (Paul, 1975).

Other Cell Lines

All neoplasms were of human origin. Five different melanomas, one nasopharyngeal carcinoma, a renal adenocarcinoma, an adenocarcinoma of the breast, a Schwannoma, and a chondrosarcoma were obtained from the laboratory of one of the authors (JS). These various human neoplastic cell lines were subcultured using methods similar to those used for culture of the pulmonary epidermoid carcinoma cells. Human fibroblasts were obtained from adult human skin specimens as described elsewhere (Martin, 1973). Fetal human endothelial cells were obtained from umbilical cord veins. Fibroblast and endothelial cell cultures were established in our laboratory. Hep-2 and HeLa cell lines were purcahsed from American Type Culture Collection, Rockville, MD.

Blood Coagulation Assays

The activity from human PEC cell lines that produced clotting of heparinized plasma was assayed by the following procedure: 0.5 units of heparin (in 0.1 ml of 0.154 M NaCl) were mixed with 0.3 ml of citrated platelet-poor plasma (CPPP) to produce heparinized, citrated platelet poor plasma (HCPPP). After 5 min incubation at 23°C, 100 µl of the test solution (or phosphate buffered saline (PBS) as a control) were added to the HCPPP. This reaction mixture was incubated at 23°C for either 30 or 60 min. Following this incubation, the amount of anticoagulant activity of heparin remaining in the reaction mixture was assayed as described (Bowie, et al., 1971); this latter assay is an activated partial thromboplastin time (APTT) test. Unless stated otherwise, all blood coagulation assays were carried out by use of a Fibrometer (BioQuest, Cockeysville, MD).

Cells grown in tissue culture flasks were rinsed gently with 10 ml of PBS 3 times to remove culture medium. Washed cells then were removed by use of a rubber policeman and suspended in 5 ml of PBS. Suspended cells were subjected to centrifugation at 100 g for 5 min. The supernatant was discarded and sedimented cells were resuspended in PBS (approximately 10^7 cells/ml) and either used as such or processed further by subcellular fractionation techniques.

Subcellular Fractionation Studies

Washed cells suspended in a small volume of PBS were subjected to 5 cycles of freezing and thawing. The disrupted cells (approximately 88% disruption by trypan blue dye exclusion studies) next were subjected to nongradient centrifugation at 40,000 g for 30 min. The supernatant was recovered and will be referred to as the "soluble fraction".

The soluble fraction from disrupted cells was studied directly or after exclusion chromatography on Sephadex G-50 (60 x 1.5 cm column, Pharmacia Fine Chemicals, Piscataway, NJ). Fractions of 2.5 ml were collected and monitored for the presence of the activity that produced clotting of heparinized plasma. Fractions that contained such activity will be referred to as "active fractions". In some experiments, fractions containing this activity were pooled, lyophilized, and dialyzed extensively against PBS before assay of the activity in blood coagulation tests; dialysis appeared to have little effect on active fractions.

Ultrafiltration

The procedure for ultrafiltration has been described previously (Lovette et al., 1976). Ultrafiltration membranes XM-100A and XM-50 (Amicon Corp., Lexington, MA) were used at a filtration pressure of 60 PSI. The filtrate was collected and examined for effects on blood coagulation tests as described above.

Thermal and pH Stability Studies

Soluble fractions from disrupted cells were incubated in silicone-coated tubes at various temperatures in a thermostatically controlled water bath. At 30 min intervals, aliquots were removed and transferred immediately to an ice bath (4°C). These aliquots later were examined for their heparin neutralizing activity.

The pH of active preparations was altered by addition of either 0.01 N HCl or 0.01 N NaOH. The test solutions were allowed to stay at the required pH for 60 min at 23°C. Following this incubation period, the pH of each solution was readjusted to 7.3. After compensating for volume change, the effect of pH on the stability of an active fraction was analyzed.

Trypsin Treatment

Fifty units of immobilized trypsin (polyacrylamide, Sigma Chemical Co., St. Louis, MO) in PBS was added either to the soluble fraction obtained from disrupted cells or to the semipurified fractions eluted from Sephadex G-50 (Pharmacia Fine Chemicals, Piscataway, NJ) columns. Immobilized trypsin was incubated with the active fractions at 23°C with gentle mixing. Aliquots of these mixtures were removed every 30 min and examined for heparin neutralizing activity. Two control experiments were carried out simultaneously also. In one case, the immobilized trypsin was incubated with PBS alone (suspension medium), whereas in the other case, immobilized trypsin was inhibited by previous exposure to purified α_1-antitrypsin (Sigma Chemical Co.) and then incubated with active fractions to serve as a control.

Adsorption of the Activity to Container Walls

Glass or plastic tubes that had contained the active fractions were rinsed 5 times with PBS. After such rinsing, HCPPP was added to the tubes and incubated at 37°C for 60 min. Following this incubation, the anticoagulant activity of heparin remaining in HCPPP was determined as described earlier.

RESULTS

Examination of nonchromatographed soluble fractions of both non-neoplastic and neoplastic human cells grown in tissue culture revealed that only the cells from 10 different lines of PEC possessed an activity that produced clotting of heparinized but not of citrated plasma (Table

1). Such an activity was not demonstrable with any of the other human normal or neoplastic cell lines examined under similar conditions. The soluble fraction from PEC cells neutralized the anticoagulant activity of heparin in a dose dependent fashion (Figure 1). Only a small amount of a similar type of heparin neutralizing activity was recovered from fibroblasts or other normal or neoplastic cell lines.

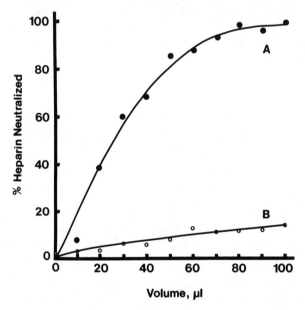

Figure 1.
Concentration dependent neutralization of heparin by the soluble fraction obtained from PEC cells (A) or human fibroblasts (B). A representative experiment. 100 µl of the soluble fraction in A neutralized approximately 1 unit of heparin in 30 min at 37°C.

When the soluble fraction from PEC cell lines was chromatographed on Sephadex G-50, the heparin neutralizing activity eluted from the column immediately after the void volume (Figure 2). Ultrafiltration studies showed that the heparin neutralizing activity passed through an XM-100A membrane filter but not through an XM-50 membrane filter. Exclusion chromatography and the ultrafiltration studies, therefore, suggested that the molecular weight of the active component was between 50,000 and 100,000 daltons. The strong affinity of this heparin neutralizing activity for surfaces, including the ultrafiltration filters, resulted in considerable loss of activity in these studies.

Treatment of fractions containing heparin neutralizing activity with immobilized trypsin resulted in eventual loss of this activity (Figure 3). Incubation of the active fractions with immobilized trypsin previously inhibited with antitrypsin did not produce significant loss of the activity.

Table 1.

Human Cell Lines Carried in Tissue Culture	Number of Cell Lines Examined	Clot Formation Induced by the Soluble Fraction from Cultured Cells‡	
		Heparinized Plasma	Citrated Plasma
Fibroblasts	4	–	–
Endothelial (fetal)	5	–	–
Pulmonary Epidermoid Carcinoma	10	+@	–
Other Malignancies*	10	–	–
HeLa	2	–	–
Hep-2	2	–	–

‡ 0.1 ml of the cell soluble fraction (obtained from approximately equal number of cells) was mixed with 0.2 ml of HPPP (containing 1 unit heparin/ml) or 0.2 ml CPPP. Reaction mixtures were incubated at 37°C and monitored visually every 5 min for the presence of clot until 60 min.
@ Clotting was observed in tests with soluble extracts from all 10 cell lines.
* Listed in Materials and Methods

The heparin neutralizing activity of soluble fractions was relatively unstable at storage temperatures above 10°C (Figure 4). Complete loss of activity occurred when preparations were heated to 100°C for 15 min.

The heparin neutralizing activity was relatively stable at 0-4°C or at -20°C for up to 4 weeks. The heparin neutralizing activity was relatively stable at pH values near physiologic pH. However, this activity was lost progressively below pH 5.0, and rapid loss of activity occurred above pH 8.0 (Figure 5).

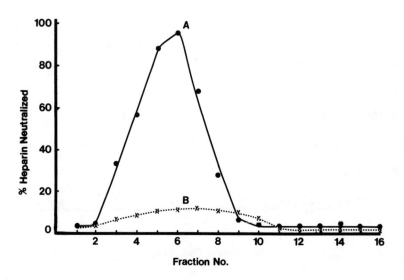

Figure 2
Elution profile of heparin neutralizing activity from Sephadex G50 column. The column was equilibrated with PBS at 10°C. 3.5 ml of the soluble fraction from PEC (A) or from normal human fibroblasts (B) were layered on the column and fractions were collected after discarding the void volume. Each fraction (2.5 ml) was analyzed for the presence of heparin neutralizing activity.

The heparin neutralizing activity appeared to have a strong affinity for glass and plastic surfaces. A gradual but significant loss of this activity from the soluble phase was noticed when active fractions were stored at 4°C. However, clotting of heparinized plasma was observed when HPPP was placed in rinsed containers in which active fractions had been stored previously. Desorption of this heparin neutralizing activity from surfaces and its effect on other coagulation assays has not yet been studied. Preliminary attempts to desorb the adsorbed activity by use of high salt or low pH conditions, or both, have resulted in complete loss of activity.

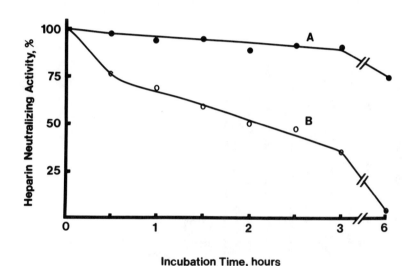

Figure 3
Loss of heparin neutralizing activity following exposure of active fractions to immobilized trypsin. The soluble fraction from PEC cells was treated with immobilized trypsin previously inhibited with α_1 antitrypsin (A) or active immobilized trypsin (B) as described in methods section. Only active trypsin caused a time dependent decrease in the heparin neutralizing activity.

DISCUSSION

The presence of an activity in human PEC cell lines capable of clotting HPPP but not CPPP was unexpected. Preliminary observations presented in this study suggest that clotting of heparinized plasma was probably due to the presence of a heparin neutralizing activity in PEC cells grown in tissue culture. The absence of this heparin neutralizing activity in a number of different normal human and non-PEC neoplastic cell lines was an intriguing finding. The present observations suggest that the heparin neutralizing activity is probably unique to PEC, but the full range of human neoplasms has not yet been tested. Nevertheless, a number of different normal human cells, such

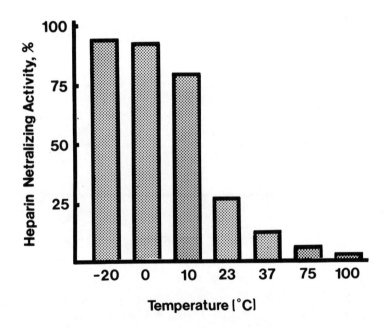

Figure 4
Effect of temperature. Aliquots stored at -20°C or 0°C were examined after 4 weeks. Aliquots incubated at 10°C, 23°C, or 37°C were tested after 24 hours, whereas those incubated at 75°C or 100°C were tested after 30 min.

as platelets and leukocytes, have been shown to possess an activity that produces slow clotting of heparinized plasma. In addition to an antiheparin protein (platelet factor 4) (Paul et al., 1980), platelets have been shown also to possess an enzyme, endoglycosidase, that degrades heparin (Oldberg et al., 1980).

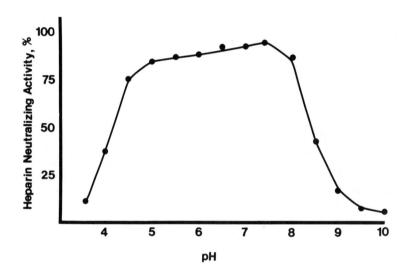

Figure 5
Effect of pH on the stability of heparin neutralizing activity. Active soluble fraction was divided into several 1 ml aliquots. Desired pH in each aliquot was adjusted with 0.01 N HCl or 0.01 N NaOH. After 60 min at 23°C, pH was readjusted to 7.4. After compensations for volume changes, aliquots were examined for heparin neutralizing activity.

Similarly, leukocytes contain cationic proteins (Zeya and Spitznagel, 1963) that may inactivate heparin. Since the active fractions obtained from PEC did not shorten APTT test values or clot CPPP, these findings support the assumption that the active moiety(s) manifests its effect through its ability to neutralize the anticoagulant activity of heparin. It is not known at this time whether the PEC neutralization of heparin occurs due to the degradation of heparin by an endoglycosidase(s) (e.g., heparinase) or due to complex formation as is observed in the case of platelet factor 4. Gradual loss of heparin neutralizing activity due to absorption on container walls suggests that the active component may be a highly charged moiety with relatively strong affinity for surfaces. That this activity is somewhat stable at slightly acidic pHs further supports this assumption (Zeya and Spitznagel, 1963).

The HCPPP-clotting activity appears to be a protein of moderate size. Both ultrafiltration and exclusion chromatographic studies suggest that the molecular weight of the active species is in the range of 50,000 to 100,000 daltons. The susceptibility of the heparin neutralizing activity to digestion by trypsin supports the conclusion that this activity is associated with a protein moiety. The possibility that the active molecular species may have enzymatic activity has not been ruled out.

Other workers have illustrated that certain human and nonhuman neoplasms possess procoagulant or thromboplastic activities as well as activities that aggregate platelets (Granlick, 1981; Gasic et al., 1978; Karpatkin et al., 1980; Paschen et al., 1979). To our knowledge, this is the first reported finding of an activity in a human neoplastic cell line that clots HPPP but not CPPP. The clinical relevance of this finding is considerable, since release of such activity from neoplastic cells could produce thrombosis in heparinized patients. Release of an activity of the type described here from neoplastic cells could be spontaneous or might be brought about by death of the cells. The presence of circulating levels of such heparin neutralizing activity from neoplastic cells would increase the levels of heparin required to produce a desired clinical anticoagulant effect or possibly render use of this anticoagulant unfeasible. Moreover, it is possible that necrotic areas with inadequate circulation may possess high concentrations of this activity that could cause thrombosis in immediately adjacent viable areas despite the presence of circulating heparin.

Efforts are underway to purify and characterize further this activity from human PEC lines. A promising approach appears to be the use of affinity chromatography using surface bound heparin.

ACKNOWLEDGEMENTS

This work was supported in part by a grant No. PDT79 from American Cancer Society. Part of this work was carried out at University of South Florida, Department of Pathology, Tampa, FL

REFERENCES

Born GVR (1981). Malignancy and hemostasis. In Donati MB, Davidson JF, Garattini S (eds): "Malignancy and the Hemostatic System". New York: Raven Press, p 1.

Bowie EJW, Thompson JH, Didisheim P, Owen CA (1971). "Mayo Clinic Laboratory Manual of Hemostasis". Philadelphia: WB Saunders Co., p 123.

Gasic GJ, Boettiger D, Catalfamo JL, Gasic TB, Stewart J (1978). Aggregation of platelets and cell membrane vesiculation by rat cells transformed in vitro by Rous Sarcoma virus. Cancer Res 38:2950.

Granlick H (1981). Cancer cell procoagulant activity. In Donati MB, Davidson JF, Garattini S (eds): "Malignancy and the Hemostatic System". New York: Raven Press, p 57.

Harris H (1971). Cell fusion and the analysis of malignancy. Proc Royal Soc B 179:1.

Karpatkin S, Smerling A, Pearlstein E (1980). Plasma requirement for the aggregation of rabbit platelets by an aggregating material derived from SV40-transformed 3T3 fibroblasts. J Lab Clin Med 96:994.

Lovette KM, Chuang HYK, Mohammad SF, Mason RG (1976). The subcellular distribution and partial characterization of cholinesterase activities of canine platelets. Biochim Biophys Acta 428:355.

Martin GM (1973). Human skin fibroblasts. In Kruse PF, Patterson MK (eds): "Tissue Culture: Methods and Application". New York: Academic Press, p 39.

Oldberg A, Heldin CH, Wasteson A, Busch C, Hook M (1980). Characterization of a platelet endoglycosidase degrading heparin-like polysaccharides. Biochemistry 19:5755.

Paschen W, Patscheke H, Worner P (1979). Aggregation of activated platelets with Walker 256 Carcinoma cells. Blut 38:17.

Paul D, Niewiarowski S, Varma KG, Rucinski B, Rucker S, Lange E (1980). Human platelet basic protein associated with antiheparin and mitogenic activities: purification and partial characterization. Proc Natl Acad Sci USA 77:5914.

Paul J (1975). "Cell and Tissue culture", New York, Churchill Livingstone, p 234.

Zeya HI, Spitznagel (1963). Antibacterial and enzymic basic proteins from leukocyte lysosomes: Separation and identification. Science 142:1085.

PLASMA MEMBRANE VESICLES AS MEDIATORS OF INTERACTIONS
BETWEEN TUMOR CELLS AND COMPONENTS OF THE HEMOSTATIC AND
IMMUNE SYSTEMS

Gabriel J. Gasic and Tatiana B. Gasic

Department of Pathology, School of Medicine,
University of Pennsylvania, Philadelphia PA 19104

Almost twenty years ago we reported that neuraminidase decreases metastasis if given to the host 6-18 hr before tumor inoculation (Gasic and Gasic, 1962). Subsequently, we demonstrated that this antimetastatic effect was mediated by thrombocytopenia and that transfusion of platelets restored the number of metastases to levels observed before neuraminidase treatment (Gasic et al., 1968). This experiment clearly suggested that platelets were involved in cancer spread. Further observations showed that cells from certain tumors can interact with platelets in vitro and that tumors capable of aggregating platelets under these conditions produce more metastases, usually located in the lung (Gasic et al., 1973; Gasic et al., 1976). This enhancing effect was presumably mediated by the release of platelet products probably affecting vascular permeability, cell motility, growth of tumor cells, or a combination of these activities. Depending on the number of tumor cells inoculated i.v., thrombocytopenia may ensue.

Other tumors do not aggregate platelets in vitro but may have the capacity to damage the vascular endothelium (Warren, 1981), decreasing the production of prostacyclin (Honn et al., 1981), and expose the thrombogenic surface of the underlying basement membrane (Baumgartner et al., 1977).

Various attempts have been made to decrease the number of metastases by the use of inhibitors of platelet aggregation (PA). The more promising of these attempts has been the employement of agents that can increase the production of endothelial prostacyclin (Honn et al., 1981).

The capacity to aggregate platelets is a property that appears during cell transformation. This has been verified by two independent groups: Pearlstein et al. (1979), using SV-40 virus-transformed mouse fibroblasts; and Gasic et al. (1978), employing Rous sarcoma virus-transformed rat embryo cells. The platelet-aggregating activity of these transformed cell lines is shed spontaneously into the medium (Gasic et al., 1978) or can be extracted with urea (Pearlstein et al., 1979). Using spontaneous sheddings, we have shown that the activity resides in plasma membrane vesicles (Gasic et al., 1978). Since vesicles are easily collected and purified and whole cells could be damaged or contaminated by other cells during the process of cell separation, vesicles were preferred to study the events mediating tumor cell-platelet interactions. The vesicles were obtained from 15091A ascitic mammary adenocarcinoma cells, a highly immunogenic neoplasia.

Shedding of vesicles is preceded by the formation of blebs at the cell surface which detach by a process of pinching off. Vesicles from the medium contain ribosomes and microfilaments. After isolation and purification by sucrose density gradient ultracentrifugation (Gasic et al., 1978), vesicles lose their content and appear as hollow bodies surrounded by a clean membrane (Stewart GL, Gasic GJ, Gasic TB, and Catalfamo JL, unpublished).

This paper will focus on the interactions of vesicles isolated from 15091A ascitic cells with blood platelets and macrophages.

The study of tumor vesicle-platelet interactions were carried out both in heparinized platelet-rich plasma (PRP) and in suspension of gel-filtered platelets (Tangen and Berman, 1972) reconstituted with heparinized plasma from rats or mice.

Initially, rat PRP was preferred because larger quantities of blood could be obtained. In addition, responses to tumor vesicles were much more consistent than in the mouse system. Figure 1 illustrates a typical tracing of PA by 15091A tumor vesicles added to rat PRP. This aggregation

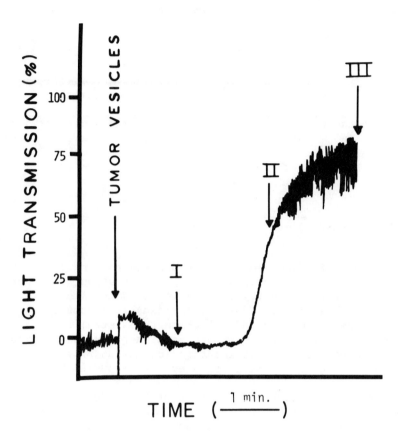

FIGURE 1

is preceded by a <u>characteristic</u> lag of 1 min or more and can be divided into three stages: the lag period, the period of increasing aggregation, and the plateau.

Studies with radioiodinated vesicles and a percoll cushion (1.039 g/ml) which permit separation of free from platelet-bound vesicles by centrifugation (8,000 g for 1 min) established that PA is preceded by binding of vesicles to platelets (Table 1, lines 1, 2, 3) and it occurs even when PA is blocked by prostacyclin (Table 1, line 4). This binding depends, almost in a linear relationship, on platelet concentration and is irreversible.

Table 1

Radioactivity bound to platelets at various stages of PA induced by ^{125}I-labeled tumor vesicles

Samples taken in stage	PGl$_2$	PA (%)	Radioactivity bound to platelets (%)		
I	–	0	59	3*	(3)
II	–	40	65		(2)
III	–	80	60	4*	(3)
IV	+	0	54		(2)

*Mean S.E.M.

Plasma is essential for the binding to occur. If platelets are freed of plasma by gel filtration, they not only lose their capacity to bind vesicles but also fail to aggregate platelets. However, if a small aliquot of heparinized plasma is added back to the system, binding takes place and aggregation proceeds in the usual fashion. This reconstitution experiment indicates the plasma cofactors are essential to both events.

Since treatment of plasma at 56°C for 30 min, a procedure that destroys the heat-labile components of the complement system, abolished the capacity of plasma to support binding, we directed our attention to complement. We confirmed its involvement by removal of its components after passage through a column of Cibacron blue on Sepharose (Gee et al., 1979). Addition of this plasma to gel-filtered platelets did not support binding of the tumor vesicles.

In a next step, we investigated which of the two pathways of complement activation was responsible for the phenomenon under study. We found that binding did not take place when C4 is absent by genetic reasons (Gasic et al., 1981) or when C4 and C2 are blocked by the use of salivary gland extract from Haementeria ghilianti (see below). Binding did not occur either in the presence of EDTA, EGTA (Fine et al., 1972), or a magnesium salt of EGTA (Fine, 1975) at the final concentration of 9 mM, or when the plasma was heated at 50°C for 30 min (Götze and Müller-Eberhard, 1971) or decomplemented by zymosan (Zucker and Grant, 1974) or cobra venom factor (Cochrane et al., 1970). These results indicate that treatments which affect early components of the complement system or activate complement at the level of C3, bypassing early components, block binding. As certain endotoxins do (Kane et al., 1973), vesicles will be capable of activating both the classical and the alternative pathway of complement.

Although complement appears to be required for vesicle binding, complement alone does not support PA. Evidence for this proposal was obtained by adding back to gel-filtered platelets rat plasma rendered deficient in proteins of the prothrombin complex by treatment of plasma donors with Coumadin. Under these conditions, binding occurred but not PA, suggesting that vitamin K-dependent clotting factors are involved in PA by tumor vesicles.

To test the possibility that PA is mediated by the generation of thrombin via activation of the clotting cascade by bound or unbound vesicles, hirudin, a specific and irreversible inhibitor of thrombin (Markwardt, 1970), was added to rat PRP prior to the addition of tumor vesicles. It was found in dose-response experiments that increasing doses of hirudin prolonged the lag period until reaching a level (80 units/ml) that completely abolished PA. The hirudin results were also confirmed by the use of synthetic antithrombins, such as DAPA (Nesheim et al., 1979) and compound No. 805 (see below).

In mice, which are widely used as experimental models in cancer research, the platelet response to tumor vesicles is variable and less intense than in the rat system. We attributed the variability to genetic factors affecting complement and the clotting cascade, as well as the documented instability of complement in the mouse system.

In a first approach, in collaboration with Dr. Arturo Ferreira of the New York University School of Medicine, we investigated whether genetic variations of the complement component 4 (C4) and the related sex-limited protein, which is encoded by a gene closely linked to that producing C4 (Ferreira et al., 1978; Michaelson et al., 1981), may be responsible for the variable PA induced by tumor vesicles.

For this purpose, more than 20 strains of mice with different levels of C4 (Staats, 1978), as determined by quantitation of the C4 antigen or the hemolytic activity, were used. Table 2 shows that most of the strains (14 out of 15) with high levels of C4 responded positively in tests of PA. Results were different, however, in mice with low C4 levels. Most of them (5 out of 7) responded negatively in tests of PA (Table 3). Although antigenic levels of C4 are low in these strains, their partial deficiency in complements can still support some hemolytic activity (data not shown) and may even permit PA by tumor vesicles as observed in AKR and C57Br mice (Table 3). It is very likely that in these strains, factors related to the clotting cascade cooperate to achieve this effect.

Table 2

Aggregation of platelets induced by tumor vesicles in heparinized mouse PRP with high levels of C4

Strain	PA
A, A.CA, A.SW	+
C57Bl/6, C57Bl/10	+
B10.A, B10.D2o, B10.D2n, B10.M	+
C3H.SW	+
C3H-H-2°	−
DBA/1, DBA/2	+
LP, SWR	+

Table 3

Aggregation of platelets induced by tumor vesicles in heparinized mouse PRP with low levels of C4

Strain	PA
AKR, C57Br	+
CBA, C3H/He, C3H/Icr	−
C3H/Bi, C3H/jk	−

The sex-limited protein, which has no hemoltyic activity (Ferreira et al., 1978), did not influence platelet aggregation at all. As shown in Table 4, there is no correlation between presence or absence of this protein and reactivity of platelets to tumor vesicles. Thefore, we believe that the sex-limited protein cannot substitute for C4 in supporting PA induced by vesicles.

Table 4

Aggregation of platelets induced by tumor vesicles in the mouse in the presence and absence of sex-limited protein

Strain	Sex	Slp	PA
A, BALB/c, B10.A, B10.D2n	M	+	+
	F	−	+
DBA/2, SJL, SWR	M	+	+
	F	−	+
AKR, A.CA, B10.Br, B10.M	M	−	+
	F	−	+
C57Bl/6, C57Br	M	−	+
	F	−	+

Some of the strains used (A, AKR, DBA/2, B10.D2o) completely lack C5 (Cinader et al., 1964). It is interesting to note that their platelets responded with normal aggregation to vesicles, indicating that the first four components of the complement system (C1, C4, C2, and C3) are the only ones participating in the PA under study.

The clotting system in the mouse is also subject to genetic variations. For example, certain strains of mice are prone to suffer from single or multiple deficiencies of the prothrombin complex (Meier et al., 1962). This condition should be taken into consideration when platelets in PRP do not respond to the aggregating activity of tumor cells. This variability in the mouse system has also been observed in humans, as described by Bastida et al. and Pearlstein et al. in papers communicated at this conference.

To further study the importance of the clotting cascade and complement in our system, we investigated whether agents interfering with their activity may inhibit PA by tumor vesicles. In this search, we found two promising new agents. One was a synthetic antithrombin, identified as compound No. 805, developed by Okamoto and the Mitsubishi Chemical Corporation; the other was a natural product present in salivary glands of the South American leech Haementeria ghiliani (Sawyer et al., 1981).

Compound No. 805 (MCI-9038), which is a new arginine derivative, is highly specific as an inhibitor of thrombin and has very low toxicity in mice (LD50 211 mg/kg, i.v.). In vivo tests in mice have shown it to be highly potent in preventing death by tumor thromboplastins. A plasma concentration as low as 1 uM fully protected mice against a lethal dose of tumor vesicles. Low concentrations (2 uM in rats and 9 uM in mice) were also effective in abolishing PA in PRP from treated donors.

The salivary gland extract proved a surprisingly interesting material. Initially, the extract was tested because of its fibrinolytic activity and capacity for inhibiting PA by ADP in rabbit PRP (Budzynski et al., 1981). Although the latter was not verified in rats and mice, the extract showed instead to be inhibitory of PA by collagen and tumor vesicles (see below). In investigating its mechanism of action, we discovered that the extract is a unique anticomplementary agent. In collaboration with Dr. Ernest Marquez of Penn State Hershey Medical College and Dr. Andrei Budzynski of the Specialized Center for Thrombosis Research at Temple University, we demonstrated that the extract can abolish in vivo the total hemolytic activity of mouse complement (Table 5) and that this effect is due to selective inhibition of two components of the classical pathway, as shown in Table 6.

Earlier we showed that low levels of C4 in mice were associated with diminished or no response of platelets to tumor vesicles. Since the Haementeria extract inhibits C4 almost specifically, we carried out experiments to determine whether its in vivo administration would suppress PA induced by tumor vesicles in mouse PRP. This was the case, as shown by Table 7. The i.v. injection of 200 ug of the extract rendered platelets, removed 2-4 hr later, unresponsive to PA induced by tumor vesicles.

Table 5

In vivo effect of salivary gland extract
on total hemolytic complement activity*

Treatment	% of cell lysis	(# of assays)
Control†	46	(2)
Extract, 1 injection	51	(2)
Extract, 2 injections	0	(2)

*B6A F_1 female mice, 4 months old, were injected i.v. with 200 ul salivary gland extract (400 ug protein) at 2 hr intervals. Prior to the first injection, plasma samples were obtained by retroorbital bleeding in the presence of heparin (5 units/ml). Thirty min after the first injection, plasma was obtained from the other eye in a similar manner. Thirty min after the second injection, mice were bled from the heart and plasma was prepared from the heparinized blood. Plasmas were immediately assayed for total hemolytic complement activity by the method of Rapp and Borsos (1970) and the degree of inhibition by salivary gland extract was determined.
†Prior to the first injection.

Table 6

In vitro effect of salivary gland extract
on mouse complement components*

Component	% decrease in lysed cells	(# of assays)
C1	6	(2)
C4	100	(2)
C2	68	(2)
C3	0	(2)
C3-C9	0	(2)

*Fresh heparinized plasma (obtained within 30 min) was incubated with buffer or salivary gland extract (40 ug for 100 ul plasma) at 30°C for 10 min, after which C1, C4, C2, and C3-C9 were determined by modification of the procedure of Rapp and Borsos (1970). The degree of hemolysis was determined by measuring hemoglobin concentration in the supernatant after removal of intact red blood cells by centrifugation. Using a spectrophotometer, supernatants were assayed at a wavelength of 412 mm and the degree of inhibition of cell lysis was calculated.

Table 7

Aggregration of platelets induced by tumor vesicles in PRP from mice treated with salivary gland extract

Strain	Salivary gland extract (ug)	Duration of treatment	PA
CF #1	0	--	+
	200	2-4 hr	-
	200	6 hr	+
B6 F_1	0	--	+
	200	2-4 hr	-
	200	6 hr	+

Based on the work of Tiffany and Penner (1980), who have studied PA by collagen in C4-deficient guinea pigs, it is likely that the capacity of the extract to abolish aggregation of rat platelets by collagen is also related to its anticomplementary activity.

In summary, our results indicate that agents which interfere with thrombin and complement were capable of abolishing PA by tumor vesicles. Further studies are required to determine whether these agents may be useful in controlling tumor spread.

The second part of this paper will focus on tumor vesicle-macrophage interactions, which we have studied in collaboration with Dr. Carol Cowing, of the University of Pennsylvania School of Medicine. First, we wanted to determine whether tumor vesicles added to a suspension of macrophages would bind to these cells, and, if so, whether these macrophages would acquire the capacity of inducing tumor immunity when injected i.v.

To achieve the first aim, we incubated 5×10^6 resting peritoneal macrophages suspended in 1 ml of RPMI 1640 medium supplemented with fresh syngenic mouse serum with radioiodinated tumor vesicles (^{125}I) from 15091 tumor containing 10 ug of protein. After 24 hr of incubation at 37°C with slow rotation, cells were pelleted, then washed thrice with PBS, and radioactivity was measured with a gamma radiation counter. Results indicated that 8% of the added radioactivity was present in the pellet, suggesting that a portion of the vesicles in the system had adhered to macrophages or were incorporated into these cells.

To investigate in a syngeneic system whether the macrophages incubated in the presence of vesicles, as described above, may confer protection against tumor growth, 5×10^6 of macrophages, exposed or unexposed to vesicle interaction, were injected i.p. into each mouse. After two weeks, these recipients were challenged with a minimal lethal dose of 15091A tumor cells, injected s.b.c.

Results of several experiments indicate that the injected vesicle-modified macrophages improve the host resistence to the tumor inoculum. The results of a representative experiment are shown in Table 8. While tumors grew in all mice receiving macrophages incubated with PBS alone, recipients treated with macrophages exposed to vesicles in tissue culture showed varying degrees of resistance to the growth of tumor inocula. In six out of 15 there was no tumor; in the remaining mice the growth was significantly retarded in some animals, while it approached control levels in others.

Table 8

Growth of 15091 tumor in CA F_1 mice treated with 15091A tumor vesicle-activated peritoneal macrophages

Amount of vesicles in macrophage incubation medium*	Number of mice without tumor/ Total number of mice	
0	0/5	
1 ug	2/5	
3 ug	1/5	(6/15)
10 ug	3/5	

*Incubated at 37°C for 24 hr.

Although macrophages constitute a major component of the cells which survive 24 hr of culture after removal from the peritoneum, one important question is whether the protective effect observed is due only to vesicle-modified macrophages or to a more complex interaction involving other types of cells. Another important question is whether the enhancing of host defenses against the tumor is due to vesicles alone, independent of their association with macrophages. To test this possibility, we injected syngeneic mice with various doses of vesicles. Two weeks later, we challenged them with the same minimal dose of 15091A tumor cells, injected s.b.c. This time no difference was observed in tumor growth between controls and mice treated with vesicles alone, suggesting that macrophages or other cells must first interact with vesicles before inducing a favorable anti-tumor protective effect.

The capacity of tumor cells to shed vesicles may also occur in vivo, as demonstrated by several investigators. Vesicles may be shed into the circulation (Poste and Nicolson, 1980) or body cavities (Dvorak et al., 1981), or may remain in the stroma of the growing tumor. Therefore, it is conceivable that vesicles may interact with macrophages in any of these environments and induce changes of their tumoricidal activity or capacity to interact with other cells of the immune system. Depending on the degree of immunogenicity of the parent tumor cells, it is very likely that antigens carried by the vesicles shed by these tumors may either activate or depress the macrophage-lymphocytic system, inducing immunity or tolerance.

From our in vitro findings we can assume that vesicles shed in vivo may interact both with components of the hemostatic system and contribute to extravasation of the blood-borne cancer cells, and with cells of the macrophage system and trigger or not trigger defensive responses. It remains to be established--and this is an intriguing question--whether surface antigens in vesicles responsible for immunogenicity are the same ones that activate the hemostatic system.

ACKNOWLEDGEMENTS

The authors are indebted to Dr. Gunther S. Stent for providing the anterior salivary glands of the leech <u>Haemen-</u>

teria ghiliani and to the Mitsubishi Chemical Industries Corporation for the generous supply of compound No. 805. This work was supported by USPHS grant 18450, awarded by the National Cancer Institute; by USPHS grants HL-18827-05 (subproject 3) and HL-14217, awarded by the National Heart, Lung, and Blood Institute; by NCI contract N01-CP-53516; and by grant CD-82 from the American Cancer Society.

REFERENCES

Baumgartner HR, Tschopp TB, Weiss HJ (1977). Platelet interaction with collagen fibrils in flowing blood II. Thromb Haemost 37:17.

Budzynski AZ, Olexa SA, Brizuela BS, Sawyer RT, Stent GS (1981). Anticoagulant and fibrinolytic properties of salivary proteins from the leech Haementeria ghiliani. Proc Soc Exp Biol Med (in press).

Budzynski AZ, Olexa SA, Sawyer RT (1981). Composition of salivary gland extracts from the leech Haementeria ghiliani. Proc Soc Exp Biol Med (in press).

Cinader B, Dubiski S, Wardlaw AC (1964). Distribution, inheritance, and properties of an antigen, MUB1, and its relation to hemolytic complement. J Exp Med 120:897.

Cochrane CG, Müller-Eberhard HJ, Aiken BS (1970). Depletion of plasma complement in vivo by a protein of cobra venom: its effects on various immunological reactions. J Immunol 105:55.

Dvorak HF, Quay SC, Orenstein NS, Dvorak AM, Hahn P, Bitzer AM (1981). Tumor shedding and coagulation. Science 212:923.

Ferreira A, Nussenzweig V, Gigli I (1978). Structural and functional differences between the H-2 controlled Ss and Slp proteins. J Exp Med 148:1186.

Fine DP, Marney SR, Colley DG, Sergent JS, Des Prez RM (1972). C3 shunt activation in human serum chelated with EGTA. J Immunol 109:807.

Fine DP (1975). Pneumococcal type-associated variability in alternate complement pathway activation. Infect Immun 12:772.

Gasic G, Gasic T (1962). Removal of sialic acid from the cell coat in tumor cells and vascular endothelium and its effects on metastasis. Proc Nat Acad Sci (USA) 48:1172.

Gasic GJ, Gasic TB, Stewart CC (1968). Antimetastatic effects associated with platelet reduction. Proc Nat Acad Sci (USA) 61:46.

Gasic GJ, Gasic TB, Galanti N, Johnson T, Murphy S (1973). Platelet-tumor cell interactions in mice. The role of platelets in the spread of malignant disease. Int J Cancer 11:704.

Gasic GJ, Koch PAG, Hsu B, Gasic TB, Niewiarowski S (1976). Thrombogenic activity of mouse and human tumors: effects on platelets, coagulation, and fibrinolysis, and possible significance for metastases. Z Krebsforsch 86:277.

Gasic GJ, Boettiger D, Catalfamo JL, gasic TB, Stewart GJ (1978). Aggregation of platelets and cell membrane vesiculation by rat cells transformed in vitro in Rouse sarcoma virus. Cancer res 38:2950.

Gasic GJ, Catalfamo JL, Gasic TB, Avdalovic N (1981). In vitro mechanism of platelet aggregation by purified plasma membrane vesicles shed by mouse 15091A tumor cells. In Donati MB, Davidson JF, Garattini S (eds.) "Malignancy and the Hemostatic System," New York: Raven Press, p. 27.

Gee AP, Borsos T, Boyle MDP (1979). Interaction between components of the human classical complement pathway and immobilized cibacron blue F3GA. J Immunol Methods 30:119.

Götze O, Müller-Eberhard HJ (1971). The C3-activator system and alternate pathway of complement activation. J Exp Med 134:90s.

Honn KV, Cicone B, Skoff A (1981). Prostacyclin: a potent antimetastatic agent. Science 212:1270.

Kane MA, May JE, Frank MM (1973). Interactions of the classical and alternate complement pathway with endotoxin lipopolysaccharide. J Clin Invest 52:370.

Markwardt F (1970). Hirudin as an inhibitor of thrombin. Methods Enzymol 19:924.

Meier H, Allen RC, Hoag WG (1962). Spontaneous hemorraghic diathesis in inbred mice due to single or multiple "prothrombin complex" deficiencies. Blood 19:501.

Michaelson J, Ferreira A, Nussenzweig V (1981). cis-Interacting genes in the S region of the murine major histocompatibility complex. Nature 289:306.

Nesheim ME, Prendergast FG, Mann KG (1979). Interactions of a fluorescent-active site-directed inhibitor of thrombin: dansylarginine N-(3-ethyl-1,5-pentanedyl) amide. Biochemistry 18:996.

Pearlstein E, Cooper LB, Karpatkin S (1979). Extraction and characterization of a platelet-aggregating material from SV40-transformed mouse 3T3 fibroblasts. J Lab Clin Med 93:332.

Poste G, Nicolson GL (1980). Arrest and metastasis of blood-borne tumor cells are modified by fusion of plasma membrane vesicles from highly metastatic cells. Proc Nat Acad Sci (USA) 77:399.

Rapp HJ, Borsos T (1970). "Molecular Basis of Complement Action," New York: Appleton Century Crofts, p. 75.

Sawyer RT, Lepont F, Stuart DK, Kramer AP (1981). Growth and reproduction of the giant glossiphonid leech Haementeria ghiliani. Biol Bull 160:322.

Staats J (1978). Standardized nomenclature for inbred strains of mice; sixth listing. Cancer Res 36:4333.

Tangen O, Berman HJ (1972). Gel filtratiuon of blood platelets: a methodological report. Adv Exp Med Biol 34:235.

Tiffany ML, Penner JA (1980). Effect of complement on collagen-induced platelet aggregation. J Lab Clin Med 96:796.

Warren BA (1981) Cancer cell-endothelial reactions: the micro-injury hypothesis and localized thrombosis in the formation of micrometastases. In Donati MB, Davidson JF, Garattini S (eds.) "Malignancy and the Hemostatic System," New York: Raven Press, p. 5.

Zucker MB, Grant RA (1974). Aggregation and release reaction in human blood platelets by Zymosan. J Immunol 112:1219.

DISCUSSION

Dr. S. Karpatkin (New York University Medical Center, New York). I would like to comment on the role of complement components and the clotting system in tumor cell-induced platelet aggregation. Dr. Gasic has just presented evidence that both situations can apply with vesicles, in particular in the tumor cell line he has employed. However, our experience is different with the different tumor cell lines which we employ. What we find is that certain tumor cell lines do not involve thrombin aggregation, but work by the complement system alone. Other tumor cell lines have nothing to do with complement but just work only by the thrombin system.

Dr. D.S. Rappaport (University of Minnesota Medical School, Minneapolis). I would like to point out some tumors have receptors for complement on the surface of the tumor and these may also be present on the vesicles you describe.

Dr. S. A. Leon (Albert Einstein Medical Center, Philadelphia). In your vesicles, in addition to ribosome and other components of the endoplasmic reticulum, are there any other components from nuclei? Do you have any DNA?

Dr. G. J. Gasic (University of Pennsylvania, Philadelphia). No.

Dr. Leon. I ask you this for one reason, we have found that cancer patients have abnormally high levels of DNA in their blood so if you can get pieces of cells floating you may also get DNA.

PLATELET AGGREGATING MATERIAL (PAM) OF TWO VIRALLY-TRANSFORMED TUMORS: SV3T3 MOUSE FIBROBLAST AND PW20 RAT RENAL SARCOMA. ROLE OF CELL SURFACE SIALYLATION

Simon Karpatkin[1], Edward Pearlstein[2], Peter L. Salk[3] and Ganesa Yogeeswaran[4]

Departments of Medicine[1,2] and Pathology[2], and Irvington House Institute[2], New York University Medical School, New York, NY 10016; and Autoimmune and Neoplastic Disease Laboratory[3] and Department of Cancer Biology[4], The Salk Institute for Biological Studies, San Diego, CA 92138

SUMMARY

Platelets are required for certain experimental tumor metastases and several lines of tumor cells have been shown to aggregate platelets. We have extracted a sedimentable sialolipoprotein, platelet aggregating material (PAM) from the cell surface of SV40 transformed Balb C3T3 fibroblasts which aggregates heparinized PRP at 2.5 µg/ml via the release reaction, following a one minute lag period. A similar extract from non-transformed 3T3 cells has barely measurable activity at 40 µg/ml. Gel-filtered platelets (GFP) do not aggregate with PAM. However, PAM aggregation can be restored by addition of 5% plasma but not by fibrinogen. Two plasma components are required: a heat-labile complement component which is activated during the lag period; and a heat-stable factor which is required for platelet aggregation.

The pathophysiologic significance of PAM has been examined in ten variant cell lines derived from a spontaneously metastatic renal cell sarcoma of rats, initially induced with polyoma virus (PW20 Wistar-Furth parental lines). These lines were selected in vitro and in vivo from a single line and differed in their capacity to form distant tumors in various organs after subcutaneous injection. These cells were examined for cell surface sialylation, PAM and PAM sialic acid content, since cell surface sialic acid is increased in a variety of tumor tissues and PAM is inhibited by neuraminidase. A good correlation was obtained between in vivo metastatic potential and cell surface sialic acid, r=0.83, p<0.003; cell surface sialic acid and PAM, r=0.85, p<0.002; in vivo metastatic potential and sialic acid content of PAM, r=0.69, P<0.03; and in vivo metastatic potential and PAM, r=0.68, p<0.03. We conclude that platelets may play a role in hematogenous metastasis via the ability of tumor cells to aggregate platelets by cell surface constituents containing sialic acid. The platelet-tumor cell interaction requires activation of the alternate complement pathway and a heat stable plasma factor.

INTRODUCTION

It has previously been demonstrated that certain animal tumor cells or media conditioned by these cells will induce platelet aggregation in vitro (Gasic et al, 1973; Gasic et al, 1976; Gasic et al, 1977). The significance of this result is apparent in light of evidence from several laboratories indicating that tumor cells may also aggregate platelets in vivo (Jones et al, 1971; Warren, Vales, 1972; Hilgard, 1973; Warren, 1973; Hilgard, Gordon-Smith, 1974; Sindelar et al, 1975; Gastpar, 1977), that blood-borne metastases induce thrombocytopenia in the host (Gasic et al, 1973) and that thrombocytopenia impairs the development of metastasis (Gasic et al, 1968). Furthermore, a rough correlation can be made between in vitro induction of platelet aggregation by certain tumor cells and their propensity for lung metastasis (Gasic et al, 1973; Hilgard, Gordon-Smith, 1974). These observations imply a direct role for the platelet in the pathogenesis of tumor cell metastasis.

In order to define the mechanism of platelet-tumor cell interaction, an in vitro system was developed employing the normal mouse fibroblast cell line 3T3 as a control and the virally transformed SV3T3 cell line as the tumor cell. A technique has been established for extracting PAM from the transformed cell. This PAM has been partially characterized and its mechansim of action studied.

A spontaneously-metastatic, polyoma-induced PW20 Wistar-Furth rat renal sarcoma was also examined in vivo and in vitro for degree of metastasis, cell surface sialylation, and in vitro platelet aggregating ability. Ten cell lines from the PW20 transformed parent cell line were selected by tissue and animal passage for their varying ability to develop metastases following subcutaneous injection (Salk, Yogeeswaran, 1978; Pearlstein et al, 1980).

MATERIALS AND METHODS

Cell Culture

A low passage Balb C/3T3 fibroblast cell line and its SV40 virally-transformed derivative (SV3T3) were obtained from stocks at the Imperial Cancer Research Fund, London, and were maintained in Dulbecco's modified Eagle's medium (E4) supplemented with 10% fetal calf serum, 2 mM glutamine, 100 U/ml penicillin, and 100 μg/ml streptomycin (Pearlstein, Seaver, 1976). Cells were passaged twice weekly.

The PW20 family of tumor cell lines (Table 1) was derived from a culture of the polyoma virus-induced PW20 Wistar-Furth rat renal sarcoma (obtained from Dr. H.O. Sjögren, University of Lund, Sweden), by varying conditions of passage in tissue culture and passage through syngeneic animals (Pearlstein et al, 1980).

Table 1

DERIVATION OF PW20 TUMOR CELL LINES

Derivative Lines	Number of Animal Passages*	In Vitro Subcultures Tested+
RO(L)	0	11-19
RO(H)	0	276-283
RO(L)R1	1	18
RO(H)R1	1	3
R1	1	88-92
R2	2	80-84
R2R1	3	6-16
R2L1	3	5-7
R2N1	3	5-6
R2R2	4	6

*Number of animal passages since receipt of the original PW20 cell line from Dr. Sjögren.

+Number of subcultures after establishment of each new derivative line. For RO(L) and RO(H) cell line the number represents the number of subcultures after receipt of the original cell line. The numbers listed represent the number of the subculture used for testing of metastatic properties.

The RO(L) and RO(H) cell lines represent low and high tissue culture passages, respectively, of the original cell culture. The Rl cell line was obtained by re-establishing a subcutaneous RO(L) tumor in tissue culture, and maintaining the culture for a large number of passages. The R2 cell line was similarly produced by reculturing a subcutaneous Rl tumor and maintaining the resultant culture for a prolonged period. The RO(L)Rl, RO(H)Rl, R2Rl and R2R2 cell lines were produced by reculturing subcutaneous RO(L), RO(H), R2 and R2Rl tumors, respectively; the R2Ll and R2Nl cell lines were obtained by culturing explanted lung (L) and lymph node (N) metastases which developed from subcutaneous R2 tumors. These latter lines were all tested at low passage numbers. In vitro assays were performed using cells within several passages of those listed in Table 1, obtained either from the same cells used for in vivo testing or from an aliquot of identical cells preserved in liquid nitrogen. For the in vitro assays, the cell lines were maintained as above for the 3T3 fibroblast cell line.

Evaluation of Metastatic Properties

The spontaneous metastatic behavior of the tumor cell lines was evaluated in syngeneic Wistar-Furth rats of both sexes obtained from Microbiological Associates, Inc. Animals were injected subcutaneously in the loose skin of the midback with tumorigenic doses of cultured cells. Tumors were measured at regular intervals following injection and excised, using a closed surgical technique, upon reaching a weight of 2-6 grams. The animals were then observed for the development of palpable regional lymph node metastases and/or clinical signs suggestive of internal metastases to the lung or other organs. Animals with large external metastases or with signs of visceral metastatic growth were sacrificed and a thorough search was conducted for macroscopic metastases to regional lymph nodes, intrathoracic and intra-abdominal sites. Asymptomatic animals were similarly sacrificed after a prolonged period of observation. In some experiments the lungs were injected with india ink and fixed according to the method of Wexler (1966) in order to facilitate visualization of the metastatic nodules.

Percent Sialylation of Cell Surface Glycoconjugates

The percent sialylation of exposed cell surface glycoconjugates was determined using the galactose oxidase-sodium borotritide labelling technique, as previously described (Yogeeswaran et al, 1979). This procedure results in the tritiation of terminal galactosyl (Gal) and N-acetyl galactosaminyl (GalNAc) groups on exposed cell surface carbohydrates. Gal and GalNAc groups which are further substituted with sialic acid are protected from labelling. It is therefore possible to determine the percent of the exposed terminal Gal and GalNAc groups which are sialylated by labelling cells with and without prior incubation with neuraminidase, which cleaves terminal sialic acid residues from the saccharide chains. Galactose oxidase was obtained from A.B. Kabi (Stockholm, Sweden), Vibrio cholera neuraminidase was obtained from Calbiochem/Behring (La Jolla, CA), and sodium borotritide (4-6 Ci/mMole) was obtained from Amersham/Searle Corp. (Arlington Heights, IL).

Extraction of PAM

Cells were grown to confluency on 10 cm tissue-culture dishes and washed twice with Veronal buffer (11.75 gm of sodium diethyl barbiturate, 14.67 gm of sodium chloride, 430 ml of 0.1 N HCl, H_2O to 2 L, final pH 7.4). Three milliliters of 1M urea, dissolved in Veronal buffer, were then added. Dishes were shaken at 30°C for 1 hr in a 5% CO_2 atmosphere, the supernatants were spun at 2000 x g for 5 min to remove any floating cells, and the cell-free supernatant was dialyzed against several changes of Veronal buffer for 2 days at 4°C. (Viability of the adherent cells was greater than 90% as judged by trypan blue exclusion following this treatment). Following dialysis, the supernatant was concentrated in an Amicon chamber with the use of an XM100 membrane (Amicon Corp., Lexington, MA) to 1/100 the volume of 1M urea used in the original extraction (approximately 100 $\mu g/ml$ protein). In parallel cultures, cells were extracted with E4 or 0.3 mM KCl, and extracts were processed in an identical fashion.

Total Sialic Acid Content of PAM

The sialic acid content of PAM was determined by the thiobarbituric acid method of Warren (1959), using N-acetyl neuraminic acid (Sigma Chemical Corp., St. Louis, MO) as a standard.

Protein Content of PAM

The protein content of PAM was determined by the method of Lowry et al (1951), using bovine serum albumin as a standard.

Centrifugation of PAM

To determine whether PAM was sedimentable, 0.5 ml of extract was spun at 100,000 x g for 1 hr at 4°C in a Beckman L2-65B ultracentrifuge equipped with an SW40Ti rotor (Beckman Instruments, Inc., Fullerton, CA). Following centrifugation the supernatant was removed, and the pellet was resuspended in 0.5 ml of Veronal buffer.

Enzyme Treatment of PAM

Ten microliters of an enzyme solution were added to 100 μl of PAM (100 μg/ml protein) to give the final enzyme concentration and the mixture was incubated for 1 hr at 37°C. Enzymes were shown to be active with their respective substrates prior to use. A 50 μl amount of either treated PAM, or enzyme alone diluted 1:11 in Veronal buffer was used in the aggregation assay. No enzyme induced aggregation by itself, nor did enzyme treatment of platelets render them unresponsive to 2×10^{-5}M ADP or to untreated PAM added simultaneously. To further ensure that the presence of enzyme was not making the platelets specifically unresponsive to PAM, treated PAM was centrifuged as previously described to pellet activity, and the pellet was resuspended in Veronal buffer prior to reacting with the platelets.

Phospholipase-A_2 was boiled for 10 min prior to use in order to inactivate possible contaminants. Neuraminidase treatment was performed in the presence of 2 mM phenylmethyl sulfonyl fluoride (PMSF) to inhibit any contaminating protease activity.

Treatment with Complement Inactivators

Forty microliters of purified cobra venom factor (100U/ml)

from Naja naja Kaouthia (Cordis Laboratories, Miami, FL) was incubated with 0.4 ml of PRP for various lengths of time at 37°C, followed by the addition of 50 μl of PAM. Cobra venom factor did not affect platelet aggregation induced in PRP with 10^{-6}M ADP. Furthermore, platelets isolated from cobra venom factor-treated PRP by gel filtration were responsive to aggregation by PAM in the presence of fresh PFP.

Washed zymosan (Sigma Chemical Co., St. Louis, MO) was resuspended in fresh plasma at a final concentration of 1.35 mg/ml, incubated for 1 hr at 37°C, and removed by centrifugation at 1000 x g for 5 min at 4°C. The supernatant plasma was assayed for its ability to support platelet aggregation with PAM.

Binding of PAM to CNBr-activated Sepharose 4-B

PAM, bovine serum albumin, or fibronectin (at 1 mg/ml) was covalently bound to CNBr-activated Sepharose 4-B and the beads were washed extensively with Veronal buffer prior to use. For aggregation experiments, 1 ml of packed derivatized beads were suspended in an equal volume of buffer. A 50 μl volume of this suspension was added to 0.4 ml of PRP or 0.4 ml of GFP containing 50 μl of PFP. Aggregation could be followed turbidometrically, and platelet-bead interaction viewed microscopically.

Preparation of PRP and PPP

Rabbits or humans were bled by venipuncture directly into a plastic syringe containing a final concentration of 5 U/ml heparin (preservative-free, Connaught Laboratories, Willowdale, Ont., Canada). PRP was obtained by centrifugation at 150 x g for 5 min at room temperature. PPP was prepared by centrifuging the remaining blood at 2000 x g for 15 min. The PRP was allowed to stand at room temperature for 30 min in tightly capped plastic tubes prior to testing.

Preparation of Washed Human Platelets

Eighteen milliliters of human blood were drawn into 2 ml of 3.8% tri-sodium citrate. PRP was prepared as described and incubated with 5 mg of apyrase at 37°C for 30 min. The platelets were then sedimented as above and resuspended in Ardlie's buffer (Mustard et al, 1972) containing 0.2 U/ml heparin and 0.5 mg/ml apyrase, except that $MgCl_2$ and $CaCl_2$ were omitted from the buffer and 2 mM EDTA was added. The resuspended platelets were again centrifuged, resuspended in half their original PRP volume of Ardlie's buffer, free of $MgCl_2$, $CaCl_2$, heparin, apyrase, and EDTA, and then incubated at 37°C for 45 min. In some experiments washed platelets were prepared by gel filtration (Tangen, Berman, 1972). These platelets aggregated with 10^{-5}M ADP or 10^{-5}M epinephrine when 2 mg/ml human fibrinogen was added to the suspension. Collagen-induced aggregation did not require fibrinogen but was enhanced by its addition.

In Vitro Platelet Aggregation

Platelet aggregation was measured turbidometrically (Karpatkin, 1978) with a Bio-Data aggregometer (Bio-Data Corp., Willow Grove, PA). In a typical experiment, 0.4 ml of PRP or washed platelets was warmed to 37°C for 3 min in a flat-bottomed cylindrical cuvette. The platelets were then stirred in the aggregometer at 1200 rpm, a baseline was obtained on the recorder, and 0.05 ml of PAM was added. Aggregation was recorded as an increase in light transmission, with PPP representing 100% transmission. In experiments where inhibitors of aggregation were assayed, 0.05 ml of inhibitor or 0.05 ml of Veronal buffer was followed by 0.05 ml of PAM.

Quantitation of the Release with ^{14}C-Serotonin

Serotonin release was determined essentially as described (Gasic et al, 1976).

RESULTS

Studies on Non-transformed 3T3 and Transformed SV40 3T3 Mouse Fibroblasts

Whole Cells. Intact SV3T3 transformed cells were capable of inducing platelet aggregation after a lag period of 1 to 2 min (Fig. 1). When 50 μl of transformed cells at an initial concentration of 10^7 cells/ml were added to 0.4 ml of PRP, the final maximal level of aggregation (optical density change) reached was 80% of that achieved by the addition of 2×10^{-5}M ADP. The normal 3T3 line was also capable of inducing

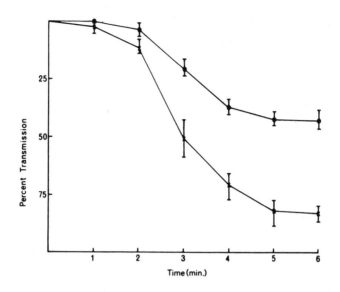

Figure 1. Induction of rabbit platelet aggregation in heparinized PRP by 10^7 intact 3T3 (●—●) or SV3T3 (x—x) cells. Each point (± SEM) represents the average of four experiments.

moderate aggregation following a slightly longer lag period than that observed with the transformed cell. However, the final percentage of aggregation induced by the normal cell line was approximately 40% of the ADP control (Fig. 1).

Urea Extract. A urea extract of SV3T3 cells induced platelet aggregation at a final concentration of 40 μg/ml protein (Fig. 2). In contrast, medium conditioned by the same cell type for an identical length of time required approximately five-fold to ten-fold higher concentration of protein to induce a similar degree of aggregation (data not shown). Dilution of the extracted SV3T3 cell line PAM by a factor of 20 (from 40 to 2 μg/ml) resulted in a loss of activity (Fig. 2) demonstrating the sensitivity of the system to dilution. Normal 3T3 cell extracts, at 40 μg/ml, were capable of inducing only slight aggregation after a prolonged lag period (Fig. 2).

The transformed cell extract was more effective in heparin-PRP, as were the intact cells, in agreement with the results of others (Gasic et al, 1976).

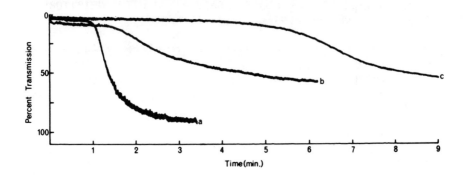

Figure 2. Induction of rabbit platelet aggregation in heparinized PRP by 1M urea extracts from SV3T3 cells at a final protein concentration of (a) 40 μg/ml or (b) 2 μg/ml and from 3T3 cells (c) at 40 μg/ml. Data taken from one of 20 experiments with similar results.

Inhibitors of PAM. As shown in Table 2, several compounds which inhibit the secondary wave of platelet aggregation (Weiss 1975) also prevented PAM induction of platelet aggregation. Thus, 5 mM EDTA, 0.05 mM indomethacin, 0.1 mM adenosine and 0.1 mM $N6,O2'$-dibutyrl cyclic AMP all inhibited PAM activity.

Release of ^{14}C-Serotonin. Direct measurement of the release reaction was performed by quantitating ^{14}C-serotonin release from aggregated platelets. The results, shown in Table 3, indicate that PAM accomplishes irreversible aggregation by inducing platelet release.

Non-Inhibitors of PAM. Certain proteases induce platelet aggregation (Davey, Luscher, 1966) and tumor cells frequently synthesize higher levels of proteases than do their normal counterparts (Ossowski et al, 1973; Goldberg, 1974). Therefore an attempt was made to inhibit PAM with several specific protease inhibitors known to interfere with transformed cell plasminogen activation (Hynes, Pearlstein, 1976). All inhibitors were assayed for activity on appropriate substrates prior to use. For example, 2.5 mM DFP inhibited thrombin-induced platelet aggregation at a concentration of 0.1 μg/ml. The

Table 2

INHIBITION OF PAM-INDUCED AGGREGATION WITH INHIBITORS
OF THE PLATELET RELEASE REACTION

Inhibitor	Aggregating Agent	Percent Aggregation
Veronal	PAM	100
Veronal buffer	2×10^{-5}M ADP	100
5 mM EDTA	PAM	0
0.1 mM dBcAMP[b]	PAM	1-30
0.1 mM dBcAMP	2×10^{-5}M ADP	0
0.1 mM adenosine	PAM	0
0.1 mM adenosine	2×10^{-5}M ADP	0
0.05 mM indomethacin	PAM	10

All inhibitors were incubated with heparinized rabbit PRP for 6-9 min at 37°C prior to the addition of PAM, 10 μg/ml. Inhibition concentration is that achieved following dilution of the inhibitor in PRP. Each experiment was performed 3 or more times.
dBcAMP = $N^6, O^{2'}$-dibutyrl cyclic AMP.

Table 3

RELEASE OF ^{14}C-SEROTONIN FOLLOWING ADDITION OF
HEPARINIZED RABBIT PRP

Addition	CPM in Supernatant	Percent Release
Veronal buffer	399	2
2×10^{-5}M ADP	387	2
PAM	10,255	53
Sonication	19,350	100

Percent release was calculated as radioactivity in PAM-containing supernatant minus background radioactivity (buffer) divided by total radioactivity taken up by the platelet suspension. Results are an average of two determinations.

results given in Table 4 clearly indicate that PAM activity is not related to any increase in proteolytic activity associated with transformed cell plasma membranes. Thus EACA (0.8 mM), soybean trypsin inhibitor (50-500 μg/ml), trasylol (50 units/ml), PMSF (2 mM), and DFP (2.5 mM) had no significant effect on platelet aggregation.

Table 4

EFFECT OF PROTEASE INHIBITORS ON PAM-MEDIATED
PLATELET AGGREGATION

Inhibitor	Final Concentration	% Aggregation
Veronal buffer	--	100
EACA (Epsilon-amino-caproic acid)	0.8 mM	95
Soybean Trypsin Inhibitor	50-500 μg/ml	98
Trasylol	50 U/ml	95
PMSF	2 mM	90
DFP (Diisopropylfluorophosphate)	2.5 mM	82

Inhibitors were incubated with PAM for 4-5 min at 20°C prior to addition to PRP. Each experiment was performed 2 or more times.

Centrifugation of PAM. The urea-extracted PAM from SV3T3 cells can be pelleted by centrifugation at 100,000 x g for 1 hr at 4°C. The results shown in Fig. 3 clearly demonstrate that all the activity was recoverable in the pellet, with no residual PAM remaining in the supernatant. The ability to pellet the material provided a means of removing enzyme from the enzyme-treated PAM in order to discriminate between the effect of enzyme on PAM and on the platelet surface.

Effect of Boiling, Enzymes, Non-Ionic Detergents, or Sonication on PAM and PRP. The following treatment, listed in Table 5, completely destroyed PAM activity: 1) boiling for 15 min (partial activity could be restored, following boiling, by storage for several weeks at 4°C); 2) crystalline trypsin at 1 mg/ml; 3) neuraminidase at 2 mg/ml in the presence of 2 mM PMSF; 4) boiled phospholipase-A_2 at 0.1 μg/ml; 5) non-ionic detergents, NP-40 (0.1%) and Tween-80 (0.1%); 6) sonication for 15 sec at 0°C.

β-Galactosidase at 5 mg/ml had no effect on PAM.

PAM was washed by centrifugation and resuspended in

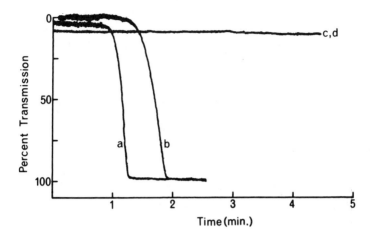

Figure 3. Platelet aggregation following sedimentation of PAM by ultracentrifugation. (a) original SV3T3 extract; (b) resuspended extract pellet in original volume of Veronal buffer following centrifugation at 100,000 x g for 1 hr; (c) supernatant following centrifugation; and (d) 3T3 extract pellet following centrifugation and resuspension of pellet in Veronal buffer. The difference in lag period between (a) and (b) is not significant.

fresh buffer prior to mixing with platelets in order to ensure that these enzymes or compounds did not affect platelets directly, making them unresponsive to PAM. Furthermore, platelet aggregation could be induced by addition of fresh PAM to PRP containing inactivated PAM. Therefore the effect was on PAM directly, not on platelet receptors for PAM.

PAM Synergism in Rabbit PRP

Dilutions of PAM were prepared which were below the threshold concentration required for platelet aggregation. Epinephrine, ADP, and collagen were also diluted to levels which were incapable of promoting platelet aggregation. By mixing these subthreshold concentrations of aggregating agents with dilute PAM (Fig. 4), we could demonstrate a synergistic effect between PAM and epinephrine at 1 $\mu g/ml$ and $10^{-5}M$, respectively, regardless of the order of addition. Neither ADP nor collagen proved to be synergistic with PAM when assayed in rabbit PRP (data not shown).

Table 5

EFFECT OF BOILING, ENZYMES, NON-IONIC DETERGENTS,
AND SONICATION ON PAM ACTIVITY

Treatment	Final Concentration or Time	Percent Aggregation	N*
Boiling	15 min	0	3
Boiling and storage at 0°C for 2 weeks	--	30-40	2
Trypsin-TPCK	1 mg/ml	0	5
Neuraminidase	2 mg/ml	0	5
Phospholipase-A_2	0.1 µg/ml	0	5
NP-40	0.1%	0	3
Tween-80	0.1%	0	3
Sonication	15 sec	0	2
β-Galactosidase	0.5 mg/ml	83	2

All enzymes, in 10 µl volume, were incubated with 100 µl of PAM, 100 µg/ml, for 1 hr at 37°C. PAM was then sedimented at 100,000 x g for 1 hr at 4°C, and the pellet was resuspended in 100 µl of Veronal buffer. Of this material, 50 µl were used for platelet-aggregation studies with heparinized rabbit PRP.
*N = number of experiments.

Effect of PAM on Washed Platelets

PAM did not aggregate washed platelets in the presence or absence of fibrinogen (a and b, Fig. 5) in most experiments (a small optical density change of 10% to 15% was occasionally noted in the presence of fibrinogen plus PAM). However, the washed human platelet system developed was responsive to platelet aggregation induced by ADP (c and d, Fig. 5) or epinephrine (a and b, Fig. 6), in the presence of purified human fibrinogen and to collagen (d and e, Fig. 6) in the absence or presence of fibrinogen. Of particular interest was the observation that PAM could replace the fibrinogen requirement for ADP-induced aggregation (e, Fig. 5) or epinephrine-induced aggregation (c, Fig. 6) but not for collagen-induced aggregation (f, Fig. 6).

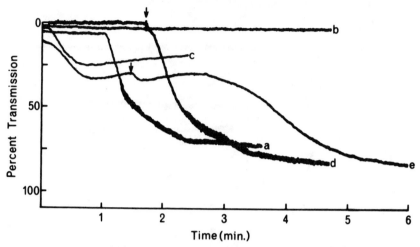

Figure 4. Synergistic effect of PAM and epinephrine on platelet aggregation in rabbit PRP. (a) 10 μg/ml PAM; (b) 1 μg/ml PAM; (c) 10^{-5}M epinephrine; (d) 1 μg/ml PAM followed by 10^{-5}M epinephrine; (e) 10^{-5}M epinephrine followed by 1 μg/ml PAM. Concentrations given are those achieved following dilution of the material in PRP during the assay. Initial additions were done at time zero. Second additions (in (d) and (e)) are indicated by arrows. Data taken from one of four experiments with similar results.

Requirement of a Plasma Factor

We have also observed that preincubation of PPP with gel-filtered (washed) platelets, prior to the addition of PAM, restores PAM's ability to induce platelet aggregation (Fig. 7). This was concentration-dependent. However, when boiled PPP was employed, no restoration of activity was observed (data not shown).

Shortening of the Lag Period

Further studies on the plasma factor have revealed that PAM interacts with a heat labile plasma factor during the lag period, to form a heat-stable product. This heat-stable product aggregates platelets with the absence of a lag period. Thus, incubation of PAM with plasma (1:1) at 37°C for 10 min

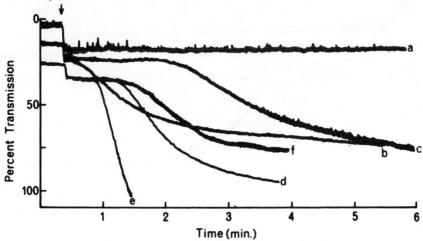

Figure 5. Effect of ADP, fibrinogen, and PAM on the aggregation of washed human platelets. Addition of PAM (10 μg/ml) in the (a) absence of, (b) presence of fibrinogen. Addition of ADP (2×10^{-5}M) in the (c) absence or (d) presence of fibrinogen (2 mg/ml) or (e) PAM. Initial incubations were started at time zero, and additions were made when indicated by arrow. Data taken from one of seven experiments with similar results.

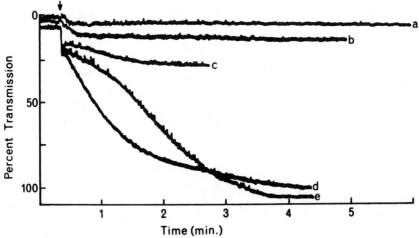

Figure 6. Effect of epinephrine, collagen, fibrinogen, and PAM on the aggregation of washed human platelets. Addition of epinephrine (10^{-5}M) in the (a) absence or (b) presence of fibrinogen or (c) PAM. Addition of collagen in the (d) absence or (e) presence of fibrinogen or (f) PAM. Initial incubations were started at time zero, and additions were made when indicated by arrow.

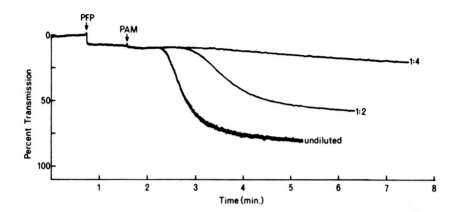

Figure 7. Aggregation of gel-filtered platelets in the presence of PPP and PAM. The addition of 50 μl of PPP (arrow) to 0.4 ml of gel-filtered platelets prior to the addition of 50 μl of PAM (arrow) reconstituted the aggregation-promoting activity of PAM. Dilution of PPP (1:2 and 1:4) with Veronal buffer resulted in a concentration-dependent loss of reconstitution activity. Data taken from one of three experiments with similar results.

produces an 'activated PAM' which abolishes the 1-2 min lag period noted in the absence of prior incubation, Fig. 8. Heating of this 'activated PAM' at 56°C for 30 min does not destroy PAM activity, whereas heating of plasma at 56°C for 30 min does destroy its ability to support PAM-induced aggregation.

Nature of the Heat-Labile Factor

The heat-labile factor(s) appears to be a protein(s) of the complement system, since treatment of PRP with cobra venom factor (Fig. 9) diminished the plasma factor activity in a time-dependent fashion. The ability of the plasma factor to support PAM aggregation of GFP was reduced to <20% of control levels by 45 min of incubation and essentially abolished by 75 min of incubation. Plasma from inbred strains of guinea pigs genetically deficient in the fourth component of comple-

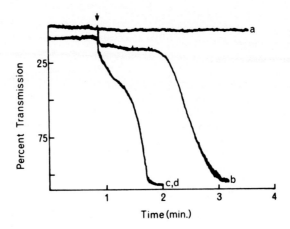

Figure 8. Aggregation of 0.4 ml of GFP with 50 μl of PAM (20 μg) after reconstitution at time zero with 50 μl of (a) PFP heated to 56°C for 30 min or (b) PFP. PAM was added at arrow, c and d, PFP was incubated 1:1 with PAM for 10 min at 37°C, with (c) or without (d) subsequent heating to 56°C for 30 min. Fifty microliters of this mixture was then added to the GFP.

ment was as capable of supporting PAM-induced aggregation of guinea pig platelets as normal guinea pig plasma (Fig. 10). This implies that the classical pathway of complement activation is not involved in PAM activity. Zymosan-treated PFP was incapable of supporting PAM-induced aggregation of GFP, strongly suggesting the requirement for an intact complement system to allow activation of the alternative pathway in the presence of platelets (Fig. 11).

Binding of PAM to Platelets by Employing PAM Coupled to Sepharose Beads

PAM-Sepharose beads can aggregate platelets in PRP as well as GFP in the presence of plasma. A 50 μl volume of a suspension of derivatized beads was added to 0.4 ml of PRP or 0.4 ml of GFP containing 50 μl of PFP. Data are presented in Fig. 12 for PRP. After platelet aggregation, binding of PAM to platelets can be visualized by the adherence of platelets and platelet clumps to the surface of PAM-coated beads (Fig. 12, A) but not to beads coated with bovine serum albumin

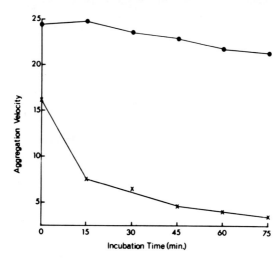

Figure 9. Inactivation of the aggregation-promoting activity of plasma by cobra venom. PRP was incubated with cobra venom for the indicated times. PAM was added immediately, and platelet aggregation was measured turbidometrically. Aggregation velocity was measured as the slope of a tangent drawn parallel to the steepest part of the aggregation curve and is plotted in arbitrary units of optical density change per unit time. Control PRP was incubated under identical conditions with an equivalent volume of buffer. X, cobra venom; ●, control.

or fibronectin (Fig. 12, B) or to underivatized beads which have been added to PRP and aggregated by exogenous PAM (Fig. 12, C).

Requirement of a Heat-Stable Plasma Factor

We have previously demonstrated that PAM activity is sedimentable at 100,000 x g for 1 hr (Pearlstein et al 1979). This physical property has enabled us to demonstrate that a multistep process occurs during PAM-plasma interaction. When PAM was activated by incubating it with PFP for 10 min at 37°C and subsequently subjected to centrifugation, platelet aggregating activity was recoverable in the pellet. However, the expression of this activity still required a plasma factor (Fig. 13). Plasma previously heated to 56°C will also restore PAM aggrega-

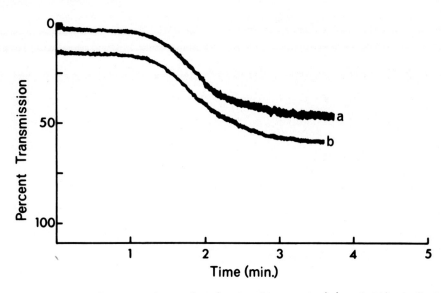

Figure 10. Aggregation of 0.4 ml of normal (a) of $C'4$-deficient (b) guinea pig PRP with 50 µl of PAM (20 µg). PAM was added at time zero.

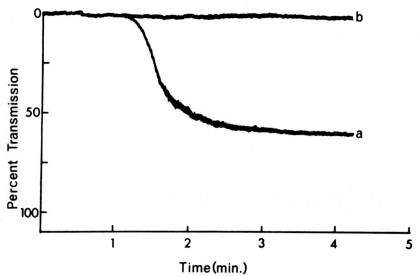

Figure 11. Aggregation of 0.4 ml of GFP with 50 µl of PAM (20 µg) in the presence of 50 µl of (a) normal plasma incubated at 37°C for 1 hr or (b) plasma treated with zymosan at 37°C for 1 hr.

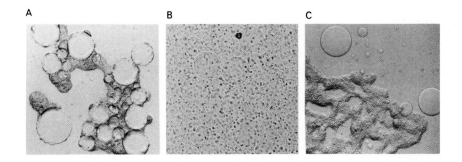

Figure 12. Aggregation of 0.4 ml of PRP with 50 μl of Sepharose 4-B beads coated with (A) PAM or (B) bovine serum albumin. (C) 50 μl of underivatized beads were added to PRP followed by 50 μl (20 μg) of PAM.

ability as long as the initial activation of PAM was induced with unheated plasma (Fig. 13, d). Rabbit fibrinogen (2 mg/ml) did not restore PAM aggregation ability (Fig. 13, e). The heat-stable factor was precipitable with 50% saturated ammonium sulfate and was stable to dialysis (data not shown).

Thus, a heat-labile factor is required to activate PAM and abolish the lag period and a heat-stable factor is required for the 'activated PAM' to be operative, Fig. 14. Preliminary studies indicate that the heat-labile factor is complement, since treatment of plasma with cobra venom abolishes its ability to activate PAM. Since C-4 deficient plasma retains the ability to activate PAM, it appears that the alternative complement pathway is involved. The requirement of complement for tumor cell-induced platelet aggregation has recently been reported by Gasic et al (1981).

Studies on PW20 Renal Cell Sarcoma Variant Cell Lines

Metastatic Properties. The percent of animals developing one or more spontaneous metastatic lesions following primary tumor excision ranged from 0-100%, with lesions appearing in the lung, mediastinal lymph nodes, abdominal viscera, and

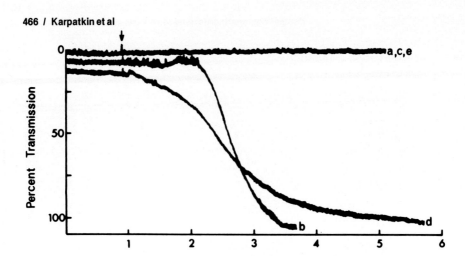

Figure 13. Requirement of a heat-stable plasma factor for PAM-induced platelet aggregation. PAM was activated by preincubation with PFP for 10 min at 37°C and then sedimented by centrifugation at 100,000 x g. The sediment was resuspended in Veronal buffer and compared to 'unactivated' fresh PAM. Aggregation of 0.4 ml of GFP with (a) 50 μl fresh PAM (20 μg); (b) 50 μl of fresh PAM in the presence of 50 μl of fresh plasma; (c) 50 μl of fresh PAM in the presence of 50 μl of 56°C-heated plasma; (d) 50 μl of activated PAM in the presence of 50 μl of 56°C-heated plasma; and (e) 50 μl of activated PAM in the presence of rabbit fibrinogen (2 mg/ml). Plasma was added at zero time and PAM was added at the arrow.

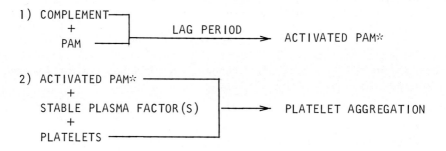

Figure 14. Proposed sequence of events during PAM-induced platelet aggregation.

external lymph nodes (Table 6).

Platelet Aggregation. The maximum average velocity of platelet aggregation induced by PAM ranged from 1.5 to 100 units of light transmission per minute. A significant correlation was observed between the platelet aggregating activity of PAM and the metastatic potential of the tumor cell lines from which the PAM was derived ($r = 0.68$, $p < 0.03$).

Sialic Acid Content of PAM. The sialic acid content of PAM ranged from 1.5 to 24.7 mg per 100 mg protein. PAM preparations from the three least metastatic cell lines, which were least active with respect to platelet aggregating ability, contained the lowest levels of sialic acid. A good correlation was observed between the platelet aggregating activity of PAM and its sialic acid content ($r = 0.60$, $p < 0.06$), and between the metastatic potential of the cell lines and the sialic content of PAM ($r = 0.69$, $p < 0.03$).

Sialylation of Cell Surface Glycoconjugates. The percent sialylation of exposed cell surface Gal and GalNAc residues ranged from 13-77%, with the three lowest values observed in the three least metastatic cell lines. A significant correlation was observed between cell surface sialylation and metastatic potential ($r = 0.83$, $p < 0.003$), cell surface sialylation and the platelet aggregating activity of PAM ($r = 0.85$, $p < 0.002$).

DISCUSSION

In agreement with Gasic and co-workers (Gasic et al 1973; Gasic et al 1976; Gasic et al 1977; Gasic et al 1978), we have demonstrated that viral transformants of established cell lines will cause platelet aggregation whereas the normal parental line is much less effective (Fig. 1). Indeed, it is conceivable that the low activity of the non-transformed cells could represent initial signs of spontaneous in vitro cell transformation. In addition, 1M urea extraction of plasma membrane components from the transformed cells permits this effect to be demonstrated by cell-free extracts (Fig. 2). The extraction of normal cells yields considerably less biologically active PAM. It is of interest that although the intact normal cells contained one-half the platelet aggregating activity of transformed cells, the urea extract of the normal

Table 6
METASTATIC PROPERTIES, PLATELET AGGREGATING ACTIVITY OF CELL SURFACE EXTRACTS (PAM), SIALIC ACID CONTENT OF PAM, AND SIALYLATION OF CELL SURFACE GLYCOCONJUGATES OF METASTATIC VARIANT PW20 TUMOR CELL LINES.

Derivative Line	Metastases from Primary Subcutaneous Tumors			Platelet Aggregating Material (PAM)		Cell Surface Sialylation
	Incidence [*]	Location [‡]	Percent	Velocity of Aggregation (Percent Transmission/min) [§]	Sialic Acid Content (Mg/100 Mg protein) [≡]	Percent Sialylation of Cell Surface Gal and GalNAc [⋈]
R2N1	13/13	T,A,N	100	73 ± 3	7.1	66 ± 7
R2L1	13/13	T,A,N	100	34 ± 5	16.4	77 ± 7
R1	45/49	T,A,N	92	70 ± 4	15.7	75 ± 3
R2	56/64	T,A,N	88	69 ± 5	24.7	73 ± 5
R2R2	5/7	T,A,-	71	9 ± 2	14.5	36 ± 2
RO(L)R1	4/6	T,-,N	67	63 ± 4	21.4	70 ± 6
RO(L)	6/9	T,A,N	67	100 ± 5	15.8	72 ± 4
R2R1	3/13	T,-,-	23	4 ± 1	5.4	13 ± 2
RO(H)	1/19	-,A,-	5	8 ± 1	5.4	31 ± 7
RO(H)R1	0/5	-,-,-	0	2 ± 1	1.5	10 ± 10

[*] Number of animals developing one or more metastases following primary tumor excision/number of animals surviving surgery.

[‡] T, intrathoracic (lung and/or mediastinal lymph nodes); A, intra-abdominal (mesenteric lymph nodes, kidney, adrenal gland, ovary and/or liver ; N, superficial lymph nodes (inguinal, axillary, cervical).

[§] Maximum average velocity of platelet aggregation induced by PAM tested at 3-5 concentrations ranging from 10-100 μg/ml. Mean ± SE of values obtained from 2-5 independent experiments.

[≡] Mean obtained from two independent determinations which were very similar.

[⋈] Mean ± SE of values obtained from 3-7 independent experiments.

cells contained 20-fold less PAM.

Both transformed cells and their cell-free extracts cause platelets in PRP to aggregate in a similar fashion: a lag period of 1 to 3 min followed by rapid platelet aggregation accompanied by the release reaction. The sensitivity of the tumor cell and the cell-free extract to trypsin was also similar. It is therefore likely that platelet aggregation induced by intact cells and extract is promoted by equivalent factors. PAM operates via the release reaction and requires divalent cation, since inhibitors of the secondary wave of platelet aggregation abolish PAM-induced aggregation of PRP (Table 2). PAM operates best in heparin-PRP rather than citrate-PRP and does not operate in the presence of EDTA (Table 2).

The mechanism of action of PAM-induced aggregation is unique because PAM aggregates PRP but does not aggregate washed platelets, which are capable of being aggregated by ADP or epinephrine in the presence of fibrinogen or by collagen in the absence of fibrinogen. However, PAM, like fibrinogen, does support ADP- and epinephrine-induced platelet aggregation and may operate via a mechanism similar to that of fibrinogen in the washed platelet system. It is conceivable that PAM has two activities or sites: one which aggregates platelets in PRP (requiring a plasma factor) and another which supports washed platelet aggregation in the absence of fibrinogen.

The plasma requirement is of interest (Karpatkin et al, 1980). A heat-labile factor (probably alternative pathway complement components) is necessary for the activation of PAM and the elimination of the lag period. PAM binds to platelets during its interaction, suggesting the requirement of physical contact prior to PAM-induced platelet aggregation. It is conceivable that activation of PAM is required for the binding of PAM to platelets. Indeed Gasic et al (1981) have recently demonstrated the requirement of complement for the binding of ^{125}I-labelled tumor vesicles to platelets. However, in our experiments it is clear that PAM* is not fully capable of inducing platelet aggregation unless a second stable plasma factor is added.

Many transformed cells have been shown to have increased levels of proteolytic enzymes (Ossowski et al, 1973; Goldberg, 1974) including the transformed lines used in these studies (Pearlstein et al, 1976). Since platelets are sensitive to aggregation by specific proteases (Davey, Luscher, 1966; Niewiarowski et al, 1973), we attempted to inhibit PAM-induced aggregation by protease inhibitors known to prevent plasminogen activation by these tumor cells (EACA and soybean trypsin inhibitor) as well as additional protease inhibitors. As shown in Table 4, none of these inhibitors prevented platelet aggregation with PAM. This finding is in agreement with a report (Gasic et al, 1976) which showed no correlation between the ability of tumor cells to induce platelet aggregation and their fibrinolytic activity in vitro.

We have demonstrated that PAM is a complex mixture of protein, lipid, and carbohydrate assembled in a macromolecular form capable of sedimentation at 100,000 × g for 1 hour. Destruction of PAM activity by trypsin, neuraminidase, or phospholipase-A_2 indicates that several components interact to produce PAM activity. This interaction may involve maintenance of a vesicular configuration, since sonication will also reduce PAM activity. Attempts to replace lipid by neutral detergents and maintain activity have been unsuccessful, and it appears that the lipid moiety may play more than a simple structural role in promoting platelet aggregation.

Since fibroblasts synthesize collagen and collagen can induce platelet aggregation in vitro (Zucker, Borelli, 1962), it was necessary to determine whether PAM activity may be related to this protein. Such does not appear to be the case because 1) treatment of tumor cells with collagenase does not interfere with their aggregation-promoting activity (Gasic et al, 1973); 2) PAM and collagen are not additive in inducing platelet aggregation; 3) transformation leads to a decrease in collagen biosynthesis (Green et al, 1966); 4) collagen is capable of aggregating washed platelets (d and e, Fig. 6) but PAM is not (Fig. 7); and 5) treatment of PAM with collagenase does not inhibit PAM-induced platelet aggregation (data not shown).

The involvement of sialic acid is interesting in light of a report demonstrating increased levels of sialic acid-containing glycolipids in serum samples obtained from mice bearing transplantable mammary carcinomas, compared to normal animals (Kloppel et al, 1977). The sialic acid components were shown to be present in the gangliosides. The sensitivity of PAM to neuraminidase but not to β-galactosidase indicates that PAM and the material present in the serum of tumor-bearing animals may be related.

Studies on PW20-Induced Renal Cell Sarcoma Variant Cell Lines

Because the SV403T3 fibroblast is not spontaneously metastatic in vivo, the pathophysiologic significance of PAM and its relationship to sialic acid was examined in 10 variant cell lines from a spontaneously metastatic PW20 renal cell sarcoma in Wistar-Furth rats. Significant correlations were observed between the metastatic potential of the cell lines, the ability of PAM to aggregate platelets, the sialic acid content of PAM and the degree of sialylation of the intact tumor cell surface. The correlation between the activity of PAM and its sialic acid content is in accord with our recent report that the activity of PAM obtained from SV3T3 cells is abolished by treatment with neuraminidase (Pearlstein et al, 1979) and suggests that the sialic acid content of PAM is important in determining its activity. The correlation between the sialic acid content of PAM and the degree of cell surface sialylation is consistent with the evidence suggesting that PAM is composed of membrane vesicles. Interestingly, the total sialic acid content of the cells (data not shown) correlates poorly with the other parameters shown in Table 6. This finding suggests that the degree of sialylation of the cell surface, which is reflected in the sialic acid content of the extracted PAM, may be a more important determinant of metastatic behavior and platelet aggregating properties than the overall sialic acid content of the cell (Pearlstein et al,1980).

One cell line (R2R2) appears to be an exception in that it shows relatively low values of PAM activity and cell surface sialylation for its degree of metastasis. Of interest is the observation that this cell line demonstrated a transient incomplete regression of the initially formed tumors, in vivo, followed by delayed progressive growth prior to excision. It is therefore possible that a further immunoselection of highly sialylated cells (Salk, Yogeeswaran,1978) with enhanced metastatic potential may have taken place from a heterogenous population of predominantly non-metastatic poorly sialylated R2R2 cells. In this regard, the reduced in vitro adhesiveness and motility of R2R2 cells is also characteristic of the poorly metastatic cell line (Salk, Lanza,1978). If this line were eliminated from the correlation, it would raise the correlation coefficients by an average of 0.06.

The correlation between the spontaneous metastatic behavior of murine tumor cells and their cell surface sialylation in vitro has recently been noted (Salk, Yogeeswaran,1978;

Yogeeswaran et al, 1979) and has now been observed in a total
of 32 murine tumor cell lines (data not shown). One mechanism
which might be proposed to account for the observed correlation is that elevated levels of sialic acid on the cell surface may reduce the immunogenicity of the tumor cells
(Ray, 1977; Salk, Yogeeswaran, 1978), thereby increasing their
ability to evade immune destruction and establish metastatic
foci (Pimm, Baldwin, 1978). Differences in cell surface sialic
acid may also bring about differences in the adhesive properties of the cells (Weiss, 1963; Salk, Lanza, 1978), thereby
altering their patterns of arrest and implantation at secondary
sites (Sinha, Goldenberg, 1974; Weiss et al, 1974). Another
mechanism, which is suggested by the results of the present
study, is that increased sialylation of the tumor cell surface
may enhance the ability of these cells to aggregate platelets,
either prior to or after their arrest in the microvasculature
(Jones et al 1971; Warren, Vales, 1972; Sindelar et al, 1975).
The resulting aggregates may prolong the survival of the tumor
cells at the site of arrest (Jones et al, 1971; Warren, Vales,
1972; Sindelar et al, 1975). The release of vasoactive substances from the aggregated platelets (Nachman et al, 1972) may
facilitate passage of the tumor cells through the vessel wall
by promoting endothelial cell damage and enhanced cell adhesion
to the subendothelial matrix.

The mechanisms of tumor cell metastasis are inherently
complex (Roos, Dingemens, 1979). The extent to which platelets
may be involved in promoting tumor cell metastasis remains to
be determined. The variant cell lines employed in this study
provide an excellent opportunity to study the role of platelets
in tumor cell metastasis. Our data are consistent with the
hypothesis that platelets play a role in certain tumor cell
metastases and lends support to the therapeutic possibility
that certain tumor metastases may be reduced by agents capable
of eradicating that property of platelets which promotes
metastasis.

REFERENCES

Davey MG, Luscher EF (1966). The action of thrombin on platelet proteins. Thromb Diath Haemorrh Suppl 20:283.
Gasic GJ, Boettiger D, Catalfamo JL, Gasic TB, Stewart GJ
(1978). Aggregation of platelets and cell membrane vesiculation by rat cells transformed in vitro by Rous sarcoma
virus. Cancer Res 38:2950.

Gasic GJ, Catalfamo JL, Gasic TB, Avdalovic N (1981). In vitro mechanism of platelet aggregation by purified plasma membrane vesicles shed by mouse 15091A tumor cells. In Donati MD (ed): "Malignancy and the Hemostatic System", New York: Raven Press, p 27.

Gasic GJ, Gasic TB, Galanti N, Johnson T, Murphy S (1973). Platelet-tumour-cell interactions in mice. The role of platelets in the spread of malignant disease. Int J Cancer 11:704.

Gasic GJ, Gasic TB, Jimenez SA (1977). Effects of trypsin on the platelet aggregating activity of mouse tumor cells. Thromb Res 10:33.

Gasic GJ, Gasic TB, Stewart CC (1968). Antimetastatic effects associated with platelet reduction. Proc Natl Acad Sci 61:46.

Gasic GJ, Koch PAG, Hsu B, Gasic TB, Niewiarowski S (1976). Thrombogenic activity of mouse and human tumors: effect of platelets, coagulation and fibrinolysis, and possible significance for metastases. Z Krebsforsch 86:263.

Gastpar H (1977). Platelet-cancer cell interaction in metastasis formation: a possible therapeutic approach to metastasis prophylaxis. J Med 8:103.

Goldberg AR (1974). Increased protease levels in transformed cells: a case in overlay assay for the detection of plasminogen activator production. Cell 2:95.

Green H, Goldberg B, Todaro GJ (1966). Differentiated cell types and the regulation of collagen synthesis. Nature 212:631.

Hagmar B, Boeryd B (1969). Disseminating effect of heparin on experimental tumour metastases. Pathol Eur 4:274.

Hilgard P (1973). The role of blood platelets in experimental metastases. Br J Cancer 28:429.

Hilgard P, Gordon-Smith EC (1974) Microangiopathic haemolytic anaemia and experimental tumour-cell emboli. Br J Haematol 26:651.

Hynes RO, Pearlstein E (1976). Investigations of the possible role of proteases in altering surface proteins of virally transformed hamster fibroblasts. J Supramolec Struct 4:1.

Jones DS, Wallace AC, Fraser EF (1971). Sequence of events in experimental metastases of Walker 256 tumor: light, immunofluorescent and electron microscopic observations. J Natl Cancer Inst 46:493.

Karpatkin S (1978). Heterogeneity of human platelets. VI. Correlation of platelet function with platelet volume. Blood 51:307.

Karpatkin S, Smerling A, Pearlstein E (1980). Plasma requirement for the aggregation of rabbit platelets by an aggregating material derived from SV40-transformed 3T3 fibroblasts. J Lab Clin Med 96:994.

Kloppel TM, Keenan TW, Freeman MJ, Morre DJ (1977). Glycolipid-bound sialic acid in serum: increased levels in mice and humans bearing mammary carcinomas. Proc Natl Acad Sci 74:3011.

Lowry OH, Rosebrough NJ, Farr AL, Randall RJ (1951). Protein measurement with the folin phenol reagent. J Biol Chem 193:265.

Mustard JF, Perry DW, Ardlie NG, Packham MA (1972). Preparation of suspensions of washed platelets from humans. Br J Haematol 22:193.

Nachman RL, Weksler B, Ferris B (1972). Characterization of human platelet vascular permeability-enhancing activity. J Clin Invest 51:549.

Niewiarowski S, Senyi AF, Gilles P (1973). Plasmin-induced platelet aggregation and platelet release reaction. Effects on hemostasis. J Clin Invest 52:1647.

Ossowski L, Unkeless JC, Tobia A, Quigley JP, Rifkin DB, Reich E (1973). An enzymatic function associated with transformation of fibroblasts by oncogenic viruses. II. Mammalian fibroblast cultures transformed by DNA and RNA tumor viruses. J Exp Med 137:112.

Pearlstein E, Cooper LB, Karpatkin S (1979). Extraction and characterization of a platelet aggregating material from SV40-transformed mouse 3T3 fibroblasts. J Lab Clin Med 93:332.

Pearlstein E, Hynes RO, Franks LM, Hemmings VJ (1976). Surface proteins and fibrinolytic activity. Cancer Res 36:1475.

Pearlstein E, Salk PL, Yogeeswaran G, Karpatkin S (1980). Correlation between spontaneous metastatic potential, platelet aggregating activity of cell surface extracts, and cell surface sialylation in 10 metastatic-variant derivatives of a rat renal sarcoma cell line. Proc Natl Acad Sci USA 77:4336.

Pearlstein E, Seaver J (1976). Non-lytic and non-ionic detergent extraction of plasma membrane constituents from normal and transformed fibroblasts. Biochim Biophys Acta 426:589.

Pimm MV, Baldwin RW (1978). In Baldwin RW (ed) "Secondary Spread of Cancer. Immunology and Immunotherapy of Experimental and Clinical Metastases", New York: Academic Press, p 163.

Ray PK (1977). Bacterial neuraminidase and altered immunological behavior of treated mammalian cells. Adv Appl Microbiol 21:227.

Roos E, Dingemans KP (1979). Mechanisms of metastasis. Biochim Biophys Acta 560:135.

Salk PL, Lanza RP (1978). In vitro growth characteristics, motility and adhesive properties of metastatic variant PW20 cell line. J Supramol Struct 9 (Suppl 3):446.

Salk PL, Yogeeswaran G (1978). Metastatic properties, immunogenicity and cell surface sialylation of tumor cells derived from a polyoma-induced rat renal sarcoma. Fed Proc 37:2686A.

Sindelar WF, Tralka TS, Ketcham AS (1975). Electron microscopic observations on formation of pulmonary metastases. J Surg Res 18:137.

Sinha BK, Goldenberg GJ (1974). The effect of trypsin and neuraminidase on the circulation and organ distribution of tumor cells. Cancer 34:1956.

Tangen O, Berman HJ (1972). Gel filtration of blood platelets: a methodological report. Adv Exp Med Biol 34:235.

Warren BA (1973). Environment of the blood-borne tumor embolus adherent to vessel wall. J Med 4:150.

Warren BA, Vales O (1972). The adhesion of thromboplastic tumor emboli to vessel walls in vivo. Brit J Exp Pathol 53:301.

Warren L (1959). The thiobarbituric acid assay of sialic acids. J Biol Chem 234:1971.

Weiss HJ (1975). Physiology and abnormalities of platelet function. N Engl J Med 293:531.

Weiss L (1963). Studies on cellular adhesion in tissue culture. V. Some effects of enzymes on cell detachment. Exp Cell Res 30:509.

Weiss L, Glaves D. Waite DA (1974). The influence of host immunity on the arrest of circulating cancer cells, and its modification by neuraminidase. Int J Cancer 13:850.

Wexler H (1966). Accurate identification of experimental pulmonary metastases. J Nat Cancer Inst 36:641.

Yogeeswaran G, Sebastian H, Stein BS (1979). Cell surface sialylation of glycoproteins and glycosphingolipids in cultured metastatic variant RNA-virus transformed nonproducer Balb/c 3T3 cell lines. Int J Cancer 24:193.

Zucker MB, Borelli J (1962). Platelet clumping produced by connective tissue suspensions and by collagen. Proc Soc Exp Biol Med 109:779.

DISCUSSION

Dr. G. J. Gasic (University of Pennsylvania, Philadelphia). How do you differentiate the complement effects?

Dr. S. Karpatkin (New York University Medical Center, New York). We have other data which indicate that you can shorten the lag time by pre-incubating PAM with plasma but not if the plasma has been heated at 56º. If you heat the plasma with zymosan or cobra venom factor you lose the shortening of the lag period. The activated PAM alone does not work only if you add plasma back. Even if you take that plasma and heat it at 56° it still works: that is the stable factor.

Dr. G. A. Jamieson (American Red Cross, Blood Services, Bethesda). Did you get PAM from all your lines?

Dr. Karpatkin. We obtained PAM from the two virally transformed lines the PW20 and 3T3. We tried getting it from lines which work by thrombin generation but were unsuccessful. These are spontaneous lines so it could be that there is a difference between spontaneous and virally-transformed. We have one animal cell line which appears to work by thrombin and we could not extract PAM from it.

Dr. Jamieson. My other question was whether the cultures of the rat renal sarcoma line synchronous since there is evidence of changes in the degree of sialylation on the surface at different phases.

Dr. Karpatkin. Yes, someone pointed that out to me at some other meeting. We checked back and found that we have isolated all of these cells at the same stage in the cycle, just before confluency.

Dr. K. Brunson (Indiana University, Gary). Have you tried to remove sialic acid from the metastatic variants and does that then change platelet aggregation or metastatic potential?

Dr. Karpatkin. We have not tried that.

Dr. Jamieson. I would like to say that in some of our preliminary work, we have done that and it increases the degree of aggregation in three of our lines.

Dr. A. Ordinas (Universidad de Barcelona, Spain). How does PAM affect aggregation?

Dr. Karpatkin. I think PAM essentially represents the tumor cell membrane. I think sialic acid has something to do with binding.

Dr. A. Scriabine (Miles Institute for Preclinical Pharmacology, New Haven). The requirement for divalent ions?

Dr. Karpatkin. We have just tried calcium, and nothing else.

Unidentified. The requirement for divalent ions may stem from a high sialic acid content on the cell membrane. Has anyone ever looked at glycosyltransferases on the cell membrane from platelets?

Dr. Jamieson. We looked at glycosyltransferases on the membrane of platelets but not in relation to the tumor cell-induced aggregation. Dr. Scialla, who was here before, has done work on sialyltransferase levels in platelets from patients with adenocarcinoma and found that there is an increase.

Dr. Karpatkin. Glycoconjugates and sialic acid-containing carbohydrates are higher in animals which have tumor metastasis, and also in tumor cells compared to the original cell. There appears to be some evidence that tumorigeneticity is associated with sialic acid or carbohydrate.

INHIBITION OF THE PLATELET-AGGREGATING ACTIVITY OF TWO HUMAN ADENOCARCINOMAS OF THE COLON AND AN ANAPLASTIC MURINE TUMOR WITH A SPECIFIC THROMBIN INHIBITOR: DANSYLARGININE N-(3-ETHYL-1,5-PENTANEDIYL)AMIDE

Edward Pearlstein,[1] Cynthia Ambrogio,[1,3] Gabriel Gasic[2] and Simon Karpatkin[3]

Departments of Medicine[1,3] and Pathology[1] and Irvington House Institute,[1] New York University Medical School, New York, NY 10016.
On sabbatical from the Department of Medicine, University of Pennsylvania[2]

SUMMARY

Platelets are required for certain experimental metastases. Several lines of animal tumor cells aggregate platelets in vitro and in vivo. Previous studies with one of these lines, an SV40 transformed 3T3 mouse fibroblast (J. Lab. Clin. Med. 93:332-44, 1979; 96:994-1001, 1980) have revealed that the platelet aggregating material (PAM) is an extractable membrane-associated sialolipoprotein which requires divalent cation, complement, and a heat-stable plasma component for activity. Little information is available on the interaction of human tumors with platelets. We now report on the ability of two human adenocarcinomas of the colon (LoVo and HCT-8) and an anaplastic mouse tumor (Hut-20) to aggregate platelets by a different mechanism, the generation of thrombin. These spontaneous cell lines aggregate human or rabbit platelet-rich plasma after a 1-2 minute lag period. This is often followed by a visible clot. Unlike SV3T3 cells, aggregation by LoVo, HCT-8, and Hut-20 cells is not inhibited by neuraminidase, trypsin, or cobra venom factor. These three cell lines markedly shorten the recalcification time of citrated plasma whereas SV3T3 cells do not. Phospholipase A_2 treatment inhibits the shortening of the recalcification time for the three tumors; this parallels its inhibitory effect on platelet aggregation. LoVo, HCT-8, and Hut-20 cells generate thrombin via the 'tissue factor' coagulation pathway (employing coagulation factor-deficient substrates). Dansylarginine-N-(3-ethyl-1,5-pentanediyl)amide (DAPA), a highly specific potent anti-thrombin antagonist, inhibits LoVo, HCT-8 and Hut-20-induced platelet aggregation at 4-15 µM, whereas its effect on SV3T3 cells is negligible. If platelets are required for certain human tumor metastases, DAPA, or other anti-thrombin agents, may prove to be valuable therapeutic agents. In addition, we have established that PRP from individual human donors demonstrates a variability in their platelet-aggregation response to different tumor cells. HCT-8 and LoVo aggregated the PRP of 8 of 8 human volunteers tested, Hut-20 cells aggregated approximately 60% of 18 subjects, and HM-29, a human melanoma (whose mechanism of action is yet to be elucidated), aggregated 3 of 28 volunteers. Preliminary evidence indicates that the ability of individuals to respond to HM-29 cells may be linked to the HLA allotype DrW5.

INTRODUCTION

A role for platelets in the hematogenous dessemination of animal tumors has been suggested by many studies (Gasic et al 1968; Jones et al 1971; Warren, Vales 1972; Gasic et al 1973; Hilgard 1973; Warren 1973; Hilgard, Gordon-Smith 1974; Sindelar et al 1975; Gasic et al 1976; Gasic et al 1977a; Gasic et al 1977b; gasic et al 1978; Fitzpatrick, Stringfellow 1979; Pearlstein et al 1979; Hara et al 1980; Karpatkin et al 1980; Marcum et al 1980; Pearlstein et al 1980). Certain tumor cells require platelets for the development of metastases (Gasic et al 1968; Gasic et al 1973; Hilgard 1973; Pearlstein et al 1980). Ultrastructural studies have demonstrated arrested tumor emboli surrounded by platelets (Jones et al 1971; Warren, Vales 1972; Hilgard, Gordon-Smith 1974; Sindelar et al 1975). Several tumor cells induce thrombocytopenia in vivo (Gasic et al 1968; Gasic et al 1973; Hilgard 1973) and aggregate platelets in vitro (Gasic et al 1973; Gasic et al 1976; Gasic et al 1977a; Gasic et al 1977b; Gasic et al 1978; Fitzpatrick, Stringfellow 1979; Pearlstein et al 1979; Hara et al 1980; Karpatkin et al 1980; Marcum et al 1980; Pearlstein et al 1980). A correlation exists between the ability of some tumor cells to aggregate platelets in vitro and their requirement for metastases (Gasic et al 1973; Pearlstein et al 1980).

We have recently studied two virally-transformed animal cell lines (Pearlstein 1979; Karpatkin et al 1980; Pearlstein et al 1980). An SV40 transformed 3T3 mouse fibroblast line was shown to have an extractable, membrane-associated platelet aggregating material (PAM) containing a sialolipoprotein which aggregates platelets following a 1-2 min lag period. The reaction requires Ca^{++}, is associated with release of serotonin and is inhibited by adenosine or indomethacin (Pearlstein et al, 1979). Two plasma factors are also required: a heat-labile complement component and a 56°C stable plasma component (Karpatkin et al, 1980). Ten metastatic-variant derivatives of a polyoma-induced PW20 renal sarcoma of Wistar-Furth rats have also been studied in vivo as well as in vitro (Pearlstein et al, 1980). They have an extractable PAM which appears to require membrane sialic acid for platelet aggregation. Excellent correlations were obtained for metastatic potential, surface sialic acid, and PAM.

The purpose of this investigation was to study whether the mechanism of action of human tumors was the same as that described for the animal tumors. Two human adenocarcinomas of the colon, (HCT-8 and LoVo) and one anaplastic murine tumor (Hut-20) were studied. In contrast to our studies with two virally-transformed lines, the three spontaneous cell lines appear to operate via the generation of thrombin and do not require trypsin-sensitive protein or neuraminidase-digestible surface sialic acid for platelet aggregation. Of particular interest is the observation that platelet aggregation induced by all three spontaneous tumors can be inhibited by a highly specific synthetic thrombin inhibitor, Dansylarginine-N-(3-ethyl-1,5-pentanediyl)amide (DAPA) (Hijikata et al 1976; Nesheim, Prendergast 1979) at approximately 10^{-6}M.

MATERIALS AND METHODS

Cell Cultures

All cell lines were grown in the presence of 2 mM glutamine, 100U/ml penicillin, and 100 µg/ml streptomycin. All tissue culture supplies were obtained from Gibco Laboratories, Grand Island, NY.

An SV40 virus-transformed Balb C/3T3 fibroblast cell line (SV3T3) was maintained in Dulbecco's modified Eagle's medium (E4) supplemented with 10% calf serum, as described previously (Pearlstein et al, 1979).

An anaplastic murine tumor cell line, Hut-20, was obtained through the courtesy of Dr. Addi Gazdar of the Veterans Administration Hospital, Washington, DC. The line was grown in RPMI-1640 containing 10% Bobby calf serum.

A human adenocarcinoma of the colon, HCT-8, was obtained through the courtesy of Dr. Edward Cadman, Yale University School of Medicine, New Haven, Conn., and was maintained in culture in RPMI-1640 containing 10% fetal bovine serum.

Another human adenocarcinoma of the colon, LoVo, was provided by Dr. Benjamin Drewinko of the M.D. Anderson Hospital, Houston, Texas, and was grown in McCoy's 5a medium containing 15% fetal bovine serum.

A human melanoma line, HM-29, was obtained from Dr. Jean-Claude Bystryn of the NYU Medical Center, and grown in RPMI-1640 containing 10% fetal bovine serum.

Cells were harvested using 5 mM EGTA in Hank's buffered salt solution, for 10 min on ice, or with trypsin treatment (0.5 mg/ml trypsin-EDTA) for 3 min at 37°C. Harvested cells were washed and suspended in Veronal buffer (0.03 M sodium barbital, 0.12 M NaCl, pH 7.4) prior to use. Cells were quantitated by counting in a hemocytometer and viability determined by trypan-blue exclusion.

The species origin of all lines was verified by karyotyping through the courtesy of Dr. M. Smith, Mt. Sinai Medical School, NY.

Platelet-rich Plasma (PRP)

Human or animal blood was anticoagulated with heparin (Liquaemin-Organon Inc., West Orange, NJ) final concentration, 5 U/ml. PRP was prepared by centrifugation at 150 g for 5 min at room temperature. Platelet-poor plasma (PPP) was prepared from the remaining blood by additional centrifugation at 2000 g for 15 min. PRP was incubated at room temperature for 30 min in tightly-capped plastic tubes prior to use. Gel-filtered platelets were prepared from PRP as described previously (Karpatkin et al 1980).

Platelet Aggregation

Aggregometry was performed in a Bio-Data aggregometer (Bio-Data, Willow Grove, PA) as previously described (Pearlstein et al, 1979; Karpatkin et al, 1980; Pearlstein et al, 1980).

Recalcification Time

Blood was anticoagulated with sodium citrate, 0.38% final concentration, and PPP prepared as described above. 0.1 ml of cell suspension was incubated with 0.1 ml of 50 mM $CaCl_2$ for 2 min at 37°C in a plastic tube. The reaction was started by the addition of 0.1 ml of platelet-poor plasma, and the clotting time recorded in seconds (monitored by inversion of the tube, every 30 seconds). Recalcification times were also performed with human plasmas deficient for coagulation factors XII, IX, X, II, V, and VII (George King Biologicals, Overland Park, KA).

Enzyme Treatment

Tumor cells removed from tissue culture dishes with EGTA were washed, suspended in Veronal buffer, and incubated with either 1U/ml neuraminidase (Type V, Sigma Chem. Co., St. Louis, MO) plus 2 mM PMSF, or boiled (10 min) phospholipase A_2, 25 μg/ml (bee venom, Sigma) for 1 hour at 37°C. The cells were then washed in Veronal buffer and resuspended to the proper cell count prior to utilization in aggregation and recalcification time studies. For trypsin treatment, cells in monolayer were treated with the enzyme (1.5 mg/ml) for 1 hour at 37°C, floating cells aspirated, and washed three times with Veronal buffer before use.

Treatment with Complement Inactivators

40-200 microliters of purified cobra venom (100 U/ml) from Naja naja Kaouthia (Cordis Laboratories, Miami, FL) was incubated with 0.4 ml of PRP for various lengths of time at 37°C, followed by addition of cell suspension. Cobra venom factor did not affect platelet aggregation induced in PRP with 10^{-6}M ADP. Furthermore, platelets isolated from cobra venom factor-treated PRP by gel filtration (Tangen, Berman 1972) were responsive to aggregation by PAM in the presence of fresh PPP.

Washed zymosan A (Sigma) was resuspended in plasma at a

final concentration of 2 mg/ml, incubated for 1 hour at 37°C, and removed by centrifugation at 1000 g for 5 min at 4°C. The supernatant was assayed for its ability to support platelet aggregation.

Dansylarginine-N-(3-ethyl-1,5-pentanediyl)amide (DAPA) was synthesized (Nesheim et al 1979) and kindly supplied by Dr. Kenneth Mann, Mayo Foundation, Rochester, MN. This synthetic inhibitor of thrombin has a Ki for fibrinogen of 10^{-7}M.

Animals

Heparinized blood was collected from female New Zealand White rabbits by cannulation of the ear artery and from C57 black mice and Wistar-Furth rats by cardiac puncture. PRP and PPP were prepared from these samples as described for human plasma.

RESULTS

Platelet Aggregation by LoVo, HCT-8, or Hut-20 Cells Is Independent of Complement

To facilitate comparisons between human lines and SV3T3-induced aggregation, rabbit PRP was often employed since SV3T3 cells do not aggregate human PRP and rabbit PRP responded in a similar fashion as responder human PRP whenever compared.

We have previously demonstrated that SV3T3 PAM-induced aggregation is impaired and the lag period prolonged (Karpatkin et al, 1980) when rabbit complement is inactivated with cobra venom factor or zymosan A. No such effect was noted when LoVo, HCT-8, or Hut-20 cells (1×10^5) were used as the aggregating agent, employing either rabbit PRP (Figure 1) or human PRP (data not shown).

Treatment of LoVo, HCT-8, Hut-20 and SV3T3 Tumor Cells With Enzymes

Platelet Aggregation. In previous studies (Pearlstein et al 1979) we noted that the PAM extract of the animal tumor cell line SV3T3 was sensitive to treatment with trypsin, neuraminidase or phospholipase-A. These experiments were repeated with intact SV3T3 cells and compared to LoVo, HCT-8, and Hut-20 cells, Figure 2, since PAM can not be prepared from the three non-virally transformed cell lines. Although SV3T3 cells lost their platelet aggregability following treatment with trypsin, neuraminidase or phospholipase-A_2, the LoVo, HCT-8, and Hut-20 cells were only affected by phospholipase-A_2 treatment (data not shown for Hut-20 cells). When compared to SV3T3 cells the sensitivity to phospholipase-A_2 was 25-fold greater for Hut-20 cells, and similar for LoVo and HCT-8 cells.

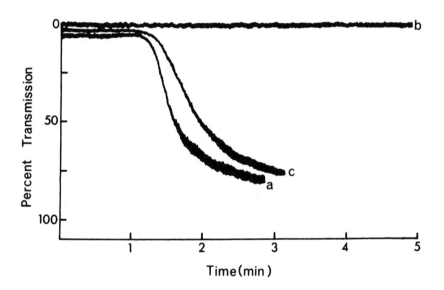

Figure 1. Aggregation of rabbit PRP with SV3T3, LoVo, HCT-8, and Hut-20 cells with or without treatment with cobra venom. Aggregation induced by SV3T3 in (a) control or (b) cobra venom treated PRP; (c) aggregation by LoVo, HCT-8, or Hut-20 cells in control or cobra venom treated PRP; these curves were superimposable.

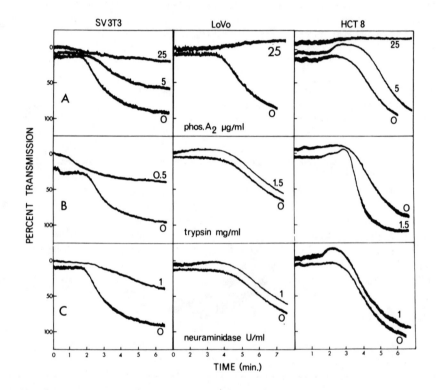

Figure 2. Platelet aggregation induced by enzyme-treated tumor cells. Tumor cells were exposed to either row (A) phospholipase-A_2; (B) trypsin; or (C) neuraminidase for 1 hr at 37°C, and washed with Veronal buffer, pH 7.4. 0.5×10^6 cells in 50 µl were added to 0.4 ml of heparinized rabbit PRP and percent transmission monitored with a Bio-Data platelet aggregometer. 0- control, untreated cells incubated in buffer for 1 hour at 37°C.

Recalcification Time. Because clot formation was occasionally noted 10-15 min after platelet aggregation with the three spontaneous tumor cells, but not with SV3T3 cells, thrombin generation was measured, employing a recalcification time. LoVo, HCT-8 and Hut-20 cells significantly shortened the recalcification time of citrated rabbit PPP. LoVo was more potent than HCT-8 or Hut-20 cells. Similar results were obtained with human PPP. This was dependent on cell concentration, Table 1. Similar results were obtained with PRP (data not shown). SV3T3 cells had a negligable effect on the recalcification time. For example, 1×10^5 cells had no effect; 1×10^6 cells shortened the recalcification time from 10.0 min to 7.8 min.

Table 1 also demonstrates the effect of trypsin, neuraminidase or phospholipase-A_2 on the ability of tumor cells to shorten the recalcification time of rabbit PPP. As with platelet aggregation results, trypsin and neuraminidase had no effect on the recalcification time whereas phospholipase-A_2 did have an effect.

Table 2 demonstrates the effect of tumor cells on the recalcification times of coagulation factor-deficient human plasmas. LoVo, HCT-8 and Hut-20 cell-induced shortening of the recalcification time was noted with factor XII and IX deficient plasma, but not with factor II, V, X or VII deficient plasma. The data suggest that these cell lines were activating the coagulation system via activation of factor VII by the extrinsic tissue pathway.

Effect of DAPA, a Specific Thrombin Inhibitor

The ability of human tumor cells to aggregate platelets via the generation of thrombin was examined by the use of a highly specific thrombin inhibitor, DAPA. Figure 3 demonstrates complete inhibition of tumor-induced rabbit platelet aggregation by DAPA at 4-15 μM, employing 5×10^5 LoVo, HCT-8 and Hut-20 cells. Similar results were obtained with human PRP. At higher concentrations, DAPA had an effect on SV3T3 cells. However, 20-fold fewer cells and 30-fold greater DAPA concentration was required to demonstrate this effect.

Variability of Human PRP to Tumor Cell-induced Aggregation

The Hut-20 murine cell line aggregated the heparinized PRP of approximately 60% of 18 human volunteer donors tested

Table 1

EFFECT OF CELL CONCENTRATION AND ENZYME TREATMENT ON RECALCIFICATION TIME OF RABBIT PLATELET-POOR PLASMA BY LoVo, HCT-8, AND Hut-20 CELLS*

	Recalcification Time (minutes)											
	Buffer			Neuraminidase			Phospholipase-A$_2$			Trypsin		
Cell Number	LoVo	HCT-8	Hut-20	LoVo	HCT-8	Hut-20	LoVo	HCT-8	Hut-20	LoVo	HCT-8	Hut-20
0	18.0	18.0	18.0	18.5	18.5	18.5	20.0	20.0	20.0	19.0	19.0	19.0
1 x 10^3	3.0	11.5	7.5									
1 x 10^4	1.5	4.5	5.0	1.5			3.5			1.0		
1 x 10^5		1.4	3.5									
5 x 10^5			2.4			2.3			>25.0			2.8
1 x 10^6			1.5									
1 x 10^6	0.4				0.4			10.0			0.5	

*0.1 ml of cells previously treated with buffer or the indicated enzymes for 1 hr at 37°C were washed and then incubated with 0.1 ml of 50 mM CaCl$_2$ for 2 min at 37°C in a plastic tube. 0.1 ml of PPP was added and the clotting time recorded.

Table 2

EFFECT OF LoVo, HCT-8, AND Hut-20 CELLS ON THE
RECALCIFICATION TIME OF COAGULATION FACTOR-DEFICIENT PLASMAS*

Test Plasma	Clotting Time (minutes)			
	Buffer	LoVo	HCT-8	Hut-20
Normal	36.3	1.3	3.2	3.5
XII-def.	>60.0	1.3	3.3	4.5
IX-def.	>60.0	1.2	4.0	7.0
V-def.	>60.0	N.D.	N.D.	40.0
VII-def.	26.0	7.0	23.0	31.0
X-def.	>60.0	3.8	19.0	19.0
II-def.	>60.0	>60.0	>60.0	>60.0

*0.1 ml of cell suspension (1×10^5 cells) in Veronal buffer was incubated with 0.1 ml of $CaCl_2$ for 2 min at 37°C in a plastic tube. The reaction was started by addition of 0.1 ml test plasma and the clotting time recorded.
Representations of 2-5 experiments
N.D. = not determined

(Table 3), whereas it aggregated 100% of numerous rabbit, rat, or mouse donors. The PRP of negative individuals was simultaneously tested with ADP or collagen to insure the presence of reactive platelets. Table 3 indicates that certain human volunteers (A,C,D,E) were persistently responsive to Hut-20 cells while subject (F) was negative on 6 of 7 occasions. Other donors tested less frequently were positive or negative. The variability in response was not due to tumor cell concentration. Increasing the number of tumor cells added, from 1×10^6 to 1×10^7 cells, did not convert a negative responder to a positive one (data not shown). Neither did the response appear to be due to heparin concentration, since collection of blood into 5, 3, or 1 unit of heparin/ml did not convert a negative responder to a positive one.

In contrast to the results with Hut-20, HCT-8 and LoVo induced aggregation in 8 of 8 individuals tested. With the human melanoma cell line HM-29, approximately 10%, or 3 of 28 human donors responded. Two of these donors were positive on 3 of 3 occasions; the third on 2 of 2 occasions. The HLA

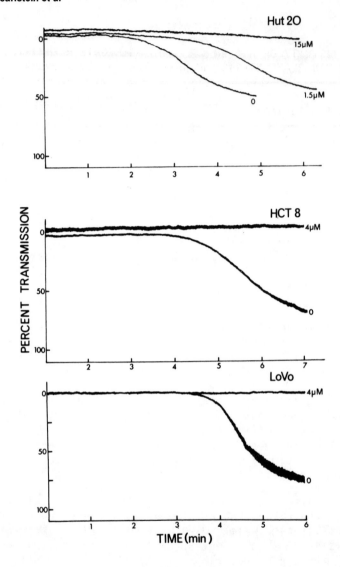

Figure 3. Effect of DAPA on platelet aggregation induced by Hut-20, HCT-8, or LoVo cells. 0.35 ml of heparinized rabbit PRP was preincubated with 50 μl of DAPA in Veronal buffer, at 37°C for 3 min prior to the addition of 50 μl of 5×10^5 tumor cells. Concentration of DAPA is given in μM.

Table 3

THE ABILITY OF Hut-20 CELLS TO AGGREGATE HEPARINIZED
HUMAN PLATELET-RICH PLASMA FROM VARIOUS DONORS*

Subject	Response Positive	Negative
A	14	0
B	5	2
C	4	0
D	3	0
E	3	0
F	1	6
G	0	3
H	1	2
I	1	1
J	1	1
K	0	1
L	0	1
M	0	1
N	0	1
O	0	1
P	0	1
Q	1	0
R	1	0

*0.4 ml of heparinized PRP was reacted with 1×10^6 Hut-20 cells of 18 different healthy subjects on various occasions

type of the responder individuals is presented in Table 4. The only allotypic determinant shared by the three individuals is the DrW5 locus, known to be present in 10% of the population.

The ability to respond to tumor cells resides with the platelet and not with the plasma factors. This was demonstrated by the results shown in Figure 4. Gel-washed platelets prepared from the PRP of a responsive individual were still capable of being aggregated by HM-29 cells in the presence of plasma from a non-responder. Conversely, platelets of a non-responder remained negative even following the addition of plasma from a responsive individual.

Table 4

HLA TYPE OF INDIVIDUALS WHOSE PLATELETS AGGREGATE
WITH THE HUMAN MELANOMA HM-29

Individual	HLA Type		
	A	B	Dr
1	28,11	W38,W49	W4,W5
2	2,3	27,12	W2,W5
3	24,11	35,44	W4,W5

Dr. M. Fotino of the New York Blood Center kindly performed HLA typing on peripheral blood lymphocytes.

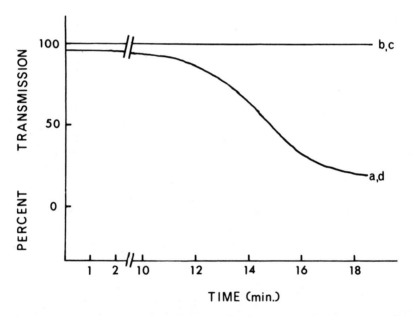

Figure 4. HM-29-induced aggregation of human PRP and GWP reconstituted with isologous plasma. (a) PRP from responder donor; (b) PRP from non-responder donor; (c) GWP from non-responder donor plus plasma from responder donor; (d) GWP from responder donor plus plasma from responder donor. 5×10^5 HM-29 cells were added at zero time. Note the unusually long lag period prior to aggregation.

DISCUSSION

Several laboratories have examined the interaction of animal tumor cells with platelets, in particular, the aggregation of platelets by tumor cells in vitro (Gasic et al 1968; Jones et al 1971; Warren, Vales 1972; Gasic et al 1973; Hilgard 1973; Warren 1973; Hilgard, Gordon-Smith 1974; Sindelar et al 1975; Gasic et al 1976; Gasic et al 1977a; Gasic et al 1977b; Gasic et al 1978; Fitzpatrick, Stringfellow 1979; Pearlstein et al 1979; Hara et al 1980; Karpatkin et al 1980; Marcum et al 1980; Pearlstein et al 1980) and the enhancement of metastasis by platelets in vivo (Gasic et al 1968; Gasic et al 1973; Pearlstein et al 1980). Little information is available on the interaction of human tumor cells with platelets. In an early report (Gasic et al 1976) Gasic and co-workers studied seven different human tumors obtained from 29 individuals. However, 6 of 7 tumors (from 25 of the 29 individuals) were fresh tissue preparations obtained at surgery. Thus the contribution of tissue collagen and contaminating fibroblasts to the platelet aggregation observed can not be excluded. Four different neuroblastoma cell lines were grown in tissue culture. All four aggregated platelets. One of these, neuroblastoma IMR, had appreciable procoagulant activity when assayed by its ability to shorten the recalcification time of platelet-poor plasma.

We have studied the mechanism of tumor-cell-induced platelet aggregation by two different virally-transformed animal cell tumors: an SV40 transformed mouse 3T3 fibroblast cell line (Pearlstein et al 1979; Karpatkin et al 1980) and a polyoma-transformed rat renal cell sarcoma (Pearlstein et al 1980). Both cell lines required surface sialic acid for platelet aggregation. The SV3T3 tumor was studied more carefully. A sedimentable platelet-aggregating material could be extracted from its cell surface, which required intact sialic acid, phospholipid and protein constituents for activity. It aggregates platelets via the release reaction following a lag period and requires Ca^{++}, complement and a stable plasma factor (56°C) for activity. PAM extracted from PW20 renal cell sarcoma cells also aggregates platelets following a lag period and requires sialic acid on the cell surface for platelet aggregating activity as well as metastases (Pearlstein et al 1980).

Since thrombocytopenia (Gasic et al 1968; Gasic et al 1973)

as well as anti-platelet agents (Kolenich et al 1972; Gasic et al 1973; Gordon et al 1979) have been reported to be capable of reducing metastases of certain animal tumors, it became important to carefully study human tumor cell lines. This was done with two human cell lines: LoVo and HCT-8. We have demonstrated that the human tumor cells aggregate platelets by a different mechanism than that obtained for the two virally-transformed animal cell lines which we had previously studied. LoVo and HCT-8 cells are not sensitive to neuraminidase or trypsin treatment, as are SV3T3 cells in their induction of platelet aggregation. Furthermore, both LoVo and HCT-8 cells appear to operate via the generation of thrombin, whereas SV3T3 cells do not. This was studied by noting the effect of LoVo and HCT-8 cells on the recalcification time of citrated platelet-poor plasma. LoVo and HCT-8 cells markedly shortened the recalcification time whereas SV3T3 cells had a negligable effect. The enhanced generation of thrombin by LoVo and HCT-8 cells could not be prevented by prior treatment of the cells with neuraminidase or trypsin. However, thrombin generation could be blocked by prior treatment with phospholipase-A_2, suggesting that phospholipid was required for thrombin generation; this was consistent with the tissue factor coagulation pathway being operative. This suggestion was supported by recalcification studies performed with coagulation factor-deficient plasmas. These indicated that LoVo and HCT-8 cells were reacting with factor VII to activate factor X to Xa leading to the activation of factor II to IIa. Preliminary results reveal a similar thrombin-induced mechanism of human platelet aggregation with two other human adenocarcinomas of the colon (SW-403 and SW-620), kindly provided by Dr. Benjamin Drewinko of M.D. Anderson Cancer Center, Houston, Texas. Tumor-induced platelet aggregation is inhibited by 1-10 μM DAPA. Similar results were obtained with a spontaneous murine tumor (Hut-20), indicating that some animal tumors may aggregate platelets in a similar fashion as that obtained for the two human cell lines.

One might ask the question: How does thrombin aggregate platelets in the presence of heparin and antithrombin III? We propose that thrombin is formed at the platelet membrane surface following tumor cell adherence to platelets (Jones et al 1971; Warren, Vales 1972; Hilgard, Gordon-Smith 1974; Sindelar et al 1975; Karpatkin et al 1980) thereby increasing the efficiency of thrombin-mediated platelet aggregation. This is consistent with the observations that VIIa (Osterund et al 1976) and tissue factor (Gitel, Wessler 1979) are not inhibited by heparin (Gitel, Wessler 1979) and antithrombin III (Osterund et al 1976), Xa is protected from heparin and antithrombin III inhibition when bound to the platelet surface (Miletich et al 1978), and binding of Xa to the platelet surface results in a 300,000-fold increase in the conversion of prothrombin to thrombin (Miletich et al 1978).

The mechanisms of induction of procoagulant activity of human tumors (Pineo et al 1973; Gordon et al 1975; Gordon et al 1979) as well as animal tumors (Gordon et al 1975; Gordon, Lewis 1978; Colucci et al 1979; Curatolo et al 1979) have been studied by other workers. However, these studies have not concerned themselves with their interaction with platelets. With most carefully studied systems, tumor cells appear to have almost exclusively, a serine or cysteine protease capable of activating factor X (Pineo et al 1973; Gordon et al 1975; Gordon, Lewis 1978; Miletich et al 1978; Colucci et al 1979; Curatolo et al 1979; Gordon et al 1979; Gordon, Cross 1981). However, activation of the tissue factor pathway via factor VII has also been reported for human leukemic cells (Gralnick, Abrell 1973; Gralnick 1979) and Toshida ascites hepatoma cells of the rat (Khato et al 1974). LoVo, HCT-8, and Hut-20 cells apparently operate via activation of the tissue factor pathway.

Marcum et al (Marcum et al 1980) recently studied the interaction of Hut-20 cells in vitro, employing platelet aggregometry, and ex vivo, employing an injured rabbit blood vessel perfusion technique. In vitro, Hut-20 cells aggregated heparinized human PRP after a 1 min lag period. Ex vivo, Hut-20 cells induced the formation of a platelet-tumor thrombus which adhered to the injured subendothelium of a rabbit aorta. The adherent platelet-thrombus formation could be prevented by administration of PGEI or use of plasma deficient in Von Willebrand's factor.

Our present data support two mechanisms for the aggregation of platelets by tumor cells: 1) Studies on the two virally-transformed animal tumor cells (Pearlstein et al 1979; Karpatkin et al 1980; Pearlstein et al 1980) indicate that cell surface-sialolipoprotein may be important and that complement activation is also required. Thrombin generation does not appear to be important because SV3T3 cells have a negligable effect on the recalcification time. Although DAPA can inhibit platelet aggregation induced by SV3T3 cells, this required 30-fold more inhibitor and 20-fold less cells than that required for inhibition of LoVo, HCT-8, or Hut-20 cells. The specificity of the reaction at this concentration of inhibitor and cells is therefore questionable. 2) Our present studies on the human LoVo and HCT-8 cell lines and the murine Hut-20 cell line indicate that thrombin generation may be important. These three cell lines are exquisitely sensitive to a highly specific synthetic thrombin inhibitor, DAPA. They are not sensitive to neuraminidase, trypsin, or complement inactivators. The two pathways for induction of platelet aggregation by tumor cells may be a reflection of spontaneous as opposed to viral transformation rather than a result of species differences.

The question therefore arises: Do most or all human tumor cell lines aggregate platelets via the generation of thrombin? If a significant number of human cell lines do aggregate human platelets via the generation of thrombin, then highly specific synthetic thrombin inhibitors (Hijikata et al 1976; Nesheim et al 1979) might be considered as possible adjuvant therapy in the treatment of human metastases. This is predicated on the supposition that platelets play a causative role in the initiation of certain tumor metastases.

Finally, the variable response of different subject's PRP's to Hut-20 and HM-29 cells is of particular interest. Some individuals are persistently or predominantly positive responders, whereas others are persistently or predominantly negative responders. These responses are not affected by tumor cell concentration or heparin concentration. Positive and negative responses have been obtained from different individuals on the same day, employing the same tumor cell suspension. It is therefore unlikely that the tumor cell line is changing its properties in tissue culture, or that we are harvesting the tumor cell line at different stages in cell growth. The ability to respond to tumor cells appears to be a property of the platelet rather than due to a constituent of plasma (Figure 4).

It is conceivable that this variable response of human PRP to Hut-20 cells represents genetic polymorphism. Indeed, positive subject C is the brother of positive subject A. Furthermore, our experience with the human melanoma cell line HM-29 was very similar, with specific individuals being positive or negative responders. Positive responders all expressed the DrW5 allotype (Table 4) and the probability of this occuring by chance is 1:1000. Although platelets express little if any D antigen, responsiveness may be closely linked with this locus. This data would suggest that if platelets are required for certain tumor cell metastases, then it is also possible that responsive individuals may have a greater potential to develop specific metastases, which may in turn parallel their in vitro aggregation response.

REFERENCES

Colucci M, Curatolo L, Donati MB, Semeraro N (1979). Direct activation of factor X by some cancer cells: Evaluation by an amidolytic assay. Thromb Haemost 42:167.

Curatolo L, Colucci M, Cambini AL, Poggi A, Morasca L, Donati MB, Semeraro N (1979). Evidence that cells from experimental tumors can activate coagulation factor X. Brit J Cancer 40:228.

Fitzpatrick FA, Stringfellow DA (1979). Prostaglandin D_2 formation by malignant melanoma cells correlates inversely with cellular metastatic potential. Proc Natl Acad Sci 76:1765.

Gasic GJ, Boettiger D, Catalfamo JL, Gasic TB, Stewart GJ (1978). Aggregation of platelets and cell membrane vesiculation by rat cells transformed in vitro by Rous sarcoma virus. Cancer Res 38:2950.

Gasic GJ, Gasic TB, Galanti N, Johnson T, Murphy S (1973). Platelet-tumor-cell interactions in mice. The role of platelets in the spread of malignant disease. Int J Cancer 11:704.

Gasic GJ, Gasic TB, Jiminez SA (1977a). Platelet aggregating material in mouse tumor cells. Removal and regeneration. Lab Invest 36:413.

Gasic GJ, Gasic TB, Jiminez SA (1977b). Effects of trypsin on the platelet aggregating activity of mouse tumor cells. Thromb Res 10:33.

Gasic GJ, Gasic TB, Stewart CC (1968). Antimetastatic effects associated with platelet reduction. Proc Natl Acad Sci USA 61:46.

Gasic GJ, Koch PAG, Hsu B, Gasic TB, Niewiarowski S (1976). Thrombogenic activity of mouse and human tumors: Effect of platelets, coagulation and fibrinolysis, and possible significance for metastases. Z Krebsforsch 86:263.

Gitel SN, Wessler S (1979). The antithrombin effects of warfarin and heparin following infusions of tissue thromboplastin in rabbits: Clinical implication. J Lab Clin Med 94:481.

Gordon SG, Cross BA (1981). A factor X-activating cysteine protease from malignant tissue. J Clin Invest 67:1665.

Gordon SG, Franks JJ, Lewis B (1975). Cancer procoagulant A: A factor activating procoagulant from malignant tissue. Thromb Res 6:127.

Gordon SG, Franks JJ, Lewis B (1979). Comparison of procoagulant activities in extracts of normal and malignant human tissue. US Natl Cancer Inst J 62:773.

Gordon SG, Lewis B (1978). Comparison of procoagulant activity in tissue culture medium from normal and transformed fibroblasts. Cancer Res 38:2467.

Gordon S, Witul M, Cohen H, Sciandra J, Williams P, Gastpar H, Murphy GP, Ambrus JL (1979). Studies on platelet aggregation inhibitors in vivo. VIII. Effect of pentoxifylline on spontaneous tumor metastasis. J Med 10:435.

Gralnick HR (1979). Cancer cell procoagulant activity. Thromb Haemostas 42:352.

Gralnick HR, Abrell E (1973). Studies on the procoagulant and fibrinolytic activity of promyelocytes in acute promyelocytic leukemia. Brit J Haematol 24:89.

Hara Y, Steiner M, Baldini MG (1980). Characterization of the platelet-aggregating activity of tumor cells. Cancer Res 40:1217.

Hijikata A, Okamoto S, Mori E, Kinjo K, Kikumoto R, Tonomura S, Tamao Y, Hara H (1976). In vitro and in vivo studies of a new series of synthetic thrombin inhibitors (OM-inhibitors). Thromb Res Suppl 2, 8:83.

Hilgard P (1973). The role of blood platelets in experimental metastases. Brit J Cancer 28:429.

Hilgard P, Gordon-Smith EC (1974). Microangiopathic haemolytic anaemia and experimental tumour-cell emboli. Brit J Haematol 26:651.

Jones JD, Wallace AC, Fraser EF (1971). Sequence of events in experimental metastases of Walker 256 tumor: light, immunofluorescent and electron microscopic observations. J Natl Cancer Inst 46:493.

Karpatkin S, Smerling A, Pearlstein E (1980). Plasma requirement for the aggregation of rabbit platelets by an aggregating material derived from SV40-transformed 3T3 fibroblasts. J Lab Clin Med 96:994.

Khato J, Suzuki M, Sato H (1974). Quantitative study on thromboplastin in various strains of Yoshida ascites hepatoma cells of rat. Gann 65(4):289.

Kolenich JJ, Mansour EG, Flynn A (1972). Hematologic effects of aspirin. Lancet 2:714.

Marcum JM, McGill M, Bastida E, Ordinas A, Jamieson GA (1980). The interaction of platelets, tumor cells, and vascular subendothelium. J Lab Clin Med 96:1046.

Miletich JP, Jackson CM, Majerus PW (1978). Properties of the factor Xa binding site on human platelets. J Biol Chem 253:6908.

Nesheim ME, Prendergast FG, Mann Kg (1979). Interactions of a fluorescent active-site-directed inhibitor of thrombin: Dansylarginine-N-(3-ethyl-1,5-pentanediyl)amide. Biochem 18:996.

Osterund B, Miller-Anderson M, Abildgaard U, Prydz H (1976). The effect of antithrombin III on the activity of the coagulation factors VII, IX and X. Thromb Haemostas 35:295.

Pearlstein E, Cooper L, Karpatkin S (1979). Extraction and characterization of a platelet-aggregating material from SV40 transformed mouse 3T3 fibroblasts. J Lab Clin Med 93:332.

Pearlstein E, Salk PL, Yogeeswaran G, Karpatkin S (1980). Correlation between spontaneous metastatic potential, platelet aggregating activity of cell surface extracts, and cell surface sialylation in ten metastatic-variant derivatives of a rat renal sarcoma cell line. Proc Natl Acad Sci USA 77:4336.

Pineo GF, Regoeczi E, Hatton MWC, Brain MC (1973). The activation of coagulation by extracts of mucous: A possible pathway of intravascular coagulation accompanying adenocarcinoma. J Lab Clin Med 82:255.

Sindelar WF, Tralka TS, Ketcham AS (1975). Electron microscopic observations on formation of pulmonary metastases. J Surg Res 18:137.

Tangen D, Berman HJ (1972). Gel filtration of blood platelets: A methodological report. Adv Exp Med Biol 34:235.

Warren BA (1973). Environment of the blood-borne tumor embolus adherent to vessel wall. J Med 4:150.

Warren BA, Vales O (1972). The adhesion of thromboplastic tumor emboli to vessel walls in vivo. Brit J Exp Pathol 53:301.

DISCUSSION

Dr. G. A. Jamieson (American Red Cross Blood Services, Bethesda). Dr. Pearlstein, did you expose your tumor cells to your antibody and then wash them? Was there any possibility of having fibronectin on your tumor cells?

Dr. E. Pearlstein (New York University Medical Center, New York). These cells do not express fibronectin; it requires an exogenous supply of fibronectin for them to adhere.

Dr. A. Scriabine (Miles Institute for Preclinical Pharmacology, New Haven). Why do you call DAPA an antithrombin antagonist? Should that not be thrombin antagonist?

Dr. Pearlstein. Thrombin antagonist.

Dr. Scriabine. Does it antagonize all the effects of the thrombin or only the vascular effects?

Dr. Pearlstein. I am not sure about that, and most of that work was done by Dr. Karpatkin.

Dr. S. Karpatkin (New York Univeristy Medical Center, New York). Dr. Mann figured out a way to synthesize the compound following reports from a Japanese company of their success with it and published the method in Biochemistry. He has since been supplying it to about fifty different investigators around the country.

Dr. L. R. Zacharski (Veterans Administration Medical Center, White River Junction). Did I understand you to say that the melanoma line was the only one showing aggregation segregated by the HLA type of the platelet.

Dr. Pearlstein. Two of the adenocarcinomas that we tried aggregated with everybody, 100% of the people. The only other line that we used for extensive studies of this nature was the Hut-20 which you have recently heard, much to our dismay, turned out to be a rodent line and that did aggregate platelets from a certain percentage of people.

Dr. Zacharski. So it is basically a melanoma line? Is the DW5 on the individual platelet?

Dr. Pearlstein. That locus is not present on the platelet to any degree. I am just saying that the individuals who are responsive contain that locus.

Subject Index

Abomino-thoracic surgery, postoperative thrombosis and, 9, 10
Acetic acid, collagen extraction and, 272
Activated coagulation products clearance, cancer and, 11
Activated partial thromboplastin or prothrombin time
  abnormality of, 2
  prolongations in cancer, 2–5
Acute non-lymphocytic leukemia, fibrinopeptide A levels and disseminated intravascular coagulation in, 114
Acute promyelocytic leukemia, hemorrhagic disseminated intravascular coagulation in, 7–8
Adenine nucleotides, intracellular levels, platelet function and, 7
Adenocarcinoma cells, platelet aggregation and, 407–410, 413
Adenocarcinomas
  and hemorrhagic disseminated intravascular coagulation, 8
  mucin-secreting, Trousseau's syndrome and, 11
Adenosine
  inhibition of platelet-aggregating material and, 454, 455
  platelet uptake, dipyridamole and, 36
Adenosine diphosphate and platelet aggregation, 460
  aspirin and, 33
  hydroxychloroquine and, 37
  sulphinpyrazone and, 38
Adenylate cyclase, RA-233 and human fibroblastic interferon and, 104
Adrenalin, aspirin and, platelet aggregation and, 33
Amaurosis fugax, sulphinpyrazone effects on transient ischemic attacks in, 46
Amnion. *See* Human amnion
Amyloidosis, immunoglobulin-type, and factor X deficiency, 3
Ancrod, metastases in tumor-injected animals and, 151
Antibiotics
  granulocytopenia after chemotherapy and, 7
  for mycoplasma elimination from cultured human tumor cells, 198–199
Anticoagulants
  and cancer cell-induced intravascular coagulation in animals, 160–161
  metastases and, 177
  prior treatment with, cancer death rates and, 116–117
  in prosthetic heart valve patients, dipyridamole or aspirin and, 43–44
  see also Antiplatelet drugs; Heparin; Warfarin
Antiplatelet drugs
  antimetastatic and antitumor effects of, 97–107, 109–111
  antithrombotic effects of, 31–53

and cancer cell-induced intravascular coagulation in animals, 160–161
clinical trial in arterial thrombosis and arteriovenous thrombosis, 51–52
clinical trial in cerebral vascular disease, 44–47
clinical trial in coronary artery disease, 39–43
clinical trial of metastasis prophylaxis in malignant human tumors, 73–75, 117–120
clinical trial in valvular heart disease and prosthetic heart valve patients, 43–44
effects on microcirculation, 111
mechanism of antithrombotic effect, 31
and metastasis, 63–64, 66–73, 81–94, 429–430
mode of action of, 33–39
and tumor cell growth, 31–53
and platelet function in tumor-bearing animals, 165–169
see also Antiplatelet drugs and cancer; Aspirin; Dipyridamole; Hydroxychloroquine; Phosphodiesterase inhibitors; Sulphinpyrazone
Antiplatelet drugs and cancer, 31–53
analogy with tissue repair, 124
biologic basis for, 113–125, 128–129
in vitro and in vivo experiments, 81–82
mechanisms of antineoplastic effects, 120–125, 128–129
and metastases in tumor-injected animal models, 148–153
types of tumors treated with, 117–118
Antithrombin, synthetic. See Compound No. 805

Antithrombotic drugs and cancer. See Antiplatelet drugs and cancer
Anturane Reinfarction Trial, 42–43
Aortocoronary bypass graft, antiplatelet drugs in prevention of arterial thrombosis and arterial venous thrombosis after, 51–52
Apyrase, platelet aggregation by human tumor cells and, 405, 407–408, 413
Arachidonate, formation of thromboxane A_2 from, 32
Arachidonic acid
DNA synthesis in mitogen-stimulated lymphocytes and, 313
metabolism, in metastatic variant cells, 169–170
prostacyclin biosynthesis and, 300
Arterial bypass surgery, antiplatelet drugs in prevention of arterial thrombosis and arterial venous thrombosis after, 51–52
Arterial catheterization, antiplatelet drugs in prevention of arterial thrombosis and arterial venous thrombosis after, 51–52
Arterial thrombosis
antiplatelet drugs and, 51–52
and dipyridamole effect on platelet survival, 37
Arteriosclerosis, platelet-tumor cell interactions and, 175
Arteriovenous shunts
aspirin effects and, sex differences in, 46
and dipyridamole effect on platelet survival, 37
Arteriovenous thrombosis, antiplatelet drugs and, 51–52
Aspirin

and bleeding time, 34
and cerebral vascular disease, 44–46
clinical trial in coronary artery disease, 40–42
dipyridamole and, platelet survival and, 37
and effects of prostacyclin in platelet adherence system, 283–284
inhibition of platelet migration by, 279
mode of action, 33–36
and mortality after myocardial infarction, 40–42
and platelet-tumor cell aggregation and attachment at subendothelial surfaces, 373–374, 379–380
and prevention of arteriovenous shunt thrombosis, 51
and prevention of thromboembolism in prosthetic heart valve patients, 43
and prevention of thrombosis in arterial catheterization patients, 51
and prostacyclin synthesis, 35–36
and prothrombin time, 34
and spontaneous metastases in tumor-transplanted animals, 148–149, 152–153
and venous thromboembolism, 47–49
see also Antiplatelet drugs
Aspirin Myocardial Infarction Study, 41
Atherosclerosis, platelets and, 25, 34, 39
Autoimmune thrombocytopenia, 21

Bacterial contamination of cultured human tumor cells, 209
Basement membranes
characterization of, 353
tumor cell interactions with, see Tumor cell -whole basement membrane interaction
Baumgartner perfusion apparatus, platelet-tumor cell interactions and attachment at subendothelial surfaces study with, 373–380
Bay g, 6575. See Nafazatrom
Bencyclane, spontaneous metastases in tumor-transplated animals and, 66, 67, 69, 149, 150, 152–153
Bleeding time
abnormalities in endothelium cell-platelet interactions in cancer, 8–9
aspirin and, 34
and platelet or endothelial cell disorders, 2
and sulphinpyrazone, 38
see also Hemorrhage
Blood. See Bleeding time; Coagulation; Hemostasis disorders in cancer; Platelets; Thrombotic disorders in cancer
Bone marrow, thrombocytopenia and, 5
see also Megakaryocytes
Boston Collaborative Drug Surveillance Group, 40–42
Bovine aortic endothelial cells, composition of, 338
Bovine capillary endothelial cells, in tumor cell-basement membrane interaction study, 360–367
Breast carcinoma, paracetamol and aspirin therapy in, 118
Breast carcinoma cells, in tumor cell-basement membrane interaction study, 360–367

Calcium, prostacyclin production by endothelial cells and, 286–287, 292

Canadian Cooperative Study Group, 45–47
Cancer
  and antiplatelet drugs, see Antiplatelet drugs
  causes of death in, 83
  diagnosis, migratory thrombophlebitis and, 11
  hemorrhage in, see Hemorrhage in cancer
  and hereditary hemorrhagic disorders, 116
  incidence in previously anticoagulated patients, 116–117
  see also Metastasis and entries following Tumor
"Cancer cell stickiness," metastasis and, 64
Carbenicillin, and acquired defects in platelet function, 7
Cardiovascular disease, anticoagulant therapy in, cancer incidence after, 116–117
Carotid transient ischemic attacks, antiplatelet drugs and, 45–46
Cell surface sialylation, platelet aggregating material of virally-transformed tumors and, 446, 449, 467, 476–477
Cerebral vascular disease, antiplatelet drugs in, 44–47
  see also Stroke; Transient cerebral ischemia
Chemotaxis, human platelet, See Platelet chemotaxis
Chemotherapy
  granulocytopenia and, 7
  microcirculation increases in tumor and, 111
  platelet-derived growth factors and, 247, 267–268
  thrombocytopenia and, 5
  and warfarin, lung carcinoma and, 119–120, 128–129
Chick embryo fibroblasts, virally transformed, growth in plasma and serum of, 249–250
Chromosome analysis of cultured human tumor cells, 195–196
  and assessment of stability, 212
  and confirmation of tumor type, 203
  and determination of extraneous cell contamination, 208
  and malignancy tests, 201
Chronic lymphocytic leukemia, plasma prekallikrein deficiency and, 3
Chronic myelogenous leukemia, factor V deficiency in, 3
α-Chymotrypsin digestion, of vessel segments in platelet-tumor cell study of attachment at subendothelial surfaces, 377
Clotting factors. See Plasma coagulation factors in cancer
CNBr-activated Sepharose 4-B, binding of platelet aggregating material to, 451
Coagulation
  assays in pulmonary epidermoid carcinoma cells, 417–418
  hemangiomata and, 23
  metastasis and, 64–66
  screening tests for classification of hemostasis disorders, 2
  see also Plasma coagulation factors in cancer
Cobra venom factor. See Complement inactivators
Collagen
  acetic acid extraction of, 272
  and biological assay for platelet-derived growth factors, 256
  -glucosal transferase, aspirin and, 35
  -induced coagulant activity, myeloproliferative disorders and, 7
  -induced platelet aggregation, aspirin and, 33–34
  -plasma interaction, low molecu-

lar substance induced human platelet chemotaxis and, 269–277, 279–280
platelet adhesion to, sulphinpyrazone and, 38
platelet aggregation and, 460
and platelet-derived growth factor, 257, 261–262
prostacyclin and, 292
type IV, tumor cell invasion of basement membrane and, 354
Collagenase
and platelet aggregating material, 470
and tumor-cell induced platelet aggregation, 392
Complement, as heat-labile factor for platelet aggregating material activity, 432–444, 463, 465, 469, 476
Complement inactivators
and DAPA inhibition of tumor-induced platelet aggregation, 483–484
and platelet aggregating material, 450–451
Compound No. 805, in tumor cell-platelet interactions, 436
Contact inhibition, morphological differentiation of cells with, 110
Coronary artery disease
clinical evaluation of antiplatelet drugs in, 39–43
dipyridamole and aspirin effect on platelet survival in, 37
Coronary artery spasm, platelet vasoconstrictor release and, 39
Coronary Drug Project Research Group, 41
Corybacterium parvum, activation of NK cells and, 180
Coumadin-resistant migratory thrombophlebitis, malignancy and, 11

Cultured human pulmonary epidermoid carcinoma cells, heparinized plasma clotting activity of soluble fractions of, 415–426
Cultured human tumor cells, 191–218, 222–223
assay for colony formation by, 253–255
assessment of variation and stability of, 211–214
bacterial, fungal or mycoplasmal contamination of, 209
cell morphology and in vitro cytopathology evaluation of, 194
chromosome analysis of, 195–196, 201
clinical data and confirmation of tumor diagnosis, 200
and clotting of heparinized plasma, 415, 417, 419–420
colony formation by various types of, 259
confirmation of malignancy of, 200–201
confirmation of tumor type of, 202–205
culture contamination tests, 197–198, 206–209
and DAPA inhibition of platelet aggregation, 479–497, 501–502
determination of characteristics of, 192
effect of platelet growth factors on, 249–264, 267–268
enzyme analysis of, 196–197
exclusion of extraneous cell contamination of, 205–209
frozen storage of, 199–200, 210
HLA typing of, 197
for human amnion in vitro invasion assay system, 360–361
in vitro cytopathology of, 201
inoculation of nude mice with, 194–195, 201, 215–217

interferon sensitivity of, 223
mechanisms of platelet aggregation by, 405–410, 413
media and serum preparation for, 193
microscopy and, 201–202
mycoplasma elimination from, 198–199
and platelet aggregation responses in normal individuals, 225–228, 229–231
poliovirus susceptibility tests of, 195
preparation of cell suspensions, 253
recording of cell line characteristics, 211
special features of cell lines, 210
see also Cultured human pulmonary epidermoid carcinoma cells
Cyclic AMP
and effect of platelet inhibitors on metastases, 71–73
and human fibroblastic interferon, 104
and inhibition of platelet aggregating material, 454, 455
intracellular, antiplatelet drugs and, 97
platelet, and dipyridamole antiplatelet action, 37
platelet, inhibition of metastases and, 115
RA-233 and, 104
Cyclic AMP phosphodiesterase
antiplatelet drugs and inhibition of, 97, 153
endothelium types and, 292
inhibitors of, metastases and, 90, 92–93, 94, 131–140
prostacyclin release and, 287
in tumor tissue and normal tissue, 94
Cyclic endoperoxidases synthesis, aspirin inhibition of, 35

Cyclo-oxygenase, sulphinpyrazone and, 39
Cyclo-oxygenase inhibitors, metastases and, 90, 94
Cyproheptacline, spontaneous metastases in tumor-transplanted animals and, 152, 153
Cytopathology, and determination of tumor type of cultured human tumor cells, 202
Cytotoxic drugs, metastases after pretreatment with, 178

Dansylarginine-N-(3-ethyl-1,5-pentanediyl) amide(DAPA), inhibition of tumor-induced platelet aggregation with, 479–497, 501–502
Detergents, non-ionic, platelet aggregating materials and, 456, 457
Dextran, inhibition of platelet function, 32
Dialysis, effects on platelet-derived growth factor activity, 258
Dipyridamole
aspirin and, platelet survival and, 34, 37
in cerebral vascular disease, 46
in coronary artery disease, 42
mechanisms of antiplatelet effects of, 36–37
metastases in tumor-cell-transplanted rats and, 66, 151
and thromboembolisms in prosthetic heart valve patients, 43
and venous thrombosis after surgery, 49
Disseminated intravascular coagulation in cancer, 10–11, 83, 113, 114
and metastases from blood-borne cancer cells, 64–66
and plasma coagulation proteins

and platelet abnormalities, 7
DNA synthesis, thromboxane A_2 and prostacyclin and, 313–314, 317
Dysfibronegenemias in cancer, 4–5

Ehrlich ascites tumor cells, intravenous transplantation into mice
  mopidamole effect on circulation time and, 70
  RA 233 and metastases from, 84–94
Endothelial disorders in cancer patients, 2, 8–9, 12–13
  see also Vascular endothelium
Enzymes
  in analysis of cultured human tumor cells, 196–197
  and DAPA inhibition of tumor-induced platelet aggregation, 482–485
  glycosaminoglycan-hydrolyzing, tumor cell-induced platelet aggregation and, 392–393
  and platelet aggregating material, 450
  and tumor cell invasiveness, 343, 353–354
  and tumor cell-vascular endothelial cell interactions, 350–351
  see also Polymorphic enzymes
Epidemiologic studies, of cancer and coagulation disorders, 116–117
Epinephrine-induced platelet aggregation abnormalities, 460
  and myeloproliferative disorders, 6
Estrogen, sex differences in antiplatelet drug clinical trials in cancer and, 82

Fibrin
  in clots associated with tumor tissue, 115

decomposition products, as nutrients for tumor cells, 83–84
deposition around tumors, platelet destruction and, 22
and platelet-tumor cell interaction, 383
Fibrinogen
  degradation products, red cell membranes and, 20
  platelet aggregation and, 460
  turnover in cancer, 114
Fibrinolytic activity, thrombosis in cancer and, 11–12
Fibrinopeptide A, disseminating intravascular coagulation in acute non-lymphocytic leukemia and, 114
Fibrinopeptides, as measure of thrombin activity, 114
Fibroblasts, separation of cultured human tumor cells from, 222
Fibronectin
  DAPA inhibition and, 501
  morphological differentiation in nude mice, 110
  platelet chemotaxis and, 279
  platelet-tumor cell interactions and, 157–158
Fungal contamination of cultured human tumor cells, 209

Glioblastoma cells, platelet aggregation and, 406–410, 413
Glucose, platelet uptake of, dipyridamole and, 36–37
Glycoconjugates, cell surface sialylation of. See Cell surface sialylation
Glycogen content of cultured human tumor cells, assessment of stability and, 213–214
Glycosaminoglycan-hydrolyzing enzymes, tumor cell-induced platelet aggregation and, 392–393
Gram-negative bacteria, superinfection in cancer and, 83

Granulocytopenia, chemotherapy and antibiotic therapy and, 7
Growth factor, platelet-derived. *See* Platelet-derived growth factor

Haementeria ghiliana salivary gland extract. *See* Salivary gland extract
Hairy cell leukemia, acquired von Willebrand's disease and, 9
Hamster embryo fibroblast cells, virally transformed, platelet derived growth factor and, 241–244
HeLa cell contamination of cultured human tumor cells, 196, 205
Hemangioendothelioma cells, prostacyclin production in response to thrombin, 284–286
Hemangiomata, thrombocytopenia and, 23
Hematopoeitic tissue malignancies, abnormal platelet function and, 5–6
Hemolytic system, interactions with tumor cells and plasma membrane vesicles, 429–440, 444
Hemorrhage, platelet abnormalities and, 5–8
  see also Hemorrhage in cancer
Hemorrhage in cancer
  and abnormalities of plasma proteins, 2–5
  as cause of death, 83
  and disseminated intravascular coagulation, 7–8
Hemostasis, plasma-collagen reactions and platelet chemotaxis and, 276–277
  see also Hemostasis disorders in cancer
Hemostasis disorders in cancer
  and abnormalities in plasma proteins, platelets, and endothelial cell function, 2

classification of, 1–13, 20
myeloma and, 4–5
see also Hemorrhage in cancer
Hemotogenous system, metastasis and, 295–296
Heparin
  metastases and, 302
  metastases in tumor-injected animals and, 149, 151
  neutralizing activity of pulmonary epidermoid carcinoma cells, 415–426
  in platelet-tumor cell interaction study, 374–380
  pork or beef, thrombocytopenia and, 230
  and postoperative venous thrombosis in cancer, 10
  and thromboembolic disorders, 20
  and tumor-injected rats, metastases and, 149, 150
  in tumor vesicle-platelet interaction study, 429–440, 444
Hepatoma, malignant, dysfibrogenemias and, 4
Hereditary hemorrhagic disorders, incidence of malignancy in, 116
Herpes simplex virus type 2, and transformation of tumor cells, platelet-derived growth factor and, 241–244
Hip replacement, aspirin therapy and venous thromboembolism in, 46, 49
Hirudin, platelet aggregation by human tumor cells and, 405, 413
HLA typing of cultured human tumor cells, 197 and determination of extraneous cell contamination, 208–209
Hodgkins disease, thrombotic thrombocytopenic purpura in, 12

Host factors, tumor meatastases and, 177-178
Human amnion in vitro invasion assay system, 356-367
Human endothelial cells
  growth on human amnion, see Human amnion in vitro invasion assay system
  in tumor cell-basement membrane interaction study, 360-367
Human fibroblastic interferon. See Interferon
Human platelet lysate
  modification of, 256
  preparation of, 255-256
Human promyleocytic leukemia cell line, RA-233-induced differentiation in, 98, 107
Human in vivo tumor metastases models, 177-184, 188-189
  and host effector mechanisms, 177-178
  role of T cell-independent immune mechanisms mediated by NK cells in, 178-184
  and use of nude mice for isolation of highly metastatic tumor cells, 183-184
  see also Cultured human tumor cells
Hut 20 tumor cells, platelet-tumor cell interactions and attachment at subendothelial surfaces in, 373-380
15-Hydroperoxyarachidonic acid, effect on metastasis, 306
Hydroxychloroquine
  mechanism of antiplatelet action of, 37
  and venous thromboembolism in surgical and orthopedic patients, 49-50
Hypersplenism, thrombocytopenia and, 5

Imidazoquinazolinone compounds, metastases prophylaxis and, 86, 88
9, 11-Iminoepoxy-prosta-5, 13-dienoic acid, inhibition of thromboxane synthetase and, 307-308
Immune response
  and host resistance against hematogenous tumor dissemination, 178-184
  and NK cells, 188-189
  nonspecific stimulation of, metastases and, 177
  and tumor vesicle-macrophage interactions, 438-440
Immune thrombocytopenia, 5
Immunoglobulins
  binding to fibrin chain in cancer patients, 4
  and prolonged prothrombin time in cancer, 12-13
Immunosuppression, metastases and, 177
Impedance plethysmography, venous thrombosis diagnosis and, 50
^{111}Indium-oxine labeled platelet studies in cancer, 25-27
  and evaluation of human platelet chemotaxis, 269-277, 279-280
  in Kasabach-Merritt syndrome, 23-25
Indomethacin
  and corticosteroids, in xeroderma pigmentosum with squamous cell carcinoma, 118
  inhibition of platelet aggregating material, 454, 455
Interferon
  inducers, activation of NK cells and, 180
  and inhibition of human neoplastic cell lines, 98, 104, 107

and RA-233, 110
  sensitivity of cultured human tumor cells, 223
Intra-tumoral platelet consumption, 21–22
  and indium-labeled platelet studies in cancer, 23, 25–27
  and Kasabach-Merritt syndrome, 23–25
  mechanisms of, 22–23
Intravascular coagulation
  and cancer cell injection into animals, 160–161
  and platelet destruction in cancer, 21–22
  see also Disseminated intravascular coagulation
Ischemic heart disease
  and sulphinpyrazone effects on platelet survival, 38
  see also Coronary artery disease
Isozymes, tissue-specific
  and confirmation of tumor type of cultured human tumor cells, 203–205
  determination in cultured human tumor cells, 197

Kanamycin, and elimination of mycoplasma from cultured human tumor cells, 198–199
Kasabach-Merritt syndrome
  and intra-tumoral platelet consumption, 27
  platelet studies in, 23–25

Leukemia, platelet function abnormalities in, 7
  see also Acute non-lymphocytic leukemia; Chronic lymphocytic leukemia; Chronic myelogenous leukemia; Hairy cell leukemia; Lymphocytic leukemia
Leukocytes, heparin-neutralizing activity of, 425

Lewis lung cancer in mice
  and antiplatelet drugs, 149, 150
  hemostatic changes and, 160–162
Leydig cell testicular tumor cells, in platelet-tumor cell interactions study, 384–400
Life Table Technique, 75
Lung cancer
  intra-tumoral platelet consumption studies in, 25
  metastases, and intravenous injection of Walker 256 tumor cells into rats, 144–152
  and RX-RA 69 phosphodiesterase inhibition of metastases in mice, 131–140
  small cell, warfarin and, 119–120
  warfarin and chemotherapy in, 128–129
  see also Lewis lung cancer
Lymphatic systems, metastases and, 177, 295–296
Lymphocytic leukemia, acquired von Willebrand's disease and, 8–9
  see also Chronic lymphocytic leukemia
Lymphoma, malignant
  antiplatelet drugs and metastases in, 63–64, 73–75
  and factor VIII inhibitors, 3
  and immune thrombocytopenia, 5
Lymphoma cells, adherence to endothelium, 288–290

Macroglobulinemia, factor VIII inhibitors and, 3–4
Macrophages, tumor vesicle interaction with, 438–440
Malignant lymphoma. See Lymphoma malignant
Mammary tumor cells
  and platelet-tumor cells interaction, 384–400

Subject Index / 513

Megakarocytes
transplanted into rats, RA-233 and, 97, 101, 103
Marantic endocarditis, chronic disseminating intravascular coagulation and, 10, 11
Megakarocytes
in cancer patients, 176
determination of production of, 175
and protein synthesis, 234–235
as source of platelet-derived growth factor, 239–241
Melanoma cells
and nafazatrom pretreatment in mice, 310–312
platelet aggregation and, 407–410, 413
Metastasis
and antiplatelet drugs, 63–64, 66–73, 83–94, 119–120, 429–430
and antitumor drugs, 97–107, 109–111
of cultured human tumor cell-produced tumors in nude mice, 216–217
defined, 295
and deuraminidase, 429
and disseminated intravascular coagulation, 64–66
and extravasation of circulating tumor cells, 366–367
host effector mechanisms and, 177–178
human in vivo models of, *see* Human in vivo tumor metastasis models
lymphatic and hemotagenous systems and, 295–296
mechanism of blood-borne, 333–334
and NK cell activity in nude mice, 179–183
pathogenesis of, 143
and platelet aggregating material, 472

and platelet-tumor cell interactions, 22–23, 63–65, 296–297, 373–380, 383–400, 479–480
prostacyclin and thromboxane synthetase inhibitors and control of, 298–321, 329–331
and RA-233 treatment in tumor-cell transplanted mice, 184–194
role of coagulation mechanism in, 115–120
role of platelets in, *see* Platelets and metastases
and RX-RA 69 phosphodiesterase inhibition in Lewis lung carcinoma-injected mice, 131–140
and thrombocytopenia, 446
and tumor cell attachment to basement membrane, 355
and tumor cell interactions with vascular endothelial cells, 333–343, 349–351
see also Platelet-tumor cell interactions
Metastatic properties of tumor cells, evaluation of, 448
Metastaic variants, and platelet-tumor cell interaction studies, 169–171
Mice
complement and tumor-platelet interaction study in, 433–440
fibroblasts, *see* Platelet aggregating material of virally-transformed tumors
intravenous tumor cell transplantation in, antiplatelet drugs and, 66, 70, 86, 89, 90, 97, 99, 103, 115, 131–140, 149, 150, 162–164, 184–194, 310–312
tumor cells from, in platelet-

tumor cells interaction
study, 384–400
see also Nude mice
Microangiopathic hemolytic anemias, disseminated intravascular coagulation and, 8, 10, 11
Microcirculation, antiplatelet drugs and, 111
Monkeys, in vivo platelet aggregation measurement in, 93–94
Monoclonal antibody techniques, confirmation of tumor type of cultured human tumor cells with, 205
Mopidamole (RA-233)
antimetastatic and antitumor effects of, 97–107, 109–111
clinical trials in human cancer patients, 98, 104, 106
-induced differentiation in human promyelocytic leukemia cell line, 98, 107
inhibition of human neoplastic cell lines and, 98, 104, 107
and metastases inhibition in rats, 66, 67, 68, 70
and metastases prevention in human tumors, 73–75
and spontaneous metastases in tumor-transplanted animals, 149, 152–153
and tumors, 118
Morphology
and assessment of stability of cultured human tumor cells, 211–214
of human cultured tumor cells, platelet-derived growth factor and, 267
Mycoplasma
contamination of cultured human tumor cells, 209
elimination from cultured human tumor cells, 198–199
tests in cultured human tumor cells for, 197–198

Myeloma
clotting factor defects in, 3–4
and dysfibrinogenemia, 4–5
factor VIII inhibitors and, 3
hemostasis defects and, 4–5
Myeloproliferative disorders
mechanisms of abnormal platelet function in, 5–7
and platelet and endothelium alterations, 12–13
and sulphinpyrazone effects on platelet survival, 38
Myocardial infarction, aspirin and, 40–42
see also Coronary artery disease

Nafazatrom, as antimetastatic agent, 310–312, 318–321
Natural killer (NK) cells, and T cell-independent immune mechanisms, metastases and, 178–184, 188–189
Neuraminidase
metastases and, 66, 429
platelet aggregating material and, 450, 456
and tumor cell-induced platelet aggregation, 390–391, 398–399
Neuroblastoma cells
in platelet-tumor cell interaction study, 384–400
transplantation into mice, antiplatelet drugs and metastases and, 86, 89, 90, 97, 99, 103
see also Wilms' tumor
NIH renal adenocarcinoma, implanted in mice, pentoxifylline and metastases study and, 90, 91
NK cells. See Natural killer cells
Nude mice, as in vivo model of human neoplasm metastatic potential, 178–184, 194–195, 203, 215–217

Occulusive coronary artery thrombobisis, platelets and, 39
Orthopedic surgery, venous thromboembolism and aspirin and, 48–49
hydroxychloroquine and, 49–50

Paracetamol, aspirin and, in breast cancer, 118
Penicillins, and acquired defects in platelet function, 7
Pentoxifylline
and circulation time of intravenously injected tumor cells in mice, 86, 87
and metastases inhibition in rats, 66, 67, 68
structural formula for, 86
Perfusion apparatus. *See* Baumgartner perfusion apparatus
Periodate oxidation, effects on platelet-derived growth factors, 258
Peripheral artery surgery, antiplatelet drugs and, 51–52
Persantine Aspirin Reinfarction Study Research Group, 42
pH, alterations of, and heparinized plasma clotting activity of pulmonary epidermoid carcinoma cells, 418–419, 425
Phenprocoumon, and metastases in tumor-injected animals, 151
Phosphodiesterase inhibitor. *See* Cyclic AMP phosphodiesterase
Phospholipases, and characterization of platelet aggregation by human tumor cells, 405, 409–410, 413, 450, 456, 457
Placentas. *See* Human amnion in vitro invasion assay system
Plasma
-collagen interaction, and low molecular substance-induced human platelet chemotaxis, 269–277, 279–280
platelet-rich, from health individuals, response to cultured human tumor cells, 225–228, 229–231
protein abnormalities in cancer, 2–5
and tumor cell-induced platelet aggregation, 389–391
Plasma coagulation factors in cancer, 2–3, 11, 21
activation mechanism, 7, 10–11, 113–115
deficiencies, immunoglobulin-type amyloidosis and, 3
hereditary deficiencies of, 116
inhibitors of, 3–4
venous thrombosis associated with, 10–11
Plasma membrane vesicles, and tumor cell interactions with components of hemostatic and immune systems, 429–440, 444
Plasma prekallikrein deficiency, in chronic lymphocytic leukemia, 3
Plasmin, conversion of plasminogen to, 33
Plasminogen, conversion to plasmin, 33
Plasminogen activator activities of tumor cells, 20
Platelet aggregating factor (PAF), 409
Platelet aggregating material (PAM) of virally-transformed tumors, 20, 22, 445–472, 477–479
binding to CNBr-activated Sepharose 4-B, 451
binding to platelets with Sepharose beads, 462–463
centrifugation of, 450, 456
determinations of sialic acid content of, 449–450

and determination of sialylation
of cell surface glycoconjugates, 449
effect on washed platelets,
458-459
enzyme treatment of, 450
and evaluation of metastatic
properties of tumor cells,
448
extraction of, 449
heat-labile factor of, 459,
461-462, 463, 465
inhibitors of, 454, 455-456
and in vitro measurement of
platelet aggregation, 452
non-inhibitors of, 454, 455
and non-transformed and transformed mouse fibroblasts,
452-457
and preparation of platelet-rich
and platelet-poor plasma
for study of, 451
and preparation of washed
human platelets for study
of, 451
protein content of, 450
quantitation of release with ^{14}C-serotonin, 452
renal cell sarcoma studies of,
465-467, 470-471
and requirement of plasma factor
for activation of, 459
synergism in rabbit platelet-rich
plasma, 457-458
treatment with complement inactivators, 450-451
treatments destroying activity of,
456-457, 458
and tumor cell culture, 447-448
Platelet aggregation
in vitro measurement of, 93-94,
452
methods, 384-385
in response to cultured human
tumor cells in normal individuals, 225-228, 229-231
tumor-cell induced, enzymes and,
392-393
tumor-cell induced, inhibition by
specific thrombin inhibitor,
479-497, 501-502
tumor cell membrane fragments-induced, 387-392
see also Platelet aggregation by
human tumor cells
Platelet aggregation by human
tumor cells, 405-410, 413
and animal models, 405-406
tumor cells used for, 406-407
Platelet aggregation inhibitors. See
Antiplatelet drugs
Platelet chemotaxis, ^{111}indium-oxine
labeled platelets and measurement of, 269-277, 279-280
Platelet consumption, extent of
malignancy and, 114
Platelet-derived growth factor
(PDGF), 233-244, 247-248
assay and characterization of,
250-251, 256-264
and cell cycle of non-transformed
mesenchymal cells, 235
chemotherapy and, 247
effect on quiescent, density-inhibited BALB/c3T3 cells, 251
effects on cultured human tumor
cell growth, 249-264,
267-268
effects of modifying procedures
on, 258
isolation of, 396-397, 399-400
and megakaryocytes, 239-241
of normal and transformed cells,
heat and, 251
purification of, 233-234,
249-250
release from platelets, 256
subcellular localization of, 234
^{3}H-thymidine incorporation into
DNA and, 234-235

Subject Index / 517

transformation of, 237–239
and tumor cells, 393–396,
    399–400
and tumor growth, 241–244
Platelet lysate
  effects on cultured human tumor
    cells, *see* Platelet-derived
    growth factors
  preparation of, 385
  tumor cell growth promoting
    activity, 394–396, 399–400
  see also Human platelet lysate
Platelet-rich plasma
  and DAPA inhibition of tumor-
    induced platelet aggrega-
    tion, 480–481
  preparation of, 451
Platelet-tumor cell interactions
  animal models for, 159–171,
    175–176
  and assays of growth-promoting
    activity, 385
  in human vs. animal models, 175
  and inhibition of platelet aggre-
    gating activity by specific
    thrombin inhibitor,
    479–497, 501–502
  and isolating of tumor cell-di-
    rected growth-promoting
    activity of platelets,
    396–397
  mechanisms of, 225
  and metastatic variants, 169–171
  and metastases, 63–75, 383–384,
    405–410, 413
  and nature of platelet-aggregat-
    ing principle of tumor cells,
    392
  in normal individuals, 225–228,
    229–231
  pharmacological modulation of,
    165–169
  and plasma membrane vesicles,
    429–440, 444
  and platelet aggregation by
    human tumor cell,
    405–410, 413
  and platelet growth-promoting
    activity for tumor cells,
    393–396, 399–400
  and release of platelet glycosami-
    noglycans by tumor cells,
    392–393
  and sensitization of platelets to
    tumor cell-induced aggre-
    gation, 389–391
  and tumor-associated hemostatic
    changes, 160–164
  and tumor-cell induced platelet
    aggregation, 386–387
  see also Platelet aggregating
    material of virally-trans-
    formed tumors; Platelets
    and metastases
Platelets
  abnormalities, classification of
    hemostasis and thrombotic
    disorders in cancer and, 2,
    5–8, 13
  adherence system, prostacyclin
    and, 282–284
  antibiotics and, 7
  aspirin inhibition of, 33–35
  collagen-glucosal transferase,
    aspirin and, 35
  destruction in cancer, 21–27
  and dipyridamole, 36–37
  growth factors secretion, 22
  heparin neutralizing activity,
    424–425
  hyperaggregable, and heparin
    therapy for thromboem-
    bolic disorders, 20
  intratumoral consumption of, *see*
    Intratumoral platelet con-
    sumption
  migration through endothelial
    lining, 277
  phosphodiesterase, *see* Cyclic
    AMP phosphodiesterase

role in thrombosis, 32–33
survival in cancer, 21
thromboxane X_2, and hemorrhagic tendency in preleukemia, 7
washed, platelet aggregating material effect on, 458–459
washed human, preparation of, 451–452
see also Antiplatelet drugs; Platelets and metastasis
Platelets and metastases, 143–153, 156–158, 296–297
and antiplatelet drugs, 148–153
and fibronectin, 157–158
historic research on, 143–144
and prostaglandins, 167–168
and route of injection of tumor cells into animals, 351–352
and spontaneously metastasizing transplantable tumors in animals, 152–153
tumor cell injections and tumor transplantation animal models of, 144–152
Poliovirus susceptibility tests, and nude mouse-grown human tumor cells, 195, 206
Polymorphic enzymes, and determination of extraneous cell contamination of cultured human tumor cells, 206–207
Prostacyclin
and adherence of lymphoma cells to endothelium, 288–290
antiplatelet activity of, 33
biosynthesis, 300
collagen and, 292
and DNA synthesis, 330
hemangioendothelioma cell production of, thrombin and, 284–286
inhibitors, and tumor growth and metastasis control, 298–321, 329–331
mechanisms of formation and release, 286–288
in melanoma-transplanted mice, metastases and, 115–116
and non-thrombogenic properties of vascular endothelium, 282–284
platelet adherence assays, 282–284
and platelet cyclic AMP, 37
role in metastasis prevention, 301–306
synthesis by vessel walls, aspirin and, 35–36
and tumor cell growth and differentiation, 313–319
Prostaglandin E
and antiplatelet drugs, 111
and platelet-tumor cell attachment to vascular subendothelial surfaces, 379
Prostaglandin I_2. See Prostacyclin
Prostaglandins
inhibitors, metastasis and, 167–168
metastases and, 90, 92–93
synthesis, 33
and tumor growth, 153, 167–168
Prostate cancer surgery, and acute disseminated intravascular coagulation, 7, 8
Prostatic hypertrophy, benign, coagulation and, 8
Prosthetic heart valves
and dipyridamole effect on platelet survival, 37
and sulphinpyrazone effect on platelet survival, 38
see also Valvular heart disease and prosthetic heart valves
Protease inhibitors, platelet aggregating material and, 455, 456, 470
Protein, platelet aggregating material content of, 450

Prothrombin
   conversion to thrombin in cancer, 114
   deficiencies, tumor-platelet interaction and, 435
Prothrombin time
   aspirin and, 34
   see also Activated partial thromboplastin time and prothrombin time
Pyrimido-pyrimidine derivatives
   and metastases prevention, 66–73, 84–94
   and potentiation of interferon production, 104
   structural formulas for, 84, 98
   see also under specific names of

RA 8. See Dipyridamole
RA 233. See Mopidamole
Rabbit
   platelet-rich plasma platelet aggregating material synergism in, 457–458
   tumor-cell injected, antiplatelet drugs and, 149
Race, and HeLa cell phenotypes, 205
Raji cells. See Lymphoma cells
Rats
   antiplatelet drug inhibition of metastases from intravenous tumor cell transplantation in, 66–73
   mammary tumor cells, and platelet-derived growth factors, 251
   platelet-rich plasma, tumor vesicle-platelet interaction study in, 429–440, 444
   renal sarcoma, see Platelet aggregating material of virally-transformed tumors
   tumor-cell injected, antiplatelet drugs and, 86, 89, 97, 100, 101, 103, 107, 144–152, 162
   tumor cells, in low NK cell activity nude mice, 182
   tumor cells, in platelet-tumor cell interactions study, 384–400
Red blood cells
   and antiplatelet drugs, 111
   and fibrinogen degradation products, 20
Red thrombi, 32
Renal adenocarcinoma cells, in platelet-tumor cell interaction study, 384–400
Renal tumor cell lines, transplanted into rats, 115
Reptilase time, 4
RX-RA 69
   and metastases inhibition in rats, 65, 67, 68
   structure of, 133

Salivary gland extract, in tumor-cell plasma membrane vesicles interactions with hemostatic and immune systems, 436–440, 444
Sarcomas
   and clinical trial of mopidamole for metastasis prevention, 73–75
   and congenital factor deficiency states, 116
Sepharose beads, plasma aggregating materials coupled to, platelet binding and, 462–463
^{14}C-Serotonin, and quantitation of platelet-aggregating material release, 452, 454
Sex differences
   in antiplatelet drug effects, 82
   in antithrombotic effects of aspirin, 46, 52
Sialic acid, in platelet aggregating material, 449–450, 467
Sialolipoprotein. See Platelet aggregating material

Sialylation. *See* Cell surface sialylation
Smoking, hyperaggregability of platelets and, 230
Sodium salicylate, platelet function and, 34
Sonication, of platelet aggregating material, 456, 457
Splenectomy, and hemorrhagic syndrome in macroglobulinemia, 4
Steroids, and immune thrombocytopenia associated with cancer, 5
Stroke, aspirin and, 45–46
Sulphinpyrazone
   clinical trial in cerebral vascular disease, 46–47
   clinical trial in coronary artery disease, 42–43
   mechanism of antiplatelet action of, 38–39
   and metastases inhibition in rats, 66
   and postoperative venous thromboembolism, 50
   and shunt thrombosis, 51–52
   and transient ischemic attacks, 45
Superinfection, cancer mortality and, 83
Surgery
   amputation of tumor-bearing leg in mice, RX-RA, 69 antimetastasis treatment and, 132–140
   and venous thromboembolism, antiplatelet drugs and, 49–50
Systemic arterial embolism, 43–44

T cell-independent immune mechanisms, and host resistance against hematogenous tumor dissemination, 176–184
Temperature
   effects on platelet-derived growth factors from normal and transformed cells, 250–251, 258
   and endothelial cell retraction, 350
   and plasma clotting activity of tumor cells, 422, 424
   and platelet binding to tumor vesicles, 431–432
Theophylline, and prostaclyclin effects on tumors, 115, 302–306
Thrombi
   composition of, 32
   platelet-tumor cell, aggregation and attachment at subendothelial surface, 373–380
Thrombin
   and biological assay for platelet-derived growth factors, 256
   conversion of prothrombin to in cancer, 114
   and platelet adherence, 282, 283
   and platelet consumption in Kasabach-Merritt syndrome, 24
   and platelet-derived growth factor, 257
   and prostacyclin production, 284–286, 292–293
   -stimulated platelet derived growth factors, 261–262
Thrombin inhibitors. *See* Dansyl-arginine-N-(3-ethyl-1, 5-pentanediyl) amide
Thrombin time, in myeloma, 4–5
Thrombocythemia
   differentiation from secondary reactive thrombocytosis, 6
Thrombocytopenia
   and hemorrhage in cancer, 5, 6
   and hemorrhagic disseminated intravascular coagulation, 8
   induced by blood-borne tumor cells, 159
   and lung tumors, 156

and metastases, 446
and neuraminidase-induced metastases decreases, 429
tumor-cell induced, 134, 135, 161-164
see also Immune thrombocytopenia; Intra-tumoral platelet consumption
Thrombocytosis
in human and animal models, 175
and myeloproliferative disorders, 6
Thromboembolism in cancer
as cause of death, 83
and circulating tumor cells in blood, 83
see also Thrombosis
Thrombophlebitis, migratory, chronic disseminated intravascular coagulation and, 10, 11
Thromboplastin time. See Activated partial thromboplastin time and prothrombin time
Thrombosis
and nafazatrom, 310
role of platelets in genesis of, 32-33
and tumor-associated disseminated intravascular coagulation, 7-8
types of, 32
see also Arterial thrombosis
Thrombotic disorders in cancer, 9-13
and abnormalities in plasma proteins, platelets, and endothelial cell function, 2
and alteration of plasma coagulation proteins, 10-12
and antiplatelet drugs, 31-53
classification of, 1-13, 20
and fibrinolytic activity, 11-12
incidence of, 9
and platelet and endothelium alterations, 12-13
screening tests for, 2

Thromboxane A_2
aspirin inhibition of, 283
and cyclo-oxygenase inhibitors, 90
synthesis, dipyridamole and, 36
synthesis from platelet membrane arachidonic acid, aspirin inhibition of, 35
and thrombosis, 32-33
and tumor cell growth and differentiation, 313-319
Thromboxane synthetase inhibitors, and tumor growth and metastasis, 306-312, 329-331
Ticarcillin, and prolongation of bleeding time, 7
Tissue thromboplastin inhibition assay, and pancoagulation inhibitors in macroglobulinemia, 4
Transient cerebral ischemia
and aspirin, 44-45
and sulphinpyrazone, 38, 46-47
Trental. See Pentoxifylline
Trousseau's syndrome, 11
Trypsin
and heparinized plasma clotting activity of pulmonary epidermoid carcinoma cells, 419, 421
and platelet aggregating material, 456, 458
and tumor-cell induced platelet aggregation, 392
Trypsin digestion
effects on platelet derived growth factors, 257-258
Tumor antigens, specific immunity against, 177
Tumor cell-platelet interactions. See Platelet-tumor cell interactions
Tumor cell-vascular endothelial cell interactions, 333-343, 349-351
and blood-borne metastases, 333-334
and cross-species reactors, 349

and endothelial cell retraction and tumor cell migration, 338–339
monolayer tissue culture techniques for study of, 335
and solubilization and invasion of endothelial basal lamina, 339, 341–342
and tumor cell adhesion, 335–338
Tumor cell-whole basement membrane interactions, 353–367
and attachment process, 354–355
and invasive process, 354, 361–367
and in vitro human amnion invasion assay system, 356–361
Tumor cells
alteration of, metastasis and, 177–178
circulating, extravasation of, 366–367
circulation in peripheral blood, metastases incidence and, 64
evaluation of metastatic properties of, 448
fibrin decomposition products as nutrients for, 83–84
growth, antiplatelet drugs and, 31–53
heterogeneity of responses, 178
-induced platelet aggregation, and thrombotic complications in cancer, 83
induced platelet aggregation, lag time in, 386–387
injection into animals, in platelet-tumor cell interaction studies, 131–140, 144–152, 160–164
invasiveness as distinguishing feature of, 355
isolation methods, 299, 301
membrane fragments, and induced platelet aggregation, 387–388

preparation of membrane fragments of, 385
selection of highly metastatic variants of, 178–184
thromboplastic activity of, 20
see also Cultured human tumor cells
Tumor growth
and assessment of stability of cultured human tumor cells, 211–214
and platelet-derived growth factors, 148, 241–244, 249–264, 267–268
prostacyclin and thromboxane synthetase inhibitors and control of, 298–321, 329–331
prostaglandins and, 153
in RA 233-treated mice, 109
role of coagulation mechanism in, 115–120
Tumor tissue, elements of clots occurring in association with, 114–115
Tumors
and antimetastatic and antitumor effect of antiplatelet drugs, 97–107, 109–111
effects on hemostatic system of host, 163–164
and fibrinopeptide A levels, 114
metastases, see Metastases
platelet consumption by, see Intra-tumoral platelet consumption
transplanted into animals, antiplatelet drugs and, 152–153
vascularization, and platelet-derived growth factor, 247–248
virally transformed, see Platelet aggregating material of virally transformed tumors
Tween-80, platelet aggregating material and, 456, 457

Urea extraction, platelet aggregating material of virally transformed tumors and, 453, 454

VA Cooperative Study Group, 118-120
Valvular heart disease and prosthetic heart valves, clinical trials of antiplatelet drugs in, 43-44
Vascular disease, platelet factors and, 22
Vascular endothelial cells, interaction with tumor cells. See Tumor cell-vascular endothelial cell interactions; Vascular endothelium
Vascular endothelium
   prostacyclin production, tumor cells and, 429
   retraction, 338-339, 349-350
   tumor cell adhesion at, 288-290, 296, 383
   see also Vascular endothelium, non-thrombogenic properties of; Vascular subendothelium
Vascular endothelium, non-thrombogenic properties of, 281-290, 292-293
   and control mechanisms of prostacyclin formation and release, 286-288
   and cultured hemangioendothelioma cells prostacyclin production in response to thrombin, 284-286
   role of prostacyclin in, 282-284, 288-290
   See also Prostacyclin
Vascular subendothelium, platelet-tumor cell interaction and attachment at, 373-380

Vascular system, metastases and, 177
Venous thromboembolism, clinical trials of antiplatelet drugs and, 47-50
Venous thrombosis, sulphinpyrazone effects on platelet survival in, 38
Vertebrobasilar transient ischemic attacks, and antiplatelet drugs, 45-46
Vesicles. See Plasma membrane vesicles
Virally-transformed tumor cells and DAPA inhibition of platelet aggregation, 480, 497, 501-502
   platelet aggregating material of, see Platelet aggregating material of virally transformed tumors
von Willebrand's disease
   acquired in cancer, 8-9
   and platelet-tumor cell interactions, 374, 379

Walker-256 tumor cells, antiplatelet drugs and, 63-73, 144-152, 383
Warfarin
   anticoagulant and antineoplastic effect of, 121
   and aortocoronary bypass graft patency rates, 51
   metastases and, 119-120, 128-129, 165, 166, 302
White thrombi, 32
Wilms' tumor, implantation into rats, antiplatelet drugs and, 86, 89, 97, 100, 103

Xeroderma pigmentosum, indomethacin and corticosteroid therapy and, 118

PROGRESS IN CLINICAL AND BIOLOGICAL RESEARCH

Series Editors
Nathan Back
George J. Brewer

Vincent P. Eijsvoogel
Robert Grover
Kurt Hirschhorn

Seymour S. Kety
Sidney Udenfriend
Jonathan W. Uhr

Vol 1: **Erythrocyte Structure and Function,** George J. Brewer, *Editor*
Vol 2: **Preventability of Perinatal Injury,** Karlis Adamsons and Howard A. Fox, *Editors*
Vol 3: **Infections of the Fetus and the Newborn Infant,** Saul Krugman and Anne A. Gershon, *Editors*
Vol 4: **Conflicts in Childhood Cancer: An Evaluation of Current Management,** Lucius F. Sinks and John O. Godden, *Editors*
Vol 5: **Trace Components of Plasma: Isolation and Clinical Significance,** G.A. Jamieson and T.J. Greenwalt, *Editors*
Vol 6: **Prostatic Disease,** H. Marberger, H. Haschek, H.K.A. Schirmer, J.A.C. Colston, and E. Witkin, *Editors*
Vol 7: **Blood Pressure, Edema and Proteinuria in Pregnancy,** Emanuel A. Friedman, *Editor*
Vol 8: **Cell Surface Receptors,** Garth L. Nicolson, Michael A. Raftery, Martin Rodbell, and C. Fred Fox, *Editors*
Vol 9: **Membranes and Neoplasia: New Approaches and Strategies,** Vincent T. Marchesi, *Editor*
Vol 10: **Diabetes and Other Endocrine Disorders During Pregnancy and in the Newborn,** Maria I. New and Robert H. Fiser, *Editors*
Vol 11: **Clinical Uses of Frozen-Thawed Red Blood Cells,** John A. Griep, *Editor*
Vol 12: **Breast Cancer,** Albert C.W. Montague, Geary L. Stonesifer, Jr., and Edward F. Lewison, *Editors*
Vol 13: **The Granulocyte: Function and Clinical Utilization,** Tibor J. Greenwalt and G.A. Jamieson, *Editors*
Vol 14: **Zinc Metabolism: Current Aspects in Health and Disease,** George J. Brewer and Ananda S. Prasad, *Editors*
Vol 15: **Cellular Neurobiology,** Zach Hall, Regis Kelly, and C. Fred Fox, *Editors*
Vol 16: **HLA and Malignancy,** Gerald P. Murphy, *Editor*
Vol 17: **Cell Shape and Surface Architecture,** Jean Paul Revel, Ulf Henning, and C. Fred Fox, *Editors*
Vol 18: **Tay-Sachs Disease: Screening and Prevention,** Michael M. Kaback, *Editor*
Vol 19: **Blood Substitutes and Plasma Expanders,** G.A. Jamieson and T.J. Greenwalt, *Editors*
Vol 20: **Erythrocyte Membranes: Recent Clinical and Experimental Advances,** Walter C. Kruckeberg, John W. Eaton, and George J. Brewer, *Editors*
Vol 21: **The Red Cell,** George J. Brewer, *Editor*
Vol 22: **Molecular Aspects of Membrane Transport,** Dale Oxender and C. Fred Fox, *Editors*
Vol 23: **Cell Surface Carbohydrates and Biological Recognition,** Vincent T. Marchesi, Victor Ginsburg, Phillips W. Robbins, and C. Fred Fox, *Editors*
Vol 24: **Twin Research, Proceedings of the Second International Congress on Twin Studies,** Walter E. Nance, *Editor*
Published in 3 Volumes:

	Part A: **Psychology and Methodology**
	Part B: **Biology and Epidemiology**
	Part C: **Clinical Studies**
Vol 25:	**Recent Advances in Clinical Oncology,** Tapan A. Hazra and Michael C. Beachley, *Editors*
Vol 26:	**Origin and Natural History of Cell Lines,** Claudio Barigozzi, *Editor*
Vol 27:	**Membrane Mechanisms of Drugs of Abuse,** Charles W. Sharp and Leo G. Abood, *Editors*
Vol 28:	**The Blood Platelet in Transfusion Therapy,** G.A. Jamieson and Tibor J. Greenwalt, *Editors*
Vol 29:	**Biomedical Applications of the Horseshoe Crab (Limulidae),** Elias Cohen, *Editor-in-Chief*
Vol 30:	**Normal and Abnormal Red Cell Membranes,** Samuel E. Lux, Vincent T. Marchesi, and C. Fred Fox, *Editors*
Vol 31:	**Transmembrane Signaling,** Mark Bitensky, R. John Collier, Donald F. Steiner, and C. Fred Fox, *Editors*
Vol 32:	**Genetic Analysis of Common Diseases: Applications to Predictive Factors in Coronary Disease,** Charles F. Sing and Mark Skolnick, *Editors*
Vol 33:	**Prostate Cancer and Hormone Receptors,** Gerald P. Murphy and Avery A. Sandberg, *Editors*
Vol 34:	**The Management of Genetic Disorders,** Constantine J. Papadatos and Christos S. Bartsocas, *Editors*
Vol 35:	**Antibiotics and Hospitals,** Carlo Grassi and Giuseppe Ostino, *Editors*
Vol 36:	**Drug and Chemical Risks to the Fetus and Newborn,** Richard H. Schwarz and Sumner J. Yaffe, *Editors*
Vol 37:	**Models for Prostate Cancer,** Gerald P. Murphy, *Editor*
Vol 38:	**Ethics, Humanism, and Medicine,** Marc D. Basson, *Editor*
Vol 39:	**Neurochemistry and Clinical Neurology,** Leontino Battistin, George Hashim, and Abel Lajtha, *Editors*
Vol 40:	**Biological Recognition and Assembly,** David S. Eisenberg, James A. Lake, and C. Fred Fox, *Editors*
Vol 41:	**Tumor Cell Surfaces and Malignancy,** Richard O. Hynes and C. Fred Fox, *Editors*
Vol 42:	**Membranes, Receptors, and the Immune Response: 80 Years After Ehrlich's Side Chain Theory,** Edward P. Cohen and Heinz Köhler, *Editors*
Vol 43:	**Immunobiology of the Erythrocyte,** S. Gerald Sandler, Jacob Nusbacher, and Moses S. Schanfield, *Editors*
Vol 44:	**Perinatal Medicine Today,** Bruce K. Young, *Editor*
Vol 45:	**Mammalian Genetics and Cancer: The Jackson Laboratory Fiftieth Anniversary Symposium,** Elizabeth S. Russell, *Editor*
Vol 46:	**Etiology of Cleft Lip and Cleft Palate,** Michael Melnick, David Bixler, and Edward D. Shields, *Editors*
Vol 47:	**New Developments With Human and Veterinary Vaccines,** A. Mizrahi, I. Hertman, M.A. Klingberg, and A. Kohn, *Editors*
Vol 48:	**Cloning of Human Tumor Stem Cells,** Sydney E. Salmon, *Editor*
Vol 49:	**Myelin: Chemistry and Biology,** George A. Hashim, *Editor*

THE LIBRARY
UNIVERSITY OF CALIFORNIA
San Francisco
666

THIS BOOK IS DUE ON THE LAST
Books not returned on time are subject to
Lending Code. A renewal may be made on certa
consult Lending Code.